CAMBRIDGE LIBRARY COLLECTION

Books of enduring scholarly value

Physical Sciences

From ancient times, humans have tried to understand the workings of the world around them. The roots of modern physical science go back to the very earliest mechanical devices such as levers and rollers, the mixing of paints and dyes, and the importance of the heavenly bodies in early religious observance and navigation. The physical sciences as we know them today began to emerge as independent academic subjects during the early modern period, in the work of Newton and other 'natural philosophers', and numerous sub-disciplines developed during the centuries that followed. This part of the Cambridge Library Collection is devoted to landmark publications in this area which will be of interest to historians of science concerned with individual scientists, particular discoveries, and advances in scientific method, or with the establishment and development of scientific institutions around the world.

The History and Present State of Electricity

When this work first appeared in 1767, electricity was seen as such a minor aspect of natural philosophy that its investigation was not considered a priority for contemporary scientists. The polymath Joseph Priestley (1733–1804) was one of the few who devoted serious effort to advancing the field. Here he charts the history of electrical study from experiments with amber in ancient Greece to the most recent discoveries. The book comprises explanations of the principal theories of electricity – both historical and contemporary – in addition to a selection of well-known experiments carried out by previous researchers. Priestley also details his own experiments, covering such topics as the colour of electric light, the effects of temperature, and even the musical tone of electrical discharges. One of his most successful works, testifying to the clarity of his explanations, the book remains an important text in the history of science.

Cambridge University Press has long been a pioneer in the reissuing of out-of-print titles from its own backlist, producing digital reprints of books that are still sought after by scholars and students but could not be reprinted economically using traditional technology. The Cambridge Library Collection extends this activity to a wider range of books which are still of importance to researchers and professionals, either for the source material they contain, or as landmarks in the history of their academic discipline.

Drawing from the world-renowned collections in the Cambridge University Library and other partner libraries, and guided by the advice of experts in each subject area, Cambridge University Press is using state-of-the-art scanning machines in its own Printing House to capture the content of each book selected for inclusion. The files are processed to give a consistently clear, crisp image, and the books finished to the high quality standard for which the Press is recognised around the world. The latest print-on-demand technology ensures that the books will remain available indefinitely, and that orders for single or multiple copies can quickly be supplied.

The Cambridge Library Collection brings back to life books of enduring scholarly value (including out-of-copyright works originally issued by other publishers) across a wide range of disciplines in the humanities and social sciences and in science and technology.

The History
and Present State
of Electricity

With Original Experiments

Joseph Priestley

CAMBRIDGE
UNIVERSITY PRESS

University Printing House, Cambridge, CB2 8BS, United Kingdom

Published in the United States of America by Cambridge University Press, New York

Cambridge University Press is part of the University of Cambridge.
It furthers the University's mission by disseminating knowledge in the pursuit of
education, learning and research at the highest international levels of excellence.

www.cambridge.org
Information on this title: www.cambridge.org/9781108064392

© in this compilation Cambridge University Press 2013

This edition first published 1767
This digitally printed version 2013

ISBN 978-1-108-06439-2 Paperback

THE
HISTORY
AND
PRESENT STATE
OF
ELECTRICITY,
WITH
ORIGINAL EXPERIMENTS,

By JOSEPH PRIESTLEY, LL.D. F.R.S.

Causa latet, vis est notissima.

OVID.

LONDON,

Printed for J. DODSLEY in Pall-Mall, J. JOHNSON and
B. DAVENPORT in Pater-noster Row, and T. CADELL
(Successor to Mr. MILLAR) in the Strand. MDCCLXVII.

THE RIGHT HONOURABLE

JAMES Earl of MORTON,

PRESIDENT OF THE ROYAL SOCIETY,

This HISTORY, &c.

IS,

WITH THE GREATEST RESPECT,

INSCRIBED,

By HIS LORDSHIP's

MOST OBEDIENT,

AND MOST HUMBLE

SERVANT,

JOSEPH PRIESTLEY.

THE
PREFACE

IN writing the *history and present state of electricity*, I flatter myself that I shall give pleasure, as well to persons who have a taste for Natural Philosophy in general, as to electricians in particular; and I hope the work will be of some advantage to the science itself. Both these ends would certainly be answered in a considerable degree, were the execution at all answerable to the design.

THE history of electricity is a field full of pleasing objects, according to all the genuine and universal principles of taste, deduced from a knowledge of human nature. Scenes like these, in which we see a gradual rise and progress in things, always exhibit a pleasing spectacle to the human mind. Nature, in all her delightful walks, abounds with such views, and they are in a more especial manner connected with every thing that relates to human life and happiness; things, in their own nature, the most interesting to us. Hence it is, that the power of association has annexed crouds of pleasing sensa-

a   tions

tions to the contemplation of every object, in which this property is apparent.

THIS pleasure, likewise, bears a considerable resemblance to that of the sublime, which is one of the most exquisite of all those that affect the human imagination. For an object in which we see a perpetual progress and improvement is, as it were, continually rising in its magnitude; and moreover, when we see an actual increase, in a long period of time past, we cannot help forming an idea of an unlimited increase in futurity; which is a prospect really boundless, and sublime.

THE pleasures arising from views exhibited in *civil*, *natural*, and *philosophical* history are, in certain respects, different from one another. Each has its advantages, and each its defects: and both their advantages and defects contribute to adapt them to different classes of readers.

CIVIL history presents us with views of the strongest passions and sentiments of the human mind, into which every man can easily and perfectly enter; and with such incidents, respecting happiness and misery, as we cannot help feeling would alarm and affect us in a very sensible manner; and therefore, we are at present alarmed and affected by them to a considerable degree. Hence the pleasure we receive from civil history arises, chiefly, from the exercise it affords our passions. The imagination is only entertained with scenes which occasionally start up, like interludes, or episodes, in the great drama, to which we are principally attentive. We are presented, indeed, with the prospect of gradual improvement during the rise of great empires; but, as we read on, we are obliged to contemplate the disagreeable reverse. And the history of

most

moſt ſtates preſents nothing but a tedious uniformity, without any ſtriking events, to diverſify and embelliſh the proſpect. Beſides, if a man have any ſentiment of virtue and benevolence, he cannot help being ſhocked with a view of the vices and miſeries of mankind; which, though they be not all, are certainly the moſt glaring and ſtriking objects in the hiſtory of human affairs. An attention, indeed, to the conduct of divine providence, which is ever bringing good out of evil, and gradually conducting things to a more perfect and glorious ſtate, tends to throw a more agreeable light on the more gloomy parts of hiſtory, but it requires great ſtrength of mind to comprehend thoſe views; and, after all, the feelings of the heart too often over power the concluſions of the head

NATURAL hiſtory exhibits a boundleſs variety of ſcenes, and yet infinitely analogous to one another. A naturaliſt has, conſequently, all the pleaſure which the contemplation of uniformity and variety can give the mind; and this is one of the moſt copious ſources of our intellectual pleaſures. He is likewiſe entertained with a proſpect of gradual improvement, while he ſees every object in nature riſing by due degrees to its maturity and perfection. And while new plants, new animals, and new foſſils are perpetually pouring in upon him, the moſt pleaſing views of the unbounded power, wiſdom, and goodneſs of God are conſtantly preſent to his mind. But he has no direct view of human ſentiments and human actions; which, by means of their endleſs aſſociations, greatly heighten and improve all the pleaſures of taſte.

THE hiſtory of philoſophy enjoys, in ſome meaſure, the advantages both of civil and natural hiſtory, whereby it is

relieved

relieved from what is moſt tedious and diſguſting in both. Philoſophy exhibits the powers of nature, diſcovered and directed by human art: it has, therefore, in ſome meaſure, the boundleſs variety with the amazing uniformity of the one and likewiſe every thing that is pleaſing and intereſting in the other. And the idea of continual riſe and improvement is conſpicuous in the whole ſtudy, whether we be attentive to the part which nature, or that which men are acting in the great ſcene.

It is here that we ſee the human underſtanding to its greateſt advantage, graſping at the nobleſt objects, and increaſing its own powers, by acquiring to itſelf the powers of nature and directing them to the accompliſhment of its own views; whereby the ſecurity, and happineſs of mankind are daily improved. Human abilities are chiefly conſpicuous in adapting means to ends, and in deducing one thing from another by the method of analogy; and where ſhall we find inſtances of greater ſagacity, than in philoſophers diverſifying the ſituations of things, in order to give them an opportunity of ſhowing their mutual relations, affections, and influences; deducing one truth and one diſcovery from another, and applying them all to the uſeful purpoſes of human life.

If the exertion of human abilities, which cannot but form a delightful ſpectacle for the human imagination, give us pleaſure, we enjoy it here in a higher degree than while we are contemplating the ſchemes of warriours, and the ſtratagems of their bloody art. Beſides, the object of philoſophical purſuits throws a pleaſing idea upon the ſcenes they exhibit; whereas a reflection upon the real objects and views of

moſt

moſt ſtateſmen and conquerors cannot but take from the pleaſure, which the idea of their ſagacity, foreſight, and com-
prehenſion would otherwiſe give to the virtuous and benevolent mind. Laſtly, the inveſtigation of the powers of nature,
like the ſtudy of Natural Hiſtory, is perpetually ſuggeſting to
us views of the divine perfections and providence, which are
both pleaſing to the imagination, and improving to the heart

But though other kinds of hiſtory may, in ſome reſpects,
vie with that of philoſophy, nothing that comes under the
denomination of hiſtory can exhibit inſtances of ſo fine a riſe
and improvement in things, as we ſee in the progreſs of the
human mind, in philoſophical inveſtigations. To whatever
height we have arrived in natural ſcience, our beginnings
were very low, and our advances have been exceeding gradual.
And to look down from the eminence, and to ſee, and compare all thoſe gradual advances in the aſcent, cannot but give
the greateſt pleaſure to thoſe who are ſeated on the eminence,
and who feel all the advantages of their elevated ſituation.
And conſidering that we ourſelves are, by no means, at the
top of human ſcience; that the mountain ſtill aſcends beyond
our ſight, and that we are, in fact, not much above the foot
of it, a view of the manner in which the aſcent has been
made cannot but animate us in our attempts to advance ſtill
higher, and ſuggeſt methods and expedients to aſſiſt us in our
further progreſs.

Great conquerors, we read, have been both animated,
and alſo, in a great meaſure, formed by reading the exploits
of former conquerors. Why may not the ſame effect be expected from the hiſtory of philoſophy to philoſophers May
not

not even more be expected in this cafe? The wars of many of thofe conquerors, who received this advantage from hiftory, had no proper connection with former wars: They were only analogous to them. Whereas the whole bufinefs of philo-fophy, diverfified as it is, is but one; it being one and the fame great fcheme, that all philofophers, of all ages and na-tions have been conducting, from the beginning of the world; fo that the work being the fame, the labours of one are not only analogous to thofe of another, but in an immediate manner fubfervient to them; and one philofopher fucceeds another in the fame field; as one Roman proconful fucceeded another, in carrying on the fame war, and purfuing the fame conquefts, in the fame country. In this cafe an intimate knowledge of what has been done before us cannot but greatly facilitate our future progrefs, if it be not abfolutely neceffary to it.

THESE hiftories are evidently much more neceffary in an advanced ftate of fcience, than in the infancy of it. At pre-fent philofophical difcoveries are fo many, and the accounts of them are fo difperfed, that it is not in the power of any man to come at the knowledge of all that has been done, as a foundation for his own inquiries. And this circumftance appears to me to have very much retarded the progrefs of difcoveries.

NOT that I think philofophical difcoveries are now at a ftand. On the other hand, as quick advances feem to have been made of late years as in any equal period of time paft whatever. Nay, it appears to me, that the progrefs is really accelerated. But the increafe of knowledge is like the in-

creafe

creafe of a city. The building of fome of the firft ſtreets makes a great figure, is much talked of, and known to every body; whereas the addition of, perhaps, twice as much building, after it has been ſwelled to a confiderable fize, is not fo much as taken no ice of, and may be really unknown to many of the inhabitants. If the additions which have been made to the buildings of the city of London, in any fingle year of late, had been made two or three centuries ago, it could not have efcaped the obfervation of hiftorians; whereas, now, they are fo fcattered, and the proportion they bear to the whole city is fo fmall, that they are hardly noticed. For the fame reafon, the improvements that boys make at fchool, or that young gentlemen make at an academy, or the univerfity, are more taken notice of than all the knowledge they acquire afterwards, though they continue their ftudies with the fame affiduity and fuccefs.

THE hiftory of experimental philofophy, in the manner in which it ought to be written, to be of much ufe, would be an immenfe work; perhaps more than any one man ought to undertake; but it were much to be wifhed, that perfons who have leifure, and fufficient abilities, would undertake it in feparate parts. I have executed it, in the beft manner I have been able, for that branch which has been my own favourite amufement; and I fhall think myfelf happy, if the attempt excite other perfons to do the like for theirs.

I CANNOT help thinking myfelf to have been peculiarly fortunate, in undertaking the hiftory of electricity, at the moft proper time for writing it, when the materials were neither too few, nor too many to make a hiftory; and when

they

they were so scattered, as to make the undertaking highly desirable, and the work peculiarly useful to Englishmen.

I LIKEWISE think myself peculiarly happy in my subject itself. Few branches of Natural Philosophy would, I think, make so good a subject for a history. Few can boast such a number of discoveries, disposed in so fine a series, all comprised in so short a space of time, and all so recent, the principal actors in the scene being still living.

WITH several of these principal actors it has been my singular honour and happiness to be acquainted; and it was their approbation of my plan, and their generous encouragement that induced me to undertake the work. With gratitude I acknowledge my obligations to Dr. Watson, Dr. Franklin, and Mr. Canton, for the books, and other materials with which they have supplied me, and for the readiness with which they have given me any information in their power to procure. In a more especial manner am I obliged to Mr. Canton, for those original communications of his, which will be found in this work, and which cannot fail to give a value to it, in the esteem of all the lovers of electricity. My grateful acknowledgments are also due to the Rev. Mr. Price F. R. S. and to the Rev. Mr. Holt, our professor of Natural Philosophy at Warrington, for the attention they have given to the work, and for the many important services they have rendered me with respect to it.

To the gentlemen above-mentioned the public is, likewise, indebted for whatever they may think of value in the *original experiments* which I have related of my own. It was from conversing with them that I was first led to entertain the

thought

thought of attempting any thing new in this way, and it was their example, and favourable attention to my experiments, that animated me in the pursuit of them. In short, without them, neither my experiments, nor this work would have had any existence.

THE historical part of the work, the reader, I hope, will find to be full and circumstantial, and at the same time succinct. Every new fact, or important circumstance, I have noted as it arose; but I have abridged all long details, and have carefully avoided all digressions and repetitions. For this purpose, I have carefully perused every original author, to which I could have recourse; and every quotation in the margin points to the authority that I myself consulted, and from which the account in the text was actually taken. Where I could not procure the original authors, I was obliged to quote them at second hand, but the reference will always show where that has been done. That I might not misrepresent any writer, I have generally given the reader his own words, or the plainest translation I could make of them; and this I have done, not only in direct quotations, but where, by a change of person, I have made the language my own.

I MADE it a rule to myself, and I think I have constantly adhered to it, to take no notice of the mistakes, misapprehensions, and altercations of electricians; except so far as, I apprehended, a knowledge of them might be useful to their successors. All the disputes which have no way contributed to the discovery of truth, I would gladly consign to eternal oblivion. Did it depend upon me, it should never be known to pofterity, that there had ever been any fuch thing as envy, jealoufy, or cavilling among the admirers of my favourite ftudy. I have, as far as my beft judgment could direct me, been juft to the merits of all perfons concerned. If any have made unjuft claims, by arrogating to themfelves the dif-coveries of others, I have filently reftored them to the right owner, and generally without fo much as giving a hint that any injuftice had ever been committed. If I have, in any cafe, given a hint, I hope it will be thought, by the offend-ing parties themfelves, to be a very gentle one; and that it will be a *memento*, which will not be without its ufe.

I THINK I have kept clear of any mean partiality towards my own countrymen, and even my own acquaintance. If Englifh authors are oftner quoted than foreign, it is becaufe they were more eafily procured : and I have found a difficulty I could not have expected, in procuring foreign publications upon this fubject.

I FIND it impoffible to write a preface to this work, with-out difcovering a little of the enthufiafm which I have con-tracted from an attention to it, by expreffing my wifhes, that more perfons of a ftudious and retired life would admit this part of experimental philofophy into their ftudies. They would find it agreeably to diverfify a courfe of ftudy, by mix-ing fomething of action with fpeculation, and giving fome employment to the hands and arms, as well as to the head. Electrical experiments are, of all others, the cleaneft, and the moft elegant, that the compafs of philofophy exhibits. They are performed with the leaft trouble, there is an amaz-ing variety in them, they furnifh the moft pleafing and fur-

prifing

prifing appearances for the entertainment of one's friends, and the expence of inftruments may well be fupplied, by a proportionable deduction from the purchafe of books, which are generally read and laid afide, without yielding half the entertainment.

THE inftruction we are able to get from books is, comparatively, foon exhaufted; but philofophical inftruments are an endlefs fund of knowledge. By philofophical inftruments, however, I do not here mean the globes, the orrery, and others, which are only the means that ingenious men have hit upon to explain their own conceptions of things to others; and which, therefore, like books, have no ufes more extenfive than the views of human ingenuity; but fuch as the air pump, condenfing engine, pyrometer, &c. (with which electrical machines are to be ranked) and which exhibit the operations of nature, that is of the God of nature himfelf, which are infinitely various. By the help of thefe machines, we are able to put an endlefs variety of things into an endlefs variety of fituations, while nature herfelf is the agent that fhows the refult. Hereby the laws of her action are obferved, and the moft important difcoveries may be made; fuch as thofe who firft contrived the inftrument could have no idea of.

IN electricity, in particular, there is the greateft room to make new difcoveries. It is a field but juft opened, and requires no great ftock of particular preparatory knowledge: fo that any perfon who is tolerably well verfed in experimental philofophy, may prefently be upon a level with the moft experienced electricians. Nay, this hiftory fhows, that feveral raw adventurers have made themfelves as confiderable, as fome who

have

have been, in other respects, the greatest philosophers. I need not tell my reader of how great weight this consideration is, to induce him to provide himself with an electrical apparatus. The pleasure arising from the most trifling discoveries of one's own far exceeds what we receive from understanding the much more important discoveries of others; and a mere reader has no chance of finding new truths, in comparison of him who now and then amuses himself with philosophical experiments.

Human happiness depends chiefly upon having some object to pursue, and upon the vigour with which our faculties are exerted in the pursuit. And, certainly, we must be much more interested in pursuits wholly our own, than when we are merely following the track of others. Besides, this pleasure has reinforcements from a variety of sources, which I shall not here undertake to trace; but which contribute to heighten the sensation, far beyond any thing else of this kind that can be experienced by a person of a speculative turn of mind.

It is a great recommendation of the study of electricity, that it now appears to be, by no means, a small object. The electric fluid is no local, or occasional agent in the theatre of the world. Late discoveries show that its presence and effects are every where, and that it acts a principal part in the grandest and most interesting scenes of nature. It is not, like magnetism, confined to one kind of bodies, but every thing we know is a conductor or non-conductor of electricity. These are properties as essential and important as any they are possessed of, and can hardly fail to show themselves wherever the bodies are concerned.

HITHERTO

HITHERTO philosophy has been chiefly converſant about the more ſenſible properties of bodies; electricity, together with chymiſtry, and the doctrine of light and colours, ſeems to be giving us an inlet into their internal ſtructure, on which all their ſenſible properties depend. By purſuing this new light, therefore, the bounds of natural ſcience may poſſibly be extended, beyond what we can now form an idea of. New worlds may open to our view, and the glory of the great Sir Iſaac Newton himſelf, and all his contemporaries, be eclipſed, by a new ſet of philoſophers, in quite a new field of ſpeculation. Could that great man reviſit the earth, and view the experiments of the preſent race of electricians, he would be no leſs amazed than Roger Bacon, or Sir Francis would have been at his. The electric ſhock itſelf, if it be conſidered attentively, will appear almoſt as ſupriſing, as any diſcovery that he made; and the man who could have made that diſcovery, by any reaſoning *a priori*, would have been reckoned a very great genius: but electrical diſcoveries have been made ſo much by accident, that it is more the powers of nature, than of human genius, that excite our wonder with reſpect to them. But if the ſimple electric ſhock would have appeared ſo extraordinary to Sir Iſaac Newton, what would he have ſaid upon ſeeing the effects of a modern electrical battery, and an apparatus for drawing lightning from the clouds! What inexpreſſible pleaſure would it give a modern electrician, were the thing poſſible, to entertain ſuch a man as Sir Iſaac for a few hours with his principal experiments!

To return from this excurſion to the buſineſs of a preface: Beſides relating the hiſtory of electrical diſcoveries, in the order

der in which they were made, I thought it neceſſary, in order to make the work more uſeful, eſpecially to young electricians, to ſubjoin a methodical treatiſe on the ſubject, containing the ſubſtance of the hiſtory in another form, with obſervations and inſtructions of my own.   The particular uſes
of theſe parts of the work are expreſſed at large in the intro-
ductions to them.   And, in the laſt place, I have given an
account of ſuch original experiments as I have been ſo fortunate as to hit upon myſelf.

I INTITLE the work the *hiſtory and preſent ſtate of electricity*;
and whether there be any new editions of the whole work or
not, care will be taken to preſerve the propriety of the title,
by occaſionally printing ADDITIONS, in the ſame ſize, as new
diſcoveries are made ; which will always be ſold at a reaſon-
able price to the purchaſers of the book ; or given *gratis*, if
the bulk be inconſiderable.

CONSIDERING what reſpectable perſons have already honoured this work with their valuable communications, I hope
it will not be deemed arrogance in me, if I here advertiſe,
that if any perſon ſhall make diſcoveries in electricity, which
he would chuſe to ſee recorded in this hiſtory, he will oblige
me by a communication of them ; and if they be really original, a proper place ſhall certainly be aſſigned to them in the
next edition, or paper of additions.   And I hope that, if
electricians in general would fall into this method, and make
either a periodical, or occaſional, but joint communication
of their diſcoveries to the public, the greateſt advantage would
thence accrue to the ſcience.

THE buſineſs of philoſophy is ſo multiplied, that all the
books

books of general philofophical tranfactions cannot be purchaf-
ed by private perfons, or read by any perfon. It is high time
to *fubdivide* the bufinefs, that every man may have an opportunity of feeing every thing that relates to his own favourite
purfuit: and all the various branches of philofophy would find
their account in this amicable feparation. Thus the numerous branches of a large overgrown family, in the patriarchal
ages, found it neceffary to feparate; and the convenience of
the whole, and the ftrength, and increafe of each branch were
promoted by the feparation. Let the youngeft daughter of
the fciences fet the example to the reft, and fhow that fhe
thinks herefelf confiderable enough to make her appearance in
the world without the company of her fifters.

But before this general feparation, let each collect together every thing that belongs to her, and march off with her
whole ftock. To drop the allufion: let hiftories be written
of all that has been done in every particular branch of fcience,
and let the whole be feen at one view. And when once the
entire progrefs, and prefent ftate of every fcience fhall be fully
and fairly exhibited, I doubt not but we fhall fee a new and
capital *æra* commence in the hiftory of all the fciences. Such
an eafy, full, and and comprehenfive view of what has been
done hitherto could not fail to give new life to philofophical
inquiries. It would fuggeft an infinity of new experiments,
and would undoubtedly greatly accelerate the progrefs of
knowledge; which is at prefent retarded, as it were, by its
own weight, and the mutual entanglement of its feveral parts.

I will juft throw out a further hint, of what, I think,
might be favourable to the increafe of philofophical knowledge.
At prefent there are, in different countries in Europe, large
incor-

incorporated focieties, with funds for promoting philofophical knowledge in general. Let philofophers now begin to fub-divide themfelves, and enter into fmaller combinations. Let the feveral companies make fmall funds, and appoint a direct-or of experiments. Let every member have a right to ap-point the trial of experiments in fome proportion to the fum he fubfcribes, and let a periodical account be publifhed of the refult of them all, fuccefsful or unfuccefsful. In this manner, the powers of all the members would be united, and increafed. Nothing would be left untried which could be compaffed at a moderate expence, and it being *one perfon's bufinefs* to attend to thefe experiments, they would be made, and reported without lofs of time. Moreover, as all incorporations in thefe fmaller focieties fhould be avoided, they would be encouraged only in proportion as they were found to be ufeful ; and fuc-cefs in fmaller things would excite them to attempt greater.

I BY no means difapprove of large, general and incorpo-rated focieties. They have their peculiar ufes too ; but we fee by experience, that they are apt to grow too large, and their forms are too flow for the difpatch of the *minutiæ* of bu-finefs, in the prefent multifarious ftate of philofophy. Let recourfe be had to rich incorporated focieties, to defray the expence of experiments, to which the funds of fmaller focie-ties fhall be unequal. Let their tranfactions contain a fum-mary of the more important difcoveries, collected from the fmaller periodical publications. Let them, by rewards, and other methods, encourage thofe who diftinguifh themfelves in the inferiour focieties ; and thus give a general attention to the whole bufinefs of philofophy.

I WISH

I wish all the incorporated philosophical societies in Europe would join their funds (and I wish they were sufficient for the purpose) to fit out ships for the complete discovery of the face of the earth, and for many capital experiments which can only be made in such extensive voyages.

Princes will never do this great business to any purpose. The spirit of adventure seems to be totally extinct in the present race of merchants. This discovery is a grand desideratum in science; and where may this pure and noble enthusiasm for such discoveries be expected but among philosophers, men uninfluenced by motives either of policy or gain? Let us think ourselves happy if princes give no obstruction to such designs. Let them fight for the countries when they are discovered, and let merchants scramble for the advantage that may be made of them. It will be an acquisition to philosophers if the seat of war be removed so far from the seat of science; and fresh room will be given to the exertion of genius in trade, when the old beaten track is deserted, when the old system of traffick is unhinged, and when new and more extensive plans of commerce take place. I congratulate the present race of philosophers on what is doing by the English court in this way; for with whatever view expeditions into the South seas are made, they cannot but be favourable to philosophy.

Natural Philosophy is a science which more especially requires the aid of wealth. Many others require nothing but what a man's own reflection may furnish him with. They who cultivate them find within themselves every thing they want. But experimental philosophy is not so independent.

c

Nature

Nature will not be put out of her way, and suffer her materials to be put into all that variety of situations which philosophy requires, in order to discover her wonderful powers, without trouble and expence. Hence the patronage of the great is essential to the flourishing state of this science. Others may project great improvements, but they only have the power of carrying them into execution.

BESIDES, they are the higher classes of men which are most interested in the extension of all kinds of natural knowledge; as they are most able to avail themselves of any discoveries, which lead to the felicity and embellishment of human life. Almost all the elegancies of life are the produce of those polite arts, which could have had no existence without natural science, and which receive daily improvements from the same source. From the great and the opulent, therefore, these sciences have a natural claim for protection; and it is evidently their interest not to suffer promising inquiries to be suspended, for want of the means of prosecuting them.

BUT other motives, besides this selfish one, may reasonably be supposed to attach persons in the higher ranks of life to the sciences; motives more exalted, and flowing from the most extensive benevolence. From Natural Philosophy have flowed all those great inventions, by means of which mankind in general are able to subsist with more ease, and in greater numbers upon the face of the earth. Hence arise the capital advantages of men above brutes, and of civilization above barbarity. And by these sciences also it is, that the views of the human mind itself are enlarged, and our common nature improved and enobled. It is for the honour

of

of the species, therefore, that these sciences should be culti-
vated with the utmost attention.

AND of whom may these enlarged views, comprehensive
of such great objects, be expected, but of those whom divine
providence has raised above the rest of mankind. Being free
from most of the cares peculiar to individuals, they may embrace the interests of the whole species, feel for the wants of
mankind, and be concerned to support the dignity of human
nature.

GLADLY would I indulge the hope, that we shall soon see
these motives operating in a more extensive manner than they
have hitherto done; that by the illustrious example of a few,
a taste for natural science will be excited in many, in whom
it will operate the most effectually to the advantage of science
and of the world; and that all kinds of philosophical inquiries
will, henceforward, be conducted with more spirit, and
with more success than ever.

WERE I to pursue this subject, it would carry me far
beyond the reasonable bounds of a preface. I shall therefore
conclude with mentioning that sentiment, which ought to be
uppermost in the mind of every philosopher, whatever be the
immediate object of his pursuit; that speculation is only of
use as it leads to *practice*, that the immediate use of natural
science is the power it gives us over nature, by means of the
knowledge we acquire of its laws; whereby human life is, in
its present state, made more comfortable and happy; but that
the greatest, and noblest use of philosophical speculation is
the discipline of the heart, and the opportunity it affords of
inculcating benevolent and pious sentiments upon the mind.

A PHI-

A PHILOSOPHER ought to be something greater, and better than another man. The contemplation of the works of God should give a sublimity to his virtue, should expand his benevolence, extinguish every thing mean, base, and selfish in his nature, give a dignity to all his sentiments, and teach him to aspire to the moral perfections of the great author of all things. What great and exalted beings would philosophers be, would they but let the objects about which they are conversant have their proper moral effect upon their minds! A life spent in the contemplation of the productions of divine power, wisdom, and goodness, would be a life of devotion. The more we see of the wonderful structure of the world, and of the laws of nature, the more clearly do we comprehend their admirable uses, to make all the percipient creation happy: a sentiment, which cannot but fill the heart with unbounded love, gratitude, and joy.

EVEN every thing painful and disagreeable in the world appears to a philosopher, upon a more attentive examination, to be excellently provided, as a remedy of some greater inconvenience, or a necessary means of a much greater happiness; so that, from this elevated point of view, he sees all temporary evils and inconveniences to vanish, in the glorious prospect of the greater good to which they are subservient. Hence he is able to venerate and rejoice in God, not only in the bright sun shine, but also in the darkest shades of nature, whereas vulgar minds are apt to be disconcerted with the appearance of evil.

NOR

NOR is the cultivation of piety useful to us only as *men*, it is even useful to us as *philosophers:* and as true philosophy tends to promote piety, so a generous and manly piety is, reciprocally, subservient to the purposes of philosophy; and this both in a direct, and an indirect manner. While we keep in view the great final cause of all the parts and the laws of nature, we have some clue, by which to trace the efficient cause. This is most of all obvious in that part of philosophy which respects the animal creation. As the great and excellent Dr. Hartley observes. " Since this world is a system of be-
" nevolence, and consequently its author the object of un-
" bounded love and adoration, benevolence and piety are our
" only true guides in our inquiries into it; the only keys
" which will unlock the mysteries of nature, and clues which
" lead through her labyrinths. Of this all branches of natural
" history, and natural philosophy afford abundant instances.
" In all these inquiries, let the inquirer take it for granted
" previously, that every thing is right, and the best that can
" be *ceteris manentibus*; that is, let him, with a pious con-
" fidence, seek for benevolent purposes, and he will be always
" directed to the right road; and after a due continuance in
" it, attain to some new and valuable truth: whereas every
" other principle and motive of examination, being foreign
" to the great plan on which the universe is constructed,
" must lead into endless mazes, errors, and perplexities. *

WITH respect to the indirect use of piety, it must be observed, that the tranquility, and chearfulness of mind, which

* Hartley's Observations on Man, Vol. 2. p. 245.

results

results from devotion forms an excellent temper for conducting philosophical inquiries; tending to make them both more pleasant, and more successful. The sentiments of religion and piety tend to cure the mind of envy, jealousy, conceit, and every other mean passion, which both disgrace the lovers of science, and retard the progress of it, by laying an undue bias upon the mind, and diverting it from the calm pursuit of truth,

LASTLY, let it be remembered, that a taste for science, pleasing, and even honourable as it is, is not one of the highest passions of our nature, and the pleasures it furnishes are even but one degree, above those of sense; and therefore that temperance is requisite in all scientifical pursuits. Besides the duties of every man's proper station in life, which ought ever to be held sacred and inviolate, the calls of piety, common friendship, and many other avocations ought generally to be heard before that of study. It is therefore only a small share of their leisure that most men can be justified in giving to the pursuit of science; though this share is more or less, in proportion to a man's situation in life, his natural abilities, and the opportunity he has for conducting his inquiries.

I SHALL conclude with another passage from Dr. Hartley to this purpose. " Though the pursuit of truth be an en-
" tertainment and employment suitable to our rational na-
" tures, and a duty to him who is the fountain of all know-
" ledge and truth, yet we must make frequent intervals and
" interruptions; else the study of science, without a view to
" God and our duty, and from a vain desire of applause,
" will get possession of our hearts, engross them wholly, and
" by

" by taking deeper root than the pursuit of vain amusements,
" become, in the end, a much more dangerous, and obsti-
" nate evil than that.   Nothing can easily exceed the vain-
" glory, self-conceit, arrogance, emulation, and envy, that
" are found in the eminent professors of the sciences, Ma-
" thematics, Natural Philosophy, and even Divinity itself.
" Temperance in these studies is, therefore, evidently re-
" quired, both in order to check the rise of such ill passions,
" and to give room for the cultivation of other essential parts
" of our natures.   It is with these pleasures as with the sen-
" sible ones ; our appetites must not be made the measure of
" our indulgence, but we ought to refer all to a higher rule.

" But when the pursuit of truth is directed by this higher
" rule, and entered upon with a view to the glory of God,
" and the good of mankind, there is no employment more
" worthy of our natures, or more conducive to their purifi-
' cation and perfection. *

* Hartley's Observations on Man, Vol. 2. p. 255, &c.

THE
CONTENTS.

PART I.
THE HISTORY OF ELECTRICITY.

PERIOD

d   SEC.

THE CONTENTS.   xxvii

PART II.

A SERIES of PROPOSITIONS, COMPRISING ALL THE GENERAL PROPERTIES OF ELECTRICITY.

PART III.

THEORIES of ELECTRICITY.

d 2   PART

VI

THE
HISTORY
AND PRESENT STATE OF
ELECTRICITY.

PART I.
THE HISTORY OF ELECTRICITY.

PERIOD I.
EXPERIMENTS AND DISCOVERIES IN ELECTRICITY PRIOR
TO THOSE OF MR. HAWKESBEE.

THE hiſtory of philoſophy contains nothing earlier than the obſervation, that yellow amber, when rubbed, has the power of attracting light bodies. Thales of Miletus, the father of the Ionic philoſophy, who flouriſhed about ſix hundred years before Chriſt, was ſo much ſtruck with this property of amber, that he imagined it was animated. But the firſt writer who expreſſly mentions this ſubſtance is Theophraſtus, who flouriſhed about the year 300 before Chriſt. He ſays, in his book concerning precious ſtones, ſect. 53, that amber (which he ſuppoſes to be a native

B   foſſil)

foſſil) has the ſame property of attracting light bodies with the lyncurium ; which, he ſays, attracts not only ſtraws, and ſmall pieces of ſticks, but even thin pieces of copper and iron. What he ſays farther of the lyncurium will be related under the article of the *tourmalin*, which Dr. Watſon has, in a manner, proved to be the ſame ſubſtance.

FROM ηλεκτρον, the Greek name for amber, is derived the term ELECTRICITY, which is now extended to ſignify, not only the power of attracting light bodies inherent in amber, but other powers connected with it, in whatever bodies they are ſuppoſed to reſide, or to whatever bodies they may be communicated.

THE attractive nature of amber is occaſionally mentioned by Pliny, and other later naturaliſts ; particularly by Gaſſen-dus, Kenelm Digby, and Sir Thomas Brown ; but excepting the electricity of the ſubſtance called *jet*, the diſcovery of which was very late (though I have not been able to find its author) no advances were made in electricity till the ſubject was undertaken by William Gilbert, a native of Colcheſter, and a phyſician at London ; who, in his excellent Latin trea-tiſe *de magnete*, relates a great variety of electrical experi-ments. Conſidering the time in which this author wrote, and how little was known of the ſubject before him, his diſ-coveries may be juſtly deemed conſiderable, though they appear trifling when compared with thoſe which have been made ſince his time.

To him we owe a great augmentation of the liſt of electri-cal bodies, as alſo of the bodies on which electrics can act ; and he has carefully noted ſeveral capital circumſtances re-
                                                                    lating

lating to the manner of their action, though his theory of electricity was very imperfect, as might be expected.

Amber and jet were, as I observed before, the only substances which, before his time, were known to have the property of attracting light bodies when rubbed; but he found the same property in the *diamond, sapphire, carbuncle, iris, amethyst, opal, vincentina, Briftol stone, beryl,* and *chryftal.* He also observes that *glass,* especially that which is clear and transparent, has the same property; likewise all *factitious gems,* made of glass or chryftal; *glass of antimony,* most *sparry substances,* and *belemnites.* Lastly, he concludes his catalogue of electric substances with *sulphur, maftic, sealing wax* made of gum lac tinged with various colours, *hard rosin, sal gem, talk,* and *roche alum.* Rosin, he said, possessed this property but in a small degree, and the three last mentioned substances only when the air was clear and free from moisture.

All these substances, he observes, attracted not only straws but all metals, all kinds of wood, stones, earth, water, oil, in short, whatever is solid, and the object of our senses But he imagined that air, flame, bodies ignited, and all matter which was extremely rare was not subject to this attraction. Grofs smoke, he found, was attracted very sensibly, but that which was attenuated very little.

Friction, he says, is, in general, necessary to excite the virtue of these substances; though he had one large and smooth piece of amber which would act without friction. But with respect to this he probably deceived himself. The most effectual friction, he observed to be that which was light and quick; and he found that electrical appearances

were

were ſtrongeſt when the air was dry, and the wind north or eaſt, at which time electric ſubſtances would act ten minutes after excitation. But he ſays, that a moiſt air, or a ſoutherly wind almoſt annihilates the electrical virtue. The ſame effect he alſo obſerved from the interpoſition of moiſture of any kind, as from the breath, and many other ſubſtances, but not always from the interpoſition of ſarſnet. He ſays that light and pure oil ſprinkled upon electrics, after excitation, did not obſtruct their virtue, but that brandy, or ſpirit of wine did. He alſo ſays, that chryſtal, talk, glaſs, and all other electrics loſt their virtue after being burned or roaſted. But this was, in ſome meaſure, a miſtake. The heat of the ſun, collected by a burning glaſs, he ſays, is ſo far from exciting amber, and other electrics, that it impairs the virtue of them all ; though, when electrics have been excited, they will retain their virtue longer in the ſun-ſhine than in the ſhade.

Moſt of the experiments of this author were made with long thin pieces of metal, and other ſubſtances, ſuſpended freely on their centers, to the extremities of which he preſented the electrics he had excited. His experiments on water were made by preſenting a round drop of water upon a dry ſubſtance to the excited electric ; and it is remarkable, that he obſerved the ſame conical figure of the electrified drops which Mr. Grey afterwards diſcovered, and which will be related more at large in its proper place. Gilbert concluded, that air was not affected by the electric attraction, becauſe the flame of a candle was not : for the flame, he ſays, would be diſturbed if the air had the leaſt motion given to it.

GILBERT

GILBERT imagined that electrical attraction was performed in the same manner as the attraction of cohesion. Two drops of water, he observed, rush together when they are brought into contact, and electrics, he says, are virtually brought into contact with the bodies they act upon, by means of their effluvia, excited by friction.

AMONG other differences between electric and magnetic attraction, some of which are very just, and others whimsical enough, he says, that magnetic bodies rush together mutually; whereas in electrical attraction it is only the electric that exerts any power. He observes also particularly, that in magnetism there is both attraction and repulsion, but in electricity only the former, and never the latter.*

SUCH were the discoveries of our countryman Gilbert, who may justly be called the father of modern electricity, though it be true that he left his child in its very infancy.

SIR FRANCIS BACON, in his Physiological Remains, gives a catalogue of bodies attractive and not attractive; but it differs in nothing worth mentioning from that of Gilbert, and he does not seem to have made any observations of his own relating to the subject.

THESE remarkable phenomena relating to amber, and other electric substances did not escape the attention of the inquisitive and sagacious Mr. Boyle, who flourished about the year 1670. He made some addition to the catalogue of electric substances, and attended to some circumstances relating to electric attraction which had escaped the observation of philosophers who lived before him.

---

* Gilbert de magnete, Lib. 2. Cap. 2.

HE

HE found that the hard cake which remains after evaporating good turpentine was electrical, as also the hard mass which remains after distilling petroleum and spirit of nitre, glass of lead, the caput mortuum of amber, and the cornelian but he could not find that property in the emerald, and he thought that glass possessed it but in a very low degree.

HE found, that the electricity of all bodies capable of having it excited in them was increased by wiping, and warming them, previous to their being rubbed. By this means he made an electric body no bigger than a pea, move a steel needle, which was freely poised, three minutes after he had left off rubbing it. He also found, that it was useful to have the surfaces of electric bodies made very smooth, except in the case of one diamond on which he tried some experiments; which, though it was rough, was, he says, possessed of a stronger electrical virtue than any polished one he had met with.

HE observed that excited electrics would attract all kinds of bodies promiscuously, whether electric or not; that excited amber, for instance, would attract both powder of amber, and small pieces of it; differing, as he takes notice, from the property of the load-stone, which acts only on one kind of matter. He found, that his electrics would attract smoke very easily, and takes some pains to account for their not sensibly attracting flame, which Gilbert excepted from the bodies attracted by electricity.

THESE attractions, he found, did not depend upon the air: for he observed that they took place in vacuo. He suspended a piece of excited amber over a light body in a glass receiver,

and

and faw, that when a vacuum was made, and the amber let down near the light body, it was attracted, as if it had been in the open air.*

Mr. Boyle made an experiment to try whether an excited electric was acted upon by other bodies as strongly as it acted upon them, and it succeeded: for, suspending his excited electric, he saw that it was sensibly moved by the approach of any other body. We should now be surprised that any person should not have concluded *a priori*, that if an electric body attracted other bodies, it must in return be attracted by them, action and reaction being universally equal to one another. But it must be considered, that this axiom was not so well understood in Mr. Boyle's time, nor until it was afterwards explained in its full latitude by Sir Isaac Newton.

These few experiments of Mr. Boyle's, we see, relate only to a few circumstances attending the simple property of electric attraction. The nearest approach that he made to the discovery of electrical repulsion was his observing, that light bodies, as feathers, &c. would cling to his fingers, and other substances, after they had been attracted by his electrics. He had never seen the electric light, and little imagined what astonishing effects would be afterwards produced by the same wonderful power, and how large a field he was opening for philosophical speculation in future time.

Mr. Boyle's theory of electrical attraction was, that the electric emitted a glutinous effluvium, which laid hold of small bodies in its way, and, in its return to the body which

* Histoire de l'electricité, p. 6.   Boyle's Mechanical production of electricity.

emitted

emitted it, carried them back with it. One James Hartman, whose account of amber is published in the Philosophical Transactions, * pretends to prove by experiment, that electric attraction was really owing to the emission of glutinous particles. He took two electric substances, viz. pieces of colophonia, and from one of them made a distillation of a black balsam, and thereby deprived it of its attractive power. He says, that the electric which was not distilled retained its fatty substance, whereas the other was, by distillation, reduced to a mere *caput mortuum*, and retained no degree of its bituminous fat. In consequence of this hypothesis, he gives it as his opinion, that amber attracts light bodies more powerfully than other substances, because it emits oily and tenacious effluvia more copiously than they do.

Contemporary with Mr. Boyle was Otto Guericke, Burgomaster of Magdebourgh, and the celebrated inventor of the air-pump, who is likewise intitled to a distinguished place among the first improvers of electricity.

This philosopher made his experiments with a globe of sulphur, made by melting that substance in a hollow globe of glass, and afterwards breaking the glass from off it. He little imagined that the glass globe itself, with or without the sulphur, would have answered his purpose as well. This globe of sulphur he mounted upon an axis, and whirled it in a wooden frame, rubbing it at the same time with his hand; and by this means he performed all the electrical experiments which were known before his time.

His was the discovery, that a body once attracted by an excited electric was repelled by it, and not attracted again

* Abridgment, Vol. 2. p. 473.

till

till it had been touched by some other body.   In this manner he kept a feather a long time suspended in the air above his sulphur globe; but he observed, that if he drove it near a linen thread, or the flame of a candle, it instantly retreated to the globe, without having been in contact with any sensible body.

NEITHER the sound, nor the light produced by the excitation of his globe escaped the notice of this accurate philosopher, though he seems not to have observed them in a very great degree: for he was obliged to hold his ear near the globe to perceive the hissing sound of the electric fire; and he compares the light which it gave in the same circumstances to that which is seen when sugar is pounded in the dark.

BUT the most remarkable experiments of this philosopher were two, which depend upon a property of the electric fluid that has not been illustrated till within these late years; viz. that bodies immerged in electric atmospheres are themselves electrified, and with an electricity opposite to that of the atmosphere.   Threads suspended within a small distance of his excited globe, he observed to be often repelled by his finger brought near them, and that a feather repelled by the globe always turned the same face towards it like the moon with respect to the earth.   This last experiment seems to have been wholly overlooked by later electricians, though it is a very curious one, and may be made with so much ease. *

A MUCH finer appearance of electric light than that which Otto Guericke's sulphur globe exhibited was observed by

* Experimenta Magdeburgica, Lib. 4. Cap. 15.

C   Dr.

Dr. Wall. The account of it is publifhed in the philofo-phical tranfactions. *

Making experiments upon artificial phofphorus, which he took to be an animal oil coagulated with a mineral acid, he was led to conjecture that amber, which he fuppofed to be a mineral oil coagulated with a mineral volatile acid, might be a natural phofphorus; and with this view he began to make experiments upon it, the refult of which, being very curious and furprifing, it will be moft agreeable to my readers to fee in the very words of the obferver himfelf.

" I found," fays he, " by gently rubbing a well polifhed
" piece of amber with my hand in the dark, that it pro-
" duced a light: whereupon I got a pretty large piece of
" amber, which I caufed to be made long and taper, and
" drawing it gently through my hand, being very dry, it
" afforded a confiderable light.

" I then ufed many kinds of foft animal fubftances, and
" found that none did fo well as wool. And now new phe-
" nomena offered themfelves: for, upon drawing the piece
" of amber fwiftly through the wollen cloth, and fqueezing
" it pretty hard with my hand, a prodigious number of little
" cracklings were heard, and every one of thefe produced a
" little flafh of light; but when the amber was drawn gently
" and flightly through the cloth, it produced only a light
" but no crackling; but by holding ones finger at a little dif-
" tance from the amber, a large crackling is produced, with
" a great flafh of light fucceding it. And, what to me is

* Abridgment, Vol. 2. p. 275.

" very

" very furprifing, upon its eruption, it ftrikes the finger very
" fenfibly, wherefoever applied, with a pufh or puff, like
" wind. The crackling is full as loud as charcoal on fire,
" and five or fix cracklings, or more, according to the
" quicknefs of placing the finger, have been produced from
" one fingle friction, light always fucceeding each of them.

" Now I make no queftion, but upon ufing a longer and
" larger piece of amber, both the cracklings and light
" would be much greater, becaufe I never yet found any
" crackling from the head of my cane, though it is a pretty
" large one. This light and crackling feems in fome degree
" to reprefent thunder and lightening."

After reciting this experiment, he gives it as his opinion,
that all, or moft bodies which have an electricity give light,
and that it is the light which is the caufe of their being
electrical. He found that light could alfo be produced by rubbing jet, red fealing wax, made with gum lac and cinnabar,
and the diamond. He even imagined he could diftinguifh
true from falfe diamonds by this teft.

Notwithstanding Dr. Wall made this beautiful difcovery, as he imagined (for he feems not to have feen what
Otto Guericke had written) of light proceeding from amber
and other electric bodies, we fee that he laboured under a
great deal of confufion and mifapprehenfion with refpect to it.
He fays, that one thing appeared ftrange to him in the courfe
of thefe experiments, which was, that though, upon friction
with wooll in the day time, the cracklings feemed to be full
as many and as large, yet that by all the trials which he made,
very little light appeared, though in the darkeft room. He

fays

says that the best time for making these experiments is when the sun is 18° below the horizon, and that when the sun was so low, though the moon shone ever so bright, the light was the same as in the darkest room, which made him chuse to call it a noctiluca.

IT is remarkable that Dr. Wall compares the light and the crackling of his amber to thunder and lightning : so early was a similarity between the effects of electricity and lightning observed. But little was it imagined that the resemblance between them extended any farther than to appearances and effects. That the cause was the same in both was a discovery reserved for Dr. Franklin in a much later period.

THE great Sir Isaac Newton, though he by no means makes a principal figure in the history of electricity, yet made some electrical observations which engaged the attention of his philosophical friends, and which, independent of their being made by him, do well deserve to be transmitted to posterity. They seem to shew that he was the first who observed, that excited glass attracted light bodies on the side opposite to that on which it was rubbed.

HAVING laid upon the table a round piece of glass, about two inches broad, in a brass ring, so that the glass might be one eight of an inch from the table, and there rubbing the glass briskly, little fragments of paper laid on the table, under the glass, began to be attracted, and move nimbly to and fro. After he had done rubbing the glass, the papers would continue a considerable time in various motions ; sometimes leaping up to the glass, and resting there a while, then leaping down and resting there, and then leaping up and

down

down again; and this fometimes in lines feemingly perpendicular to the table, fometimes in oblique ones; fometimes alfo leaping up in one arch, and leaping down in another, divers times together, without fenfibly refting between; fometimes fkipping in a bow from one part of the glafs to another, without touching the table, and fometimes hanging by a corner, and turning often about very nimbly, as if they had been carried about in the midft of a whirlwind, and being otherwife varioufly moved, every paper with a different motion. Upon his fliding his finger on the upper fide of the glafs, though neither the glafs nor the inclofed air below were moved, yet he obferved, that the papers, as they hung under the glafs, would receive fome new motion, inclining this way or that, according as he moved his finger.

Some of the motions, as that of hanging by a corner and twirling about, and that of leaping from one part of the glafs to another without touching the table happened but feldom; but, he fays, it made him take the more notice of them.*

An account of this experiment Sir Ifaac fent to the members of the R. Society in the year 1675, defiring it might be tried by them, and after fome ineffectual attempts, and receiving further inftructions, how to make it, they at length fucceeded, and the thanks of the fociety were formally returned to him.

Upon repeating the experiment with fome variety of circumftances, Sir Ifaac obferves, that rubbing varioufly, or with various things, altered the cafe. At one time he rubbed a glafs four inches broad, and one fourth thick with a

* Birch's Hift. of the R. Society, Vol. 3, p. 260, &c.   † Ib. 271.

napkin

napkin twice as much as he used to do with his gown, and nothing would stir, and yet presently after, rubbing it with something else, the motion soon began. After the glass had been much rubbed, he thought the motions were not so lasting, and the day following he found the motions fainter, and more difficult to be excited than at first. *

SIR ISAAC also mentions electricity in two queries annexed to his treatise on Optics, from which we learn, that he imagined electric bodies when excited emitted an elastic fluid, which freely penetrated glass, and that the emission was performed by the vibratory motions of the parts of the excited bodies. †

* Birch's Hist. of the R. Society, Vol. 3. p. 270
† Newton's Optics, octavo p. 314, and 327.

PERIOD

PERIOD II.

THE EXPERIMENTS AND DISCOVERIES OF MR. HAWKESBEE.

AFTER Gilbert, Mr. Boyle, and Otto Guericke, Mr. Hawkesbee, who wrote in the year 1709, distinguished himself by experiments and discoveries in electricity. He first observed the great electric power of glass, the light proceeding from it, and the noise occasioned by it, together with a variety of phenomena relating to electric attraction and repulsion. He was indefatigable in making experiments, and there are few persons to whom we are more indebted for a real advancement of this branch of knowledge. This will appear by the following succinct account of his experiments, related not exactly in the order in which he has published them, but according to their connection. This method I have chosen as best adapted to give the most distinct view of the whole.

I SHALL first relate the experiments he made concerning electric attraction and repulsion, in many of which we shall see reason to admire his ingenious contrivances, and shall see that little was added to his observations, till the capital discovery of a plus and minus electricity by Dr. Watson and Dr. Franklin, and the farther illustration of that doctrine by Mr. Canton.

THE

THE moſt curious of his experiments concerning electric attraction and repulſion are thoſe which ſhew the direction in which thoſe powers are excited.

HAVING tied threads round a wire hoop, and brought it near to an excited globe or cylinder, he obſerved, that the threads kept a conſtant direction towards the center of the globe, or towards ſome point in the axis of the cylinder, in every poſition of the hoop; that this effect would continue for about four minutes after the whirling of the globe ceaſed, and that the effect was the ſame whether the wire was held above, or under the glaſs; or whether the glaſs was placed with its axis parallel, or perpendicular to the horizon.

HE obſerved, that the threads pointing towards the center of the globe were attracted and repelled by a finger preſented to them; that if the finger, or any other body, was brought very near the threads, they would be attracted; but that if it were brought to the diſtance of about an inch, they would be repelled, the reaſon of which difference he did not ſeem to underſtand. *

HE tied threads to the axis of a globe and cylinder, and found that they diverged every way in ſtraight lines from the place where they were tied, when the globe was whirled and rubbed. In both caſes, he ſays, the threads would be repelled by the finger held on the oppoſite ſide of the glaſs, even without touching the glaſs; though they would ſometimes ſuddenly jump towards it. † He obſerved further, that by blowing with his mouth towards the the glaſs, at

* Phyſico-Mechanical Experiments, p. 75.   † Ib. p. 78.

three

three or four feet diftance, the threads would have a very confiderable motion given to them.

He found that threads, hanging freely in an unexcited globe, at reft, would be moved by the approach of any excited electric at a confiderable diftance, except in moift weather; which failure he accounts for, by fuppofing the moifture on the furface of the glafs prevented the free paffage of the electric effluvia through it. *

The varieties he obferved on the appearances and properties of the electric light, are even more curious and furprifing than his difcoveries concerning electric attraction and repulfion. It is fomething remarkable that Mr. Hawkesbee's tranfition to electric light was like that of Dr. Wall, viz. from the light of phofphorus.

Mr. Hawkesbee firft produced a confiderable quantity of light by fhaking quick-filver in a glafs veffel, out of which the air was exhaufted. Sometimes what he calls ftrange flafhes of pale light were feen darting in a variety of directions, when the mercury was put in motion within an exhaufted receiver. † But the difcovery was probably accidental, and he did not feem, at that time, to know the reafon of the appearance. He called this light the *mercurial phofphorus*, and did not confider the glafs as any way concerned in producing it.

He alfo found, that this appearance of electric light (which he ftill calls the mercurial phofphorus) did not require a very perfect vacuum, nor even a near approach to it. ‡ Nay he

* Phyfico-Mechanical Experiments, p. 160.   † Ib. 12.   ‡ Ib. 14.

fometimes

fometimes produced that appearance of light by fhaking mercury in a veffel in which the air was of the fame denfity with the external atmofphere, but ftill he had no idea of the glafs contributing to the phenomenon. *

HE obferved a ftrong light in vacuo, and a fmall one in the open air, from rubbing amber upon woollen, but feems to have confidered it as any hard body rubbing againft a foft one. † He alfo obferved a vivid purple, and afterwards a pale light produced, by rubbing glafs upon woollen in vacuo. ‡ He fays that every frefh glafs firft gave a purple, and then a pale light, and that woollen tinctured with falts or fpirits produced a new, ftrong, and fulgurating light. ‖

IN the following experiments we find his ideas of electric light much more diftinct, and the appearances the fame that are ufually exhibited by our prefent electrical machines, the ftructure of which, we fhall find to be nearly the fame with thofe which he ufed.

HE provided himfelf with a machine in which he could whirl a glafs globe; and he obferved, when the air was exhaufted out of it, that, upon applying his hand to the globe, a ftrong light appeared on the infide, and, upon letting in the air, he obferved the light on the outfide alfo; but with fome very confiderable differences in its appearances, fticking upon his fingers, and other bodies held near the globe. He alfo obferved, upon this occafion, that one fourth of an atmofphere in the globe did very little diminifh the light within. It is pleafing to obferve, that the fimilar appearance in this

* Phyfico-Mechanical Experiments, p. 18.   † Ib. 26.   ‡ Ib. 32.   ‖ Ib. 34.

experiment

experiment, and that with mercury in vacuo before-mentioned, made him fufpect, though only fufpect, that the light produced in the former cafe proceeded not from the mercury but from the glafs.

THE next experiment is of a very delicate and curious nature. It is not to be wondered at, that Mr. Hawkefbee did not underftand the circumftances which contributed to it, as the explication of it depends upon principles which were not difcovered till a much later period by Mr. Canton.

HOLDING an exhaufted globe within the effluvia of an excited one, he obferved a light in the exhaufted globe, which prefently died away if it was kept at reft, but which was revived, and continued very ftrong, if the exhaufted globe was kept in motion. Prefenting an exhaufted tube to the effluvia of an excited globe, it produced what he calls an interrupted flafhing light. He imagined that the exhaufted globe was excited by the attraction of the effluvia from the other globe, fo little did he underftand the true caufe of this curious experiment. * When he fays that light is producible by the effluvia of one glafs falling upon another, he adds; that electric, (by which he means attractive) matter, is not to be brought forth by any fuch feeble ftrokes. He had before obferved that, upon rubbing an exhaufted tube, it difcovered no attractive power, nor gave any light outwards, but only inwards.

HE found that when the friction was performed in vacuo no electricity (that is attraction) could be produced;† but that though the *attractive quality* required the prefence both of the exter-

* Phyfico-Mechanical Experiments, p. 82.   † Ib. 242.

nal

nal and internal air, in order to its fhewing itfelf, yet the *light* requires the prefence of only one of them in order to its appearance; for that either a glafs globe full of air rubbed *in vacuo*, or with its air exhaufted and rubbed *in pleno*, would, produce a very confiderable light. *

HE fays alfo, that thofe lights are lefs fenfibly affected by the return of air which are produced by the attrition of exhaufted glafs in pleno, than thofe which are produced by the attrition of glafs full of air in vacuo; for that, in the former cafe, no great alteration was found in the light or colour, until a certain quantity of air was let in to the infide of the exhaufted glafs; but that, in the latter cafe, both light and colour were fenfibly changed, upon every admiffion of air to the outfide of the full glafs. †

THE greateft electric light Mr. Hawkefbee produced, was when he enclofed one exhaufted cylinder within another not exhaufted, and excited the outermoft of them, putting them both in motion. Whether their motions confpired or not, he obferved, made no difference. When the outer cylinder only was in motion, he fays, the light was very confiderable, and fpread itfelf over the furface of the inner glafs. What furprifed him moft was, that after both glaffes had been in motion fome time, during which the hand had been applied to the furface of the outer glafs, the motion of both ceafing, and no light at all appearing, if he did but bring his hand again near the furface of the outer glafs, there would be flafhes of light, like lightning, produced in the inner glafs; as if, he

* Phyfico-Mechanical Experiments, p. 248.   † Ib. 249.

says,

says, the effluvia from the outer glass had been pushed with more force upon it by means of the approaching hand. * This experiment was similar to those which he made with the excited and exhausted globe, and with the exhausted tube; and his reasoning upon it shews, that he was still far from being fully apprized of all the circumstances attending this fact.

THE next experiments which I shall relate of Mr. Hawkesbee's are those which shew the great copiousness, and extreme subtilty of electric light. They are really amazing, and have not yet been pursued in the manner they deserve.

HE lined more than half of the inside of a glass globe with sealing wax, and having exhausted the globe, he put it in motion; when, applying his hand to excite it, he saw the shape and figure of all the parts of his hand distinctly and perfectly on the concave superficies of the wax within. It was as if there had only been pure glass, and no wax interposed between his eye and his hand. The lining of wax, where it was spread the thinest, would but just allow the sight of a candle through it in the dark; but in some places the wax was, at least, one eighth of an inch thick; yet even in those places the light and the figure of his hand were as distinguishable through it as any where else. Nay though, in some places, the sealing wax did not adhere so closely to the glass as in others, yet the light on these appeared just as on the rest. †

* Physico-Mechanical Experiments, p. 87.   † Ib. 168.

THESE

THESE experiments succeeded equally with pitch instead of sealing wax. And he observed, that when the air was let into the glass, every part of it, the lined part and the unlined, seemed to attract with equal vigour. * Melted flowers of sulphur had no such effect, but common sulphur answered as well as sealing wax, or pitch. In both these last experiments the sulphur was found to have been separated from the glass. †

USING a large quantity of common sulphur in the same manner, the light in the inside was four times as great, but the figure of his fingers was not so distinguishable as in the former cases. He likewise observed, that on the part near the axis, where the substance of the sulphur was the greatest, no light was produced ; which he attributed, chiefly, to the slowness of the motion in that place. ‡

UPON the admission of a small quantity of air into the globe, thus partially lined with sealing wax, the light wholly disappeared on the part covered with the wax, but not on the other.

HE also observed, that when all the air was let in, and the hoop of the threads before-mentioned held over the glass, the threads were attracted at greater distances by the part which was coated with the wax than by the other : when all the air was exhausted, he says, the wax would attract bodies placed near the out-side of the glass ; that even in this case, the threads preserved their central direction, though not so vigo-rously as when the air was let in ; but that they would not be

* Physico-Mechanical Experiments, p. 269. † Ib. 274. ‡ Ib. 275

attracted

attracted at all when there was no wax on the infide of the exhaufted globe.

Mr. Hawkesbee was not unattentive to the found made by the emiffion of the electric effluvia, or to the manner in which it effected the fenfe of feeling. He obferved, that when an excited tube of glafs attracted various bodies, and threw light upon them as they were held near it, a noife, which he calls a *fnapping*, was likewife heard, He alfo fays, that the rubbed tube, held near the face, gave a feeling as if fine hairs had been drawn over it ; and when he repeated the experiment of whirling and rubbing the glafs globe, he obferved the light to proceed from it with fome noife, and to make a kind of preffure upon the finger, when it was held within half an inch of it. *

Nor was Mr. Hawkefbee's attention confined to the electric power of *glafs*. He made experiments with a globe of *fealing wax*, in the center of which was a globe of wood; from which he concluded, that the electricity of fealing wax was, in general, the fame with that of glafs, but different from it in degree. He could not make any light adhere to his finger when prefented to the excited fealing wax, any more than when it was prefented to an exhaufted and excited globe of glafs.

He provided himfelf, in like manner, with a globe of fulphur, and another of rofin with a mixture of brick duft, but the fulphur could hardly be excited at all; whereas the rofin acted more powerfully than the fealing wax had done. This

* Physico-Mechanical Experiments, p. 65.

he

he afcribed to its being ufed while it was warm : for, in the fame warm ftate, it attracted leaf brafs without any attrition at all. *

He fays, that the excited rofin gave no light in the dark, and the fulphur but little. †

WITH refpect to the power of electricity in general, he obferved, that a flight friction was fufficient to excite it, and that a greater preffure, or a more violent motion did not confiderably increafe it. ‡ He fays, that all the phenomena of electricity were improved by warmth; and diminifhed by moifture; which he attributed to the refiftance that the aqueous particles gave to the effluvia; and, like Mr. Boyle, and others before him, he was confirmed in this hypothefis, by finding, that the interpofition of linen cloth prevented any effects from being obferved beyond it.

HE alfo obferved, that when the tube was filled with other matter than air, as with dry writing fand (which he actually tried) the attractive power of the effluvia was confiderably abated, but he did not know what kind of bodies would produce that effect. He himfelf obferves, that he found the electric virtue of a folid cylinder of glafs to be, not indeed quite fo ftrong, as that of a hollow tube, but more permanent. ||

THAT Mr. Hawkefbee, after all, had no clear idea of the diftinction of bodies into electrics and non electrics, appears from fome of his laft experiments, in which he attempted to produce electric appearances from metals, and from the reafons he gives for his want of fuccefs in thofe attempts. " From

* Phyfico-Mechanical Experiments, p. 154. † Ib. 156. ‡ Ib. 52. || Ib. 64.

" thefe

" thefe experiments," he fays, " I may fafely conclude, that
" if there be any fuch quality as light to be excited from a
" brafs body, under the forementioned circumftances," viz.
of whirling and rubbing, " all the attrition of the feveral bo-
" dies I have ufed for that purpofe, have been too weak to
" force it from it.   And indeed, confidering the clofenefs of
" the parts of metals, and with what firmnefs they adhere,
" entangle, and attract one another, a fmall degree of attri-
" tion is not fufficient to put their parts into fuch a motion
" as to produce an electrical quality, which quality, un-
" der the forementioned circumftances, I take to be the ap-
" pearance of light in fuch a medium.

CONSIDERING what great fuccefs Mr. Hawkefbee had with
his globe of glafs, and his machine to give motion to it, it is
furprifing that the ufe of it fhould be fo long difcontinued
after his death.   To this circumftance we may perhaps, in a
great meafure, afcribe the flow progrefs that was afterwards
made in electrical difcoveries.   Mr. Hawkefbee's fucceffors
confined themfelves to the ufe of tubes.   I fuppofe becaufe
they were lighter, more portable, and more eafily managed in the experiments which they chiefly attended to : but
the ufe of the globe would certainly have put them much
fooner in the way of making the capital difcoveries, which
were afterwards made in electricity.

E   PERIOD

PERIOD III.

THE EXPERIMENTS AND DISCOVERIES OF MR. STEPHEN GREY, WHICH WERE MADE PRIOR TO THOSE OF MONSIEUR DU FAYE, AND WHICH BRING THE HISTORY OF ELECTRICITY TO THE YEAR 1733.

NOTWITHSTANDING the important discoveries of Mr. Hawkesbee, and the promising appearance they made, as an opening to further discoveries, we find, after him, a great chasm in the history of electricity, an interruption of discoveries, and, as far as we can learn, of experiments too, for the space of near twenty years; and at a time when philosophical knowledge of every other kind was making the most rapid progress under the auspices of the great Sir Isaac Newton. But the attention of this great man happened to be engaged by other subjects, and this very circumstance might be the reason why the attention of other philosophers was also diverted from electricity.

AFTER this long interval, commences a new æra in the history of electricity; in which we shall have the works of another labourer in this new field of philosophy to contemplate, viz. Mr. Stephen Grey, a pensioner at the Charter House. No person who ever applied to this study was more assiduous in making experiments, or had his heart more entirely in the work.

work. This will appear by the prodigious number of experiments he made, and some considerable discoveries with which his perseverance was crowned; as well as by the self deceptions, which his passionate fondness for new discoveries exposed him to.

BEFORE the year 1728, Mr. Stephen Grey had often observed, in electrical experiments made with a glass tube, and a down feather tied to the end of a small stick, that, after its fibres had been drawn toward the tube, they would, upon the tube's being withdrawn, cling to the stick, as if it had been an electric body, or as if there had been some electricity communicated to the stick, or to the feather. This put him upon thinking, whether, if the feather were drawn through his fingers, it might not produce the same effect, by acquiring some degree of electricity. This experiment succeeded accordingly, upon his first trial; the small downy fibres of the feather being attracted by his finger, when held near it; and sometimes the upper part of the feather with its stem would be attracted also.

IT will be obvious to every electrician, that the success of this experiment depended upon other principles than those to which he had a view in making it. Proceeding, however, in the same manner, he found the following substances to be all electric; *hair, silk, linen, woollen, paper, leather, wood, parchment*, and *ox gut* in which leaf gold had been beaten. He made all these substances very warm, and some of them quite hot before he rubbed them. He found light emitted in the dark by the silk and the linen, but more especially by a piece of white pressing paper, which is of the same nature with

E 2   card

card paper. Not only did this fubftance, when made as hot as his fingers could bear, yield a light; but, when his fingers were held near it, a light iffued from them alfo attended with a crackling noife, like that produced by a glafs tube, though not at fo great a diftance from the fingers. *

THE preceeding experiments bring us to the eve of a very confiderable difcovery in electricity, viz. the communication of that power from native electrics to bodies, in which it is not capable of being excited; and alfo to a more accurate diftinction of electrics from non electrics. I fhall relate the manner in which thefe important difcoveries were made pretty fully, but, at the fame time, as fuccinctly as poffible.

IN the month of February 1729, Mr. Grey, after fome fruitlefs attempts to make metals attractive, by heating, rub_ bing, and hammering, recollected a fufpicion which he had fome years entertained; that, as a tube communicated its light to various bodies when it was rubbed in the dark, it might poffibly, at the fame time, communicate an electricity to them, by which had hitherto been underftood only the power of attracting light bodies. For this purpofe he pro-vided himfelf with a tube three feet five inches long, and near one inch and two tenths in diameter; and to each end was fitted a cork, to keep the duft out when the tube was not in ufe.

THE firft experiments he made upon this occafion were in-tended to try, if he could find any difference in its attraction when the tube was ftopped at both ends by the corks, and when

* Philofophical Tranfactions abridged, Vol. 8. p. 9.

left

left entirely open; but he could perceive no sensible difference. It was, however, in the course of this experiment
that, holding a down feather over against the upper end of
the tube, he found that it would fly to the cork, being attracted and repelled by it, as well as by the tube itself.
He then held the feather over against the flat end of the cork,
and observed, that it was attracted and repelled many times
together; at which, he says, he was much surprised, and concluded, that there was certainly an attractive virtue communicated to the cork by the excited tube.

He then fixed an ivory ball upon a stick of fir, about four
inches long; when, thrusting the other end into the cork, he
found, that the ball attracted and repelled the feather, even
with more vigour than the cork had done, repeating its attractions and repulsions many times successively. He afterwards fixed the ball upon long sticks, and upon pieces of brass
and iron wire, with the same success; but he observed, that
the feather was never so strongly attracted by the wire, though
it were held very near the tube, as by the ball at the end of
it.

When a wire of any considerable length was used, its vibrations, caused by the action of rubbing the tube, made it
troublesome to manage. This put Mr. Grey upon thinking whether, if the ball were hung to a packthread, and
suspended by a loop on the tube, the electricity would not be
carried down the line to the ball; and he found it to succeed
according to his expectation. In this manner he suspended
various bodies to his tube, and found all of them to be capable of receiving electricity in the same manner.

After

AFTER trying these experiments with the longest light canes and reeds that he could conveniently use, he ascended a balcony twenty six feet high; and, fastening a string to his tube, he found, that the ball at the end of it would attract light bodies in the court below.

HE then ascended to greater heights, and by putting his long canes in the end of his tube, and fastening a long string to the end of the canes, he contrived to convey the electricity to much greater distances than he had done before; till, being able to carry it no further perpendicularly, he next attempted to carry it horizontally; and from these attempts arose a discovery, of which he was not in the least aware when he began them.

IN his first trial he made a loop at each end of a packthread, by means of which he suspended it, at one end, on a nail driven into a beam, the other end hanging downwards. Through the loop which hung down, he put the line to which his ivory ball was fastened, fixing the other end of it by a loop on his tube; so that one part of the line, along which the electricity was to be conveyed, viz. that to which the ball was fastened, hung perpendicular, the rest of the line lay horizontal. After this preparation, he put the leaf brass under the ivory ball, and rubbed the tube, but not the least sign of attraction was perceived. Upon this he concluded, that when the electric virtue came to the loop of the packthread, which was suspended on the beam, it went up the same to the beam; so that none, or very little, of it went down to the ball; and he could not, at that time, think of any method to prevent it.

ON

On June the 30th 1729, Mr. Grey paid a vifit to Mr. Wheeler, to give him a fpecimen of his experiments; when, after having made them from the greateft heights which the houfe would admit, Mr. Wheeler was defirous of trying whether they could not carry the electric virtue to a greater diftance horizontally. Mr. Grey then told him of the fruitlefs attempt he had made to convey it in that direction: upon which Mr. Wheeler propofed to fufpend the line to be electrified by another of filk, inftead of packthread; and Mr. Grey told him, it might do better, on account of its fmallnefs; as lefs of the virtue would probably pafs off by it than had done by the thick hempen line, which he had ufed before. With this expedient, they fucceeded far beyond their expectations.

The firft experiment they made after this expedient occurred to them, was in a matted gallery at Mr. Wheeler's houfe July 2d. 1729, about ten o'clock in the morning, as Mr. Grey, after his ufual manner, has minutely recorded it. About four feet from the end of the gallery, they faftened a line acrofs the place. The middle part of this line was filk, the reft packthread. They then laid the line to which the ivory ball was hung, by which the electric virtue was to be conveyed to it from the tube, and which was eighty feet and a half in length acrofs this filk line, fo that the ball hung about nine feet below it. The other end of the line was, by a loop, faftened to the tube, which they excited at the other end of the room. After this preparation, they put the leaf brafs under the ivory ball, and, upon rubbing the tube, it was attracted, and kept fufpended for fome time.

The

THE gallery not permitting them to go any greater lengths with a fingle line of communication, they contrived to return the line, making the whole length of it almoft twice that of the gallery, or about one hundred and forty feven feet, which anfwered very well. But, fufpecting that the attraction would be ftronger without doubling or returning the line, they made ufe of a line one hundred and twenty four feet long, running in one direction in the barn; and, as they expected, they found the attraction ftronger than when the line had been returned in the gallery.

JULY the 3d. proceeding to make more returns of the line, the filk which fupported it happened to break, not being a-ble to bear the weight of it, when fhaken with the motion that was given to it by rubbing the tube. Upon this they endeavoured to fupport it by a fmall iron wire, inftead of the filk ftring: but this alfo breaking, they made ufe of a brafs wire a little thicker. But this brafs wire, though it fupported the line of communication very well, did not anfwer the purpofe of thefe young electrians: for, upon rubbing the tube, no electricity was perceived at the end of the line. IT had all gone off by the brafs wire which fupported it. They had had recourfe to brafs wires, as being ftronger than their filk lines, and no thicker; for the fame reafon that they had before ufed filk lines in preference to hempen ftrings; becaufe they could have them ftronger, and at the fame time fmaller. But the refult of this experiment convinced them, that the fuccefs of it depended upon their fupporting lines being *filk*, and not, as they had imagined, upon their being *fmall*. For the electric virtue went off as effectually by the fmall brafs wire, as it had done by the thick hempen cord.

BEING

Being obliged, therefore, to return to their silk lines, they contrived them to support very great lengths of the hempen line of communication; and actually conveyed the electric virtue seven hundred and sixty five feet, nor did they perceive that the effect was sensibly diminished by the distance. *

In the same manner in which *silk* was found to be a non-conductor, it is probable that, about the same time, *hair, rosin, glass,* and perhaps some other electric substances were found to have the same property, though the discovery be no where particularly mentioned: for we shall presently find Mr. Grey making use of them, to insulate the bodies that he electrified.

After this, they amused themselves with trying how large surfaces might be impregnated with the electric effluvia; electrifying a large map, table cloth, &c. They also carried the electric virtue several ways at the same time, and to a considerable distance each way.

The magnetic effluvia, they found, did not in the least interfere with the electric; for when they had electrified the load stone, with a key hanging to it, they both attracted leaf brass like other substances.

Sometime after this, Mr. Wheeler, in the absence of Mr. Grey, electrified a red hot poker, and found the attraction to be the same as when it was cold. He also suspended a live chicken upon the tube by the legs, and found the breast of it strongly electrical. †

In August 1729, Mr. Grey advanced one step further in his electrical operations. He found that he could convey the electric virtue from the tube to the line of communication without touching it, and that holding the excited tube near

* Phil. Transf. abridged, Vol. 7. p. 15.   † Ib. p. 16.

F   it

it was sufficient. Repeating his former experiments with this variation, in conjunction with Mr. Wheeler, and among others, carrying the electric virtue several ways at the same time without touching the line, they always observed, that the attraction was strongest at the place which was most remote from the tube; a fact which they might have observed, if they had attended to it, in their former experiments. *

Sometime in the same month, Mr. Wheeler and Mr. Grey in conjunction made some experiments, in order to try whether the electric attraction was in proportion to the quantity of matter in bodies; and with this view they electrified a solid cube of oak, and another of the same dimensions which was hollow; but they could not perceive any difference in their attractive power; though it was Mr. Grey's opinion, that the electric effluvia passed through all the parts of the solid cube. †

On the 13th. of August in the same year, Mr. Grey made another improvement in his electrical apparatus, by finding that he could electrify a *rod*, as well as a *thread*, without inserting any part of it into his excited tube. He took a large pole twenty seven feet long, two inches and one half diameter at one end, and one inch and one half at the other. It was a sort of wood which is called horse beach, and had its rind on. This pole he suspended horizontally by hair lines, and at the small end of the pole he hung a cork, by means of a packthread about one foot long, and put a small leaden ball upon the cork, to keep the packthread extended. Then the leaf brass being put under the cork, the tube rubbed, and held near the larger end of the pole, the cork ball at the op-

* Phil. Transf. abridged, Vol. 7. p. 17.   † Ib.

posite

posite end attracted the leaf brass strongly, to the height of an inch or more. Mr. Grey also observed, that though the leaf brass was attracted by any part of the pole, it was not near so strongly as by the cork. *

About the beginning of September, Mr. Grey made experiments, to shew that the electric effluvia might be carried in a circle, as well as along lines, and be communicated from one circle to another; and also that it might be done whether the circles were vertical, or horizontal.

About the latter end of autumn, or the beginning of winter 1729, Mr. Grey resumed his inquiries after other electrical bodies, and found many more to have the same property, but he mentions only the dry leaves of several trees; from whence he concluded, that the leaves of all vegetables had that attractive virtue. †

We are now advanced to a new scene of Mr. Grey's electrical experiments, viz. upon *fluids*, and upon *animal bodies*. Having no other method of trying whether any substances could have the electric virtue communicated to them, but by making them raise light bodies placed upon a stand under them; it may easily be imagined, that they could not well contrive to put a fluid body into that situation. The only thing that Mr. Grey could do in this way, was to make use of a bubble, in which form a fluid is capable of being held in a state of suspension. Accordingly, on March 23d, and 25th. 1730, he dissolved soap in Thames water, and suspending a tobacco pipe, he blew a bubble at the head of it; and, bringing the excited tube near the small end, he found the bubble to attract leaf brass to the height of two, and of four inches. ‡

* Phil. Transf. abridged, Vol. 7. p. 18,   † Ib. p. 19.   ‡ Ib.

APRIL the 8th. 1730, Mr. Grey suspended a boy on hair lines, in a horizontal position, just as all electricians had, before, been used to suspend their hempen lines of communication, and their wooden rods; then, bringing the excited tube near his feet, he found that the leaf brass was attracted by his head with much vigour, so as to rise to the height of eight, and sometimes of ten inches. When the leaf brass was put under his feet, and the tube brought near his head, the attraction was small; and when the leaf brass was brought under his head, and the tube held over it, there was no attraction at all. Mr. Grey does not attempt to assign any reason for these appearances. It was not till many years after this time, that the influence of *points* in receiving and emitting the electric effluvia was observed. While the boy was suspended, Mr. Grey, amused himself with making the electricity operate in several parts of his body, at the same time; and at the ends of long rods, which he made him hold in his hands, and in diversifying the experiment several other ways. *

IT is curious to observe the inference wich Mr. Grey makes from these experiments. By them, says he, we see, that animals receive a greater quantity of electrical fluid than other bodies; and that it may be conveyed from them several ways at the same time, to considerable distances. He had no idea that the bodies of animals receive electricity only by means of the moisture that is in them, – and that his hempen line of communication, and his wooden rods could not have been electrified at all, if they had been perfectly dry.

IN all these experiments Mr. Grey observed, that the leaf brass was attracted to a much greater height from the top of

* Phil. Transf. abridged, Vol. 7. p. 20.

a narrow stand than from the table; and, at least, three times higher than when it was laid on the floor of the room.

About this time Mr. Grey communicated to the Royal Society his suspicion, that bodies attracted more or less according to their colour, though the substance was the same, and the weight and size equal. He says, he found red, orange, and yellow attracted at least three or four times stronger than green, blue, or purple; but he forbore communicating a more particular account of them, till he had tried a more accurate method which, he says, he had thought of, to make the experiments. The communication, however, was never made. The thing itself was only a deception, and the cause of it will be shewn in some subsequent experiments made by Mr. Wheeler. *

Mr. Grey, having found that he could communicate electricity to a bubble of soap and water, was encouraged to attempt communicating it to water itself. In order to this, he electrified a wooden dish full of water, placed on a cake of rosin, or a pane of glass, and observed, that if a small piece of thread, a narrow slip of thin paper, or a piece of sheet glass was held over the water, in an horizontal position, at the distance of an inch or something more, they would be attracted to the surface of the water, and then repelled; but he imagined that these attractions and repulsions were not repeated so often as they would have been, if the body had been solid.

But he afterwards contrived to shew the effect of electricity upon water in another and more effectual manner. As this experiment was very curious, and exhibited an appearance

* Phil. Transf. abridged, Vol. 7. p. 22.

which

which was quite new to the electricians of those times, I shall relate the particulars of it very fully, and generally in Mr. Grey's own words. *

He filled a small cup with water higher than the brim, and when he had held an excited tube over it, at the distance of about an inch or more, he says, that if it were a large tube, there would first arise a little mountain of water from the top of it, of a conical form; from the vertex of which there proceeded a light, very visible when the experiment was performed in a dark room, and a snapping noise, almost like that which was made when the finger was held near the tube, but not quite so loud, and of a more flat sound. Upon this, says he, immediately the mountain, if I may so call it, falls into the rest of the water, and puts it into a tremulous and waving motion.

When he repeated this experiment in the sun shine, he perceived that very small particles of water were thrown from the top of the mountain; and that, sometimes, there would arise a fine stream of water from the vertex of the cone, in the manner of a fountain, from which there issued a fine stream or vapour, whose particles were so small as not to be seen; yet, he says, he was certain it must be so; since the under side of the tube was wet, as he found when he came to rub it afterwards. He adds, that he had since found, that though there does not always arise that cylinder of water, yet that there is always a stream of invisible particles thrown on the tube, and sometimes to that degree as to be visible on it.

When some of the larger cups were used (his sizes were from three fourths to one tenth of an inch in diameter) which,

* Phil. Transf. abridged, Vol. 7. p. 23.

he

he fays, were to be filled as high as could be done without running over; the middle part of the furface, which was flat, would be depreffed, upon the approach of the tube, into a concave, and the parts towards the edge be raifed; and that when the tube was held over againft the fide of the water, little conical protuberances of water iffued out from it horizontally; and, after the crackling noife, returned to the reft of the water; and that fometimes fmall particles would be thrown off from it, as from the fmall portions of it above mentioned.

THIS laft experiment he repeated with hot water, and found that it was attracted much more ftrongly, and at a much greater diftance than before. The fteam arifing from the vertex was in this cafe vifible, and the tube was fprinkled with large drops of water.

HE tried thefe experiments, in the fame manner, with quick filver, which was likewife raifed up, but, by reafon of its gravity, not to fo great a height as the water: but, he fays, that the fnapping noife was louder, and lafted much longer than it did with the water.*

IT is not eafy to know what to make of the next fet of experiments which engaged the attention of Mr. Grey, or how far he deceived himfelf in the refults of them. He fancied that he had difcovered a perpetual attractive power in all electric bodies which did not require heating, rubbing, or any kind of attrition to be excited. The following experiments, he imagined, proved the difcovery.

HE took nineteen different fubftances, which were either rofin, .gum lac, fhell lac, bees wax, fulphur, pitch, or two

* Phil. Tranf. abridged, Vol. 7. p. 24.

or

or three of thefe differently compounded. Thefe he melted in a fpherical iron ladle; except the fulphur, which was beft done in a glafs veffel. When thefe were taken out of the ladle, and their fpherical furfaces hardened, he fays, they would not attract until the heat was abated, or until they came to a certain degree of warmth; that then there was a fmall attraction, which encreafed until the fubftance was cold, when it was very confiderable. *

THE manner in which he preferved thefe fubftances in a ftate of attraction, was by wrapping them in any thing which would keep them from the external air. At firft, for the fmaller bodies, he ufed white paper, and for the larger ones, white flannel; but he afterwards found that black worfted ftockings would do as well. Being thus cloathed, he put them into a large firm box; where they remained, till he had occafion to make ufe of them.

HE obferved thefe bodies for thirty days, and found that they continued to act as vigoroufly as on the firft or fecond day; and that they retained their power until the time of his writing, when fome of them had been prepared above four months.

HE makes the moft particular mention of a large cone of ftone fulphur, covered with a drinking glafs in which it was made; and fays, that whenever the glafs was taken off, it would attract as ftrongly as the fulphur, which was kept covered in the box. In fair weather, the glafs would attract alfo, but not fo ftrongly as the fulphur, which never failed to attract, let the wind or the weather be ever fo variable; as would all the other bodies, only in wet weather, the attraction was not fo great, as in fair weather.

* Phil. Tranf. abridged, Vol. 7. p. 24.

HE

He also mentions a cake of melted sulphur, which he kept without any cover, in the same place with the body above-mentioned, and where the sun did not shine upon them; and says, that it continued to attract till the time of his writing; but that its attraction was not one tenth part of that of the cone of sulphur which was covered.

These attractions he tried by a fine thread hanging from the end of a stick. He held the electric body in one hand, and the stick in the other; and could perceive the attraction at as great a distance as he could hold them.

At the time of his writing, he was upon the subject of permanent electricity in glass, but had not then compleated his experiments.*

Great light will be thrown upon these experiments of Mr. Grey by some that will be hereafter related of Mr. Wilke. It is probable that the glass vessel in these experiments was possessed of one electricity, and the sulphur, &c. of the other. But the two electricities were not discovered till afterwards.

We are now come to a different set of electrical experiments, made by Mr. Grey and Mr. Wheeler in conjunction, similar to some of Mr. Hawkesbee's.

In the first place, Mr. Grey made some experiments, which, probably unknown to him, had been made before by Mr. Boyle, on excited glass, and several other bodies in vacuo, and found that they would attract at very near the same distance as in pleno. To determine this, he suspended the excited substance in a receiver of an air pump, and when it was

* Philosophical Transactions abridged, Vol. 6. p. 27.

G   exhausted

exhaufted, he let the electric down to a proper diftance from fome light bodies, placed on a ftand below. The event was, as near as could be judged, the fame in vacuo as in the open air, if the experiment was made in the fame receiver, and if the electric was brought to the light bodies at the fame diftance of time from the act of excitation.*

ABOUT the latter end of Auguft 1732, Mr. Grey and Mr. Wheeler fufpended, from the top of a receiver, a white thread, which hung down to the middle of it. Then exhaufting the receiver, and rubbing it, the thread was attracted vigoroufly. When it was at reft, and hung perpendicularly, the excited tube attracted it; and when the tube was taken away flowly, the thread returned to its perpendicular fituation; but the tube being removed haftily, the thread jumped to the oppofite fide of the receiver. This laft effect followed, if the hand was haftily removed from the receiver; and at firft it appeared, in both cafes, unaccountable to them; but upon further confideration, they concluded, that it proceeded from the motion of the air, made by the tube, or the hand, which took off the attraction on that fide, and not on the other. †
They alfo found, that an excited tube would attract the thread through another receiver, which was put over that in which it was fufpended. And, fometime after, Mr. Wheeler found, that the thread was attracted through five receivers put one over another, and all exhaufted: He even thought that, in this cafe, the attraction was rather ftronger than when a fingle receiver was ufed. N. B. The more ef-

* Phil. Tranf. abridged, Vol. 6. p. 27.   † Ib. Vol. 7. p. 56.

fectually

fectually to keep any thing of moisture out of the receivers, which would have been of bad consequence in this experiment, instead of wet leather, he made use of a cement made of wax and turpentine, which Mr. Boyle used in his experiments. *

These two gentlemen, about the same time, made a curious experiment which, they say, shewed, that attraction is communicated through opaque, as well as transparent bodies, not in vacuo. But a little knowledge of metal, as a conductor of electricity, would have saved them the trouble they gave themselves. They took a large hand-bell, and taking out the clapper, they suspended a cork, besmeared with honey, from the top of it; and set it on a piece of glass, on which they had put some leaf brass. The excited tube was then brought near several parts of the bell; and, upon taking it up, several pieces of leaf brass were found sticking to the cork, and others were removed from the places in which they had been left, having, probably, been attracted by the bell.†

We see by how small steps advances were made in this science by some experiments made by Mr. Grey, 16th June 1731, and which he has thought worth recording; though they contain hardly any thing which we should think new, notwithstanding the discoveries appeared pretty considerable to him.

He electrified a boy standing on cakes of rosin, as strongly as he had before electrified him when suspended on hair lines. He afterwards electrified a boy suspended on hair lines, by

* Phil. Transf. abridged, Vol. 7. p. 97.   † Ib. 96.

G 2   means

means of a line of communication from another boy who was electrified, at some feet distance from him. He varied this experiment with rods and boys several ways; and concluded from it, that the electric virtue might not only be carried from the tube, by a rod, or line, to distant bodies; but that the same rod, or line, would communicate that virtue to another rod, or line, at a distance from it; and that, by this other rod, or line, the attractive force might be carried to still more distant bodies. This experiment shews that Mr. Grey had not properly considered the line of communication, and the body electrified by it, as one and the same thing, in an electrical view, differing only in form, as they were both alike conductors of electricity.

In December following, Mr. Grey carried this experiment something further, by conveying electricity to bodies which did not touch the line of communication, making it pass through the center of hoops standing on glass. One of his hoops was twenty, another forty inches in diameter. *

Phil. Transf. abridged, Vol. 7. p. 100.

PERIOD

PERIOD IV.

THE EXPERIMENTS AND DISCOVERIES OF MR. DU FAYE.

HITHERTO the spirit of electricity seems to have been confined to England; but, about this time, we find that it had passed the seas, and that ingenious foreigners were ambitious of distinguishing themselves, and acquiring reputation in this new field of glory. Mr. Du Faye, intendant of the French king's gardens, and member of the academy of sciences at Paris, assiduously repeated the experiments above-mentioned of Mr. Grey, and likewise added to the common stock several new ones of his own. To him we are also indebted for the observation of several general properties of electricity, or rules concerning the method of its action, which had not been taken notice of before, and which reduced to fewer propositions what had been before discovered concerning it. These experiments of his were comprised in six long memoirs, inserted in the history of the academy of sciences for the years 1733 and 1734. An account of them also makes an article in the Philosophical Transactions, dated December 27th 1733.

He found that all bodies, except metallic, soft, and fluid ones, might be made electric, by first heating them, more or less, and then rubbing them on any sort of cloth. He also excepts those substances which grow soft by heat, as gum;

or

or which diffolve in water, as glue. He alfo remarked, that the hardeft ftone and marble required more chafing and heating than other bodies, and that the fame rule obtains with regard to woods; fo that box, lignum vitæ, and other kinds of very hard wood muft be chafed almoft to a degree of burning; whereas fir, lime tree, and cork require but a moderate heat. *

HE fays, that, purfuing Mr. Grey's experiments, to make water receive electricity, he found; that all bodies, without exception, whether folid or fluid, were capable of it, when they were placed on glafs flightly warmed, or only dried, and the excited tube was brought near them. He particularly mentions his having made the experiment with ice, lighted wood coal, and every thing that happened to be at hand at the time; and conftantly remarked, that fuch bodies as were of themfelves the leaft electric, had the greateft degree of electricity communicated to them by the approach of the excited tube.

HE refutes Mr. Grey's affertion concerning the different electricity of differently coloured bodies, and fhews that it proceeded not from the colour as a colour, but from the fubftance which was employed in dying it.

HAVING communicated the electricity of the tube, by means of a packthread, after Mr. Grey's manner, he obferved, that the experiment fucceeded better for wetting the line; and though he made the experiment at the diftance of one thoufand two hundred and fifty fix feet, when the wind was high, the line making eight returns, and paffing through two different walks of a garden, that the electric virtue was ftill communicated. †

* Phil. Tranf. abridged, Vol. 8. p. 393.   † Ib. p. 395.

THE

THE electric spark from a living body, and which makes a principal part of the diversion of gentlemen and ladies, who come to see experiments in electricity, was first observed by Mr. du Faye, accompanied at that time, as in most of his experiments, by the Abbe Nollet, who, afterwards, we shall find, did himself obtain a distinguished name among electricians.

MR. DU FAYE, having got himself suspended on silk lines, as Mr. Grey had done the child mentioned above, observed, that, as soon as he was electrified, if another person approached him, and brought his hand within an inch, or thereabouts, of his face, legs, hands, or cloaths, there immediately issued from his body one or more pricking shoots, attended with a crackling noise. He says this experiment occasioned to the person who brought his hand near him, as well as to himself, a little pain, resembling that of the sudden prick of a pin, or the burning from a spark of fire; and that it was felt as sensibly through his cloaths, as on his bare face, or hands. He also observes, that, in the dark, those snappings were so many sparks of fire. *

THE Abbe Nollet says he shall never forget the surprize which the first electrical spark which was ever drawn from the human body excited, both in Mr. Du Faye, and in himself.†

HE says, that those snappings and sparks were not excited, if a bit of wood, of cloth, or of any other substance than a living human body, was brought near him; except metal, which produced very nearly the same effect as the human body.

* Phil. Transf. abridged, Vol. 8. p. 395.   † Leçons de Physique, Vol. 6. p. 408.

He

He was not aware, that it was owing to points, or partial dryneſs, in the ſubſtances which he mentions, that they did not take a full and ſtrong ſpark. He ſeems alſo to have been under ſome deceptions, when he imagined, that the fleſh of dead animals gave only an uniform light,. without any ſnapping, or ſparks. *

THE two next capital obſervations of Mr. Du Faye I ſhall repeat in his own words, becauſe they are important and curious; and yet the former of them is little more that Otto Guericke had obſerved before him. " I diſcovered," he ſays, " a very ſimple " principle, which accounts for a great part of the irregula- " rities, and, if I may uſe the term, of the caprices, that " ſeem to accompany moſt of the experiments in electricity. " This principle is, that electric bodies attract all thoſe " which are not ſo, and repel them as ſoon as they are be- " come electric, by the vicinity or contact of the electric " body. Thus leaf gold is firſt attracted by the tube; ac- " quires electricity by approaching it, and, conſequently, is " immediately repelled by it; nor is reattracted, while it re- " tains its electric quality. But if, while it is thus ſuſtained " in the air, it chance to light on ſome other body, it " ſtraightway loſes its electricity, and conſequently is reat- " tracted by the tube; which, after having given it a new " electricity, repels it a ſecond time; and this repulſion con- " tinues as long as the tube keeps its power. Upon apply- " ing this principle to various experiments of electricity, one " will be ſurpriſed at the number of obſcure and puzzling

* Phil. Tranſ. abridged, Vol. 8. p. 395.

" facts

" facts which it clears up." By the help of this principle, he, particularly, endeavours to explain several of Mr. Hawkesbee's experiments. *

" Chance, he says, has thrown in my way another prin-
" ciple, more universal and remarkable than the preceeding
" one ; and which casts a new light upon the the subject of
" electricity. The principle is, that there are two distinct
" kinds of electricity, very different from one another ; one
" of which I call vitreous, the other resinous electricity.
" The first is that of glass, rock-chrystal, precious stones,
" hair of animals, wool, and many other bodies. The se-
" cond is that of amber, copal, gum lac, silk, thread,
" paper, and a vast number of other substances. The cha-
" racteristics of these two electricities is, that they repel
" themselves, and attract each other. Thus a body of the
" vitreous electricity repels all other bodies possessed of the
" vitreous, and on the contrary, attracts all those of the resi-
" nous electricity. The resinous, also, repels the resinous,
" and attracts the vitreous. From this principle, one may
" easily deduce the explanation of a great number of other
" phenomena ; and it is probable, that this truth will lead us
" to the discovery of many other things.

In order to know, immediately, to which of the two classes of electricity any body belonged, he made a silk thread
electrical, and brought it to the body, when it was excited.
If it repelled the thread, he concluded it was of the same
electricity with it, viz. resinous ; if it attracted it, he concluded it was vitreous. †

* Phil. Transf. abridged, Vol. 8. p. 396.   † Ib. p. 397.

HE alſo obſerved, that communicated electricity had the ſame property as the excited. For having electrified, by the glaſs tubes, balls of wood or ivory; he found them to repel the bodies which the tube repelled, and to attract thoſe which the tube attracted. If they had the reſinous electricity communicated to them, they obſerved the ſame rule, by attracting thoſe bodies which had the vitreous electricity communicated to them, and repelling thoſe which had received the reſinous. But, he obſerves, the experiment would not ſucceed, if the bodies were not made equally electrical; for, if one of them was weakly electrical, it would be attracted by that which was much more ſtrongly electrical, of whatever quality it was.

THIS diſcovery of the two electricities was certainly a capital one, but was, notwithſtanding, left very imperfect by Mr. Du Faye. We ſhall ſee that Dr. Franklin found, that, in all probability, the vitreous electricity was poſitive, or a redundancy of electrical matter; and the reſinous, negative, or a want of it; and that Mr. Canton has diſcovered, that it depends upon the ſurface of the electric bodies, and of the rubber, whether the electricity be poſitive or negative.

THE doctrine of two different electricities, produced by exciting different ſubſtances, conſiderable as the diſcovery of it was, ſeems to have been dropped after Mr. Du Faye, and thoſe effects aſcribed to other cauſes; which is an inſtance that ſcience ſometimes goes backwards.

MR. DU FAYE himſelf ſeems at laſt to have adopted the
opinion,

opinion, which generally prevailed to the time of Dr. Franklin; that the two electricities differed only in degree, and that the stronger attracted the weaker: not considering that, upon this principle, bodies possessed of the two electricities ought to attract one another less forcibly, than if one of them had not been electrified at all, which is contrary to fact.

It will be seen that, many years after, Mr. Kinnersley of Philadelphia, a friend of Dr. Franklin's, being at Boston in New England, made some experiments which again shewed that difference of the two electricities. He communicated those experiments to Dr. Franklin, who repeated and explained them. *

It must be added to the experiments of Mr. Du Faye, that he communicated electricity from one body to another, through the interval of ten or twelve inches, in the middle of which there was a lighted candle.† He also found, that red hot iron might be electrified very well. ‡

Mr. Du Faye was the first person who endeavoured to excite a tube in which air was condensed, and found the attempt ineffectual. Suspecting this might be owing to moisture, which he might force into the tube, in using his condensing instrument, he cemented a large copper eolipile to his tube, and compressed the air in it, by putting the eolipile upon the fire. After this, he turned a cock, which he had placed to prevent the return of the compressed air, and disengaged the tube from the eolipile; but he still found the excitation to be impossible. The Abbe Nollet, who assisted at most of this

* See his Letters.   † Nollet's Recherches, p. 203.   ‡ Ib. p. 212.

gentleman's experiments, declares himself not satisfied even with this precaution; thinking that the nonexcitation of the tube might still be owing to the moisture, which always exists in the air, and the particles of which must be drawn nearer together by condension. * In answer to this objection, Mr. Boulanger says, that a small glass full of water poured into a tube, and immediately thrown out again, will not destroy the excitability of the glass near so much as the condensed air. †

It must be observed, that Mr. Granville Wheeler, in the autumn of the year 1732, made several curious experiments, relating to the repulsive force of electricity. These he repeated to Mr. Grey in the summer following, and designed to communicate them, through his hands, to the Royal Society; but, deferring the execution of it from time to time, he was informed, that Mr. Du Faye had taken notice of the same solution of the repulsive force. Upon this he laid aside all thoughts of communicating his discovery to the public: but, finding that his experiments were different from those of Mr. Du Faye, he was persuaded to publish them in the Philosophical Transactions for the year 1739.

The experiments were made by threads of various kinds, and other substances, hanging down from silk lines, and generally made to repel one another by the approach of an excited tube. The result of them all he comprised in the three following propositions. 1st. That bodies made electrical, by communication with an excited electric, are in a state of

* Nollet's Recherches, p. 258. † Boulanger, p. 132.

repulsion

repulsion with respect to such excited bodies. 2dly. That two, or more bodies, made electrical by communicating with an excited electric, are in a state of repulsion with respect to one another. 3dly. Excited electrics do themselves repel one another. *

ONE of his experiments, to prove the second of these propositions, deserves to be mentioned for its curiosity. He tied a number of silk threads together, by a knot at each extremity; when, upon electrifying them, the threads repelled one another, and the whole bundle was swelled out into a beautiful spherical figure; so that he could with pleasure, he says, observe the knot at the bottom rising upwards, as the electricity and mutual repulsion of the threads increased; and he could not help imagining his bundle of silks to resemble a bundle of muscular fibres.

BY way of corollary to the same proposition, he observes, that it suggests, more plainly than any other known experiment, a reason for the dissolution of bodies in menstrua; viz. that the particles of the solvend, having imbibed particles of the menstrua, so as to be saturated with them, the saturated particles become repulsive of one another, separate, and fly to pieces. †

* Phil. Transf. abridged, Vol. 8. p. 411.   † Ib. p. 410.

PERIOD

PERIOD V.

THE CONTINUATION, AND CONCLUSION OF MR. GREY's
EXPERIMENTS.

MR. GREY, upon resuming his experiments, expresses
great satisfaction that his former observations had been
confirmed by so judicious a philosopher as Mr. Du Faye, who,
he acknowledges, had made several new ones of his own, particularly that important luciferous one, as he calls it; which,
put him upon making the experiments which follow,
and which were made in the months of July and August
1734. *

As Mr. Du Faye had said, that the snappings and the
sparks, he had mentioned, were strongly excited by a piece
of metal, presented to the person suspended on silk lines;
Mr. Grey concluded, that if the person and the metal should
change places, the effect would be the same. He, accordingly, suspended several pieces of metal on silk lines, beginning
with the common utensils, which were at hand, as the iron
poker, tongs, fire shovel, &c. and found, that, when they
were electrified, they gave sparks, in the same manner as

* Phil. Transf. abridged, Vol. 8. p. 397.

the

the human body had done in like circumftances. This was the origin of metallic conductors, which are in ufe to this day. *

MR. GREY did not, at that time, think of making his experiments in the dark, in order to fee the light proceeding from the iron; not imagining, that electricity, communicated to metals, would have produced fuch furprifing phenomena, as, he fays, he afterwards found it to do.

CONTINUING his experiments at Mr. Wheeler's, they found, that the flefh of dead animals exhibited, very nearly, the fame appearances as that of living animals, contrary to the affertion of Mr. Du Faye.

BUT what moft furprifed Mr. Grey, and the gentlemen then prefent, in the experiments he made upon that occafion, was the phenomenon above referred to, and what he now calls *a cone*, or *pencil of electrical light*; fuch as is commonly feen to iffue from an electrified point. As this was the firft time that this phenomenon, which is now fo common, was diftinctly feen, I fhall relate the experiment, of which it was the refult, at large.

MR. GREY, and his friends, provided themfelves with an iron rod four feet long, and half an inch in diameter, pointed at each end, but not fharp. Sufpending this iron rod upon filk lines in the night; and, applying the excited tube to one end of it, they perceived, not only a light upon that end, but another iffuing from the oppofite end, at the fame time. This light extended itfelf, in the form of a cone,

* Phil. Tranf. abridged, Vol. 8. p. 398.

whofe

whofe vertex was at the end of the rod: and Mr. Grey fays, that he and his company could plainly fee, that it confifted of feparate threads, or rays of light, diverging from the point of the rod, the exterior rays being incurvated. This light appeared at every ftroke they gave the tube.

THEY likewife obferved, that this light was always attended with a fmall hiffing noife, which, they imagined, began at the end next the tube, increafing in loudnefs till it came to the oppofite end. He fays, however, that this noife could not be heard, but by perfons who ftood near the rod, and attended to it. *

MR. GREY, repeating thefe experiments in the September following, after his return to London, obferved an appearance, which he fays, furprifed him very much. After the tube had been applied to the iron rod, as before, when the light, which had been feen at both ends, had difappeared; it was vifible again, upon bringing his hand near the end of the rod; and, upon repeating this motion of his hand, the fame phenomenon appeared for five or fix times fucceffively; only the rays were, at each time, fhorter than the other. He alfo obferved, that thefe lights, which were emitted by the tube upon the approach of his hand, were, like the others, attended with a hiffing noife.

HE took notice, that the light which appeared on the end next the tube, when it was held oblique to the axis of the rod, had its rays tending towards it; and that, all the time he was rubbing the tube, thofe flafhes of light appeared upon

* Phil. Tranf. abridged, Vol. 8. p. 398.

every

every motion of his hand up or down the tube, but that the largest flashes were produced by the motion of his hand downwards. *

When he used two or three rods, laying them, either in a right line, or so as to make any angle with each other, and applied the tube to any one of their ends; he observed, that the farthest end of the farthest rod exhibited the same phenomena as one single rod. †

Using a rod pointed only at one end, he observed, that the other end gave but one single snap, but that it was much louder than the greatest of those which were given by the point of the rod; also that the pain, resembling pricking or burning, was more strongly felt, and that the light was brighter and more contracted.

Connecting a pewter plate with the iron rod, and filling the plate with water, he observed the same light, the same pushing of the finger, as he calls it, and the same snapping, as when the experiment was made with the empty plate. And when the experiment was made with water, in the day light, it appeared to rise in a little hill, under the finger which was presented to it; and, after the snapping noise, fell down again, putting the water into a waving motion near the place where it had risen.

These effects were the same which he had before observed to proceed from the immediate action of the tube, but by these experiments, he says, that he found (what, no doubt, appeared a real advance in the science to him) that an actual

* Phil. Transf. abridged, Vol. 8. p. 399.   † Ib. p. 400.

I   flame

flame of fire, together with an explosion, and ebullition of cold water, might be produced by communicative electricity. What he adds is so remarkable, that I shall repeat his own words. " And although these effects are at present but *in* " *minimis*, it is probable, in time, there may be found out " a way to collect a greater quantity of the electric fire, and " consequently to increase the force of that power; which, " by several of these experiments, *si licet magnis componere* " *parva*; seems to be of the same nature with that of thunder " and lightning. " *

How exactly has this prophecy been fulfilled in the discoveries of the Leyden electricians, and Dr. Franklin; the former having discovered the amazing accumulation of the electric power, in what is called the Leyden phial; and the latter having proved the matter of lightning to be the very same with that of electricity; though Mr. Grey might possibly mention thunder and lightning only by way of common comparison.

On February the 18th. 1735, Mr. Grey, repeating the experiments of the iron rods with wooden ones, found all the effects to be similar, but much weaker; as it is now well known must have been the case; wood being so imperfect a conductor, and only in proportion to the moisture it contains.

At the same time, he relates, that repeating the electrification of water, he found, that the phenomena before mentioned were produced, not only by holding the tube near the water, but when it was removed, and the finger afterwards brought near it. †

* Phil. Transf. abridged, Vol. 8. p. 401.   † Ib. p. 402.

MAY

May the 6th. of the fame year, he again fufpended a boy on filk, and found that this boy was able to communicate the electric fire, firft to one, and then to feveral perfons ftanding upon electric-bodies.

Mr. Grey feems ftill to have imagined, that electricity depended, in fome meafure, upon colour. The boy fufpended on blue lines, he fays, retained his power of attraction fifty minutes; on fcarlet lines, twenty five minutes; and on orange coloured lines, twenty one minutes. By thefe experiments, he fays, we fee the efficacy of electricity upon bodies fufpended upon lines of the fame fubftance, but of different colour. *

But the greateft deception which this ingenious gentleman feems to have lain under, was occafioned by the experiments which he made with balls of iron, to obferve the revolution of light bodies about them. The paragraph relating to thefe experiments, being the laft which Mr. Grey wrote, I fhall give it at length, as a curiofity.

" I have lately made," fays he, " feveral new experi-
" ments upon the projectile and pendulous motion of fmall
" bodies by electricity; by which fmall bodies may be made
" to move about larger ones, either in circles, or in ellipfes;
" and thofe either concentric, or eccentric to the center of
" the larger body, about which they move, fo as to make
" many revolutions about them. And this motion will con-
" ftantly be the fame way that the planets move about the
" fun; viz. from the right hand to the left, or from weft to

* Phil. Tranf. abridged, Vol. 8. p. 403.

I 2   " eaft.

" eaft. But thefe little planets, if I may fo call them, move
" much fafter in their *apogeon*, than in the *perigeon* parts of
" their orbits; which is directly contrary to the motion of
" the planets about the fun. *

THESE experiments Mr. Grey had thought of but a very
little while before his laft illnefs, and had not time to com-
pleat them; but the progrefs he had made in them he
revealed, on the day before his death, to Dr. Mortimer, then
fecretary to the Royal Society. He faid they ftruck him with
new furprife every time he repeated them, and hoped that, if
God would fpare his life a little longer, he fhould, from
what thefe phenomena pointed out, bring his electrical experiments to the greateft perfection. He did not doubt but, in a
fhort time, he fhould be able to aftonifh the world with a new
fort of planetarium, never before thought of; and that, from thefe
experiments, might be eftablifhed a certain theory, to account
for the motions of the grand planetarium of the univerfe.
Thefe experiments, fallacious as they are, deferve to be
briefly related, together with thofe which were made in confequence of them after Mr. Grey's death. I fhall relate them
in Mr. Grey's own words, as they were delivered to Dr.
Mortimer, on his death bed.

PLACE a fmall iron globe, faid he, of an inch or an inch
and a half in diameter, on the middle of a circular cake of
rofin, feven or eight inches in diameter, gently excited; and
then a light body fufpended by a very fine thread, five or fix
inches long, held in the hand over the center of the table,

* Phil. Tranf. abridged, Vol. 8. p. 404.

will,

will, of itfelf, begin to move in a circle round the iron globe, and conftantly from weft to eaft. If the globe be placed at any diftance from the center of the circular cake, it will defcribe an ellipfe, which will have the fame eccentricity, as the diftance of the globe from the center of the cake.

If the cake of rofin be of an elliptical form, and the iron globe be placed in the center of it, the light body will defcribe an elliptical orbit, of the fame eccentricity with the form of the cake.

If the iron globe be placed in, or near, one of the foci of the elliptical cake, the light body will move much fwifter in the apogee, than in the perigee of its orbit.

If the iron globe be fixed on a pedeftal, an inch from the table, and a glafs hoop, or a portion of a hollow glafs cylinder, excited, be placed round it; the light body will move as in the circumftances mentioned above, and with the fame varieties.

He faid, moreover, that the light body would make the fame revolutions, only fmaller, round the iron globe placed on the bare table, without any electrical body to fupport it; but he acknowledged he had not found the experiment fucceed, if the thread was fupported by any thing but a human hand; though, he fancied, it would have fucceeded, if it had been fupported by any animal fubftance, living or dead. *

Mr. Grey went on to recite to Dr. Mortimer other experiments ftill more fallacious; which, out of regard to his

* Phil. Tranf. abridged, Vol. 8. p. 404, 405.

memory,

memory, I shall forbear to quote. Let the chimeras of this great electrician teach his followers in the same, and still but newly opened field of philosophy, a proper degree of caution in their reasonings from induction. Let not the example, however, discourage any person from trying what may appear improbable; but let it induce a man to delay the publication of his discoveries, till they have been perfectly ascertained, and performed in the presence of others. In experiments of great delicacy, a strong imagination will have great influence even upon the external senses; of which we shall have frequent instances in the course of this history.

DR. MORTIMER himself seems to have been deceived by these experiments of Mr. Grey. He says, that, in trying them after his death, he found, that the light body would make revolutions round bodies of various shapes and substances, as well as round the iron globe; and that he had actually tried the experiment, with a globe of black marble, a silver standish, a small chip box, and a large cork. *

THESE experiments of Mr. Grey's were tried by Mr. Wheeler, and other gentlemen, at the Royal Society's house, and with a great variety of circumstances; but no conclusion could be drawn from what they at that time observed. Mr. Wheeler himself took a great deal of pains to verify them, with various success; and at last he gave it as his opinion, that a desire to produce the motion from west to east was the secret cause that determined the pendulous body to move in

* Phil. Transf. abridged, Vol. 8. p. 405.

that

that direction, by means of some impression from Mr. Grey's hand, as well as his own; though he was, at the same time, persuaded, that he was not sensible of giving any motion to his hand himself. *

* Phil. Transf. abridged, Vol. 8. p. 418.

PERIOD

PERIOD VI.

THE EXPERIMENTS OF DR. DESAGULIERS.

WE are now come to the labours of that indefatigable experimental philosopher, DR. DESAGULIERS, in this new field of science. The reason which he gives why he had avoided entertaining the Royal Society upon this subject before, and why he had not pursued it so far as he might have done; considering, as he says, that he could excite as strong an electricity in glass, by rubbing with his hand, as any body could, is worth mentioning for its curiosity, and for the light that it throws upon the temper and manner of Mr. Grey. He says, that he was unwilling to interfere with the late Mr. Stephen Grey, who had wholly turned his thoughts to electricity; but was of a temper to give it entirely over, if he imagined that any thing was done in opposition to him. *

DR. DESAGULIERS begins with observing very sensibly (and the observation is still true) that the phenomena of electricity are so odd, that, though we have a great many experiments upon that subject, we have not yet been able, from their comparison, to settle such a theory as will lead us

* Philof. Tranf. abridged, Vol. 8. p. 419.

to

to the caufe of that property in bodies, or even to judge of all its effects, or find out what ufeful influence electricity has in nature; though certainly, from what we have feen of it, we may conjecture that it muft be of great ufe, becaufe it is fo extenfive.

His firft experiments, of which an account is given in the Philofophical Tranfanctions, dated July, 1739, were made with a hempen ftring, extended upon cat gut. To the end of the hempen ftring, he fufpended various fubftances; and fays, that all thofe which he tried, amongft which were feveral *electrics per fe*, as fulphur, glafs, &c. without exception, received electricity. *

He changed one of the cat gut ftrings, on which his hempen line of communication was extended, and put various other fubftances in its place, to try what bodies would tranfmit electricity to the fufpended body, and what would not; and from the refult of his experiments, partly concluded, that bodies in which electricity could not be excited intercepted the electric effluvia; and that thofe in which electricity could be excited, did not intercept it, but permitted it to go on to the extremity of the hempen ftring. But ftill he had no juft idea, that, except metals, it was the moifture in the bodies he tried which intercepted the electric effluvia, and his ideas of the manner in which they were intercepted were very imperfect.

To Dr. Defaguliers we are indebted for fome technical terms which have been extremely ufeful to all electricians to

* Phil. Tranf. abridged, Vol. 8. p. 420.

K   this

this day, and which will probably remain in ufe as long as the fubject is ftudied. He firft applied the term *conductor* to that body to which the excited tube conveys its electricity; which term has fince been extended to all bodies which are capable of receiving that virtue. And he calls thofe bodies in which electricity may be excited by heating or rubbing *electrics per fe*.

IN the writings of this author we find many axioms relating to electric experiments, fome of which are expreffed in a more clear, and diftinct manner than they had been before, but the real improvements which he made, were very few and immaterial.

ON feveral occafions, and particularly in a paper delivered to the Royal Society, in the month of January 1741, he lays down, among others, the following general rules, which feem to be more accurate than any which had been delivered before upon the fubject. *

" AN electric per fe will not receive electricity from another " electric per fe, in which it has been excited, fo as to run a- " long its whole length : but will only receive it a little way, " being, as it were, faturated with it.

" AN electric per fe will not lofe all its electricity at " once, but only the electricity of thofe parts near which " a non electric has been brought. It, confequently, lofes its " electricity the fooner, the more of thofe bodies are near it. " Thus, in moift weather, the excited tube holds its virtue but " a little while, becaufe it acts upon the moift vapours which

* Phil. Tranf. abridged, Vol. 8. p. 430.

" float

" float in the air.   And if the excited tube be applied to leaf
" gold laid upon a ftand, it will act upon it much longer, and
" more ftrongly, than if the fame quantity of leaf gold be laid
" upon a table, which has more non electric furface than the
" ftand." * This, however, feems not to be the whole reafon;
for if the leaf gold were laid upon a broad furface of glafs, it
would not be acted upon fo powerfully; as if it were placed
upon a narrow ftand of any kind of matter.

" A NON-ELECTRIC, when it has received electricity, lofes
" it all at once, upon the approach of another non electric."
This however could only be the cafe when the approaching
electric was not infulated, but had a communication with the
earth.   It muft alfo be brought into contact with the electrified body.

" ANIMAL fubftances are non electrics by reafon of the
" fluids they contain. †

" EXCITED electricity exerts itfelf in a fphere round the
" electric per fe, or rather in a cylinder, if the body be cy-
" lindrical." ‡

FEW of the many experiments which were made by Dr.
Defagulicrs (accounts of which were publifhed in the Philofophical Tranfactions) had, as I obferved before, any thing
new in them.   Thofe which were the moft fo are the following.

ENDEAVOURING to communicate electricity to a burning
tallow candle, he obferved, that the candle attracted the
thread of trial, but not within two or three inches of the

* Phil. Tranf. abridged, Vol. 8. p. 427.   † Ib. p. 429.   ‡ Ib. p. 431.

flame;

flame; but that, as soon as the candle was blown out, the thread was attracted by every part of it, and even by the wick, when the fire was quite extinguished. He electrified a wax candle in the same manner, and the experiment succeeded as well, only the electricity came not so near the flame in the wax, as in the tallow candle.

HE says, that only warming a glass receiver, without any rubbing, would cause the threads of a down feather, tied to an upright skewer to extend themselves, as soon as it was put over the feather; and that sometimes rosin and wax would exert their electricity by being only exposed to the open air.

HE observed that if a hollow glass tube, supporting the line of communication, were moistened by blowing through it, it would intercept the electricity.

HE says, that when an excited tube has repelled a feather, it will attract it again, after being suddenly dipped into water, but in fair weather it will not attract it unless it hath been dipt pretty deep into the water, a foot of its length at least; whereas, in moist weather, an inch or two will suffice. *

HE shewed the attraction of water by an excited tube in a better manner than it had been shewn before, viz. by bringing the tube to a stream issuing from a condensing fountain, which, thereupon, was evidently bent towards it.

DR. DESAGULIERS seems to have been the first who expressly said, that pure air might be ranked amongst electrics

* Phil. Transf. abridged, Vol. 8. p. 429.

per

per se, and that cold air in frosty weather, when vapours rise least of all, is preferable, for electrical purposes, to warm air in summer, when the heat raises the vapours. * He also supposed that the electricity of the air was of the vitreous kind; and he accounted for the electricity appearing on the inside only of an exhausted glass vessel, by its going where it met with the least resistance from so electrical a body as the air. †

He endeavoured to account for the fixing of air by the steams of sulphur, according to the experiment of Dr. Hales; by supposing that the particles of sulphur, and those of air, being possessed of different kinds of electricity, attracted one another, whereby their repulsive power was destroyed. He also proposed the following conjecture concerning the rise of vapour. The air at the surface of water being electrical, particles of water, he thought, jumped to it, then, becoming themselves electrical, they repelled both the air and one another, and consequently ascended into the higher regions of the atmosphere. ‡

The last paper of Dr. Desaguliers in the Philosophical Transactions, upon the subject of electricity is dated, June 24th. 1742, in which year he published a dissertation on electricity, by which he gained the prize medal of the academy at Bourdeaux. The dissertation is excellently drawn up, and comprizes all that was known of the subject till that time.

* Phil. Transf. abridged, Vol. 8. p. 437.   † Ib. p. 438.   ‡ Ib. p. 437.

PERIOD

PERIOD VII.

EXPERIMENTS OF THE GERMANS, AND OF DR. WATSON, BEFORE THE DISCOVERY OF THE LEYDEN PHIAL IN THE YEAR 1746.

ABOUT the time that Dr. Defaguliers had concluded his experiments in England, 1742, feveral ingenious Germans began to apply themfelves to the fame ftudies with great affiduity, and their labours were crowned with confiderable fuccefs.

To the Germans we are indebted for many capital improvements in our electrical apparatus within this period, without which the bufinefs would have gone on very flowly and heavily; but, by the help of their contrivances, we fhall fee aftonifhing effects were foon produced.

MR. BOZE, a profeffor of philofophy at Wittemburgh, fubftituted the *globe* for the tube, which had been ufed ever fince the time Hawkefbee. He likewife added a *prime conductor*, which confifted of a tube of iron or tin, at firft fupported by a man ftanding upon cakes of rofin, and afterwards fufpended on filk horizontally before the globe. *

* Hiftoire de l'electricité, p. 27.

To

To prevent the tube from doing any harm to the globe, he put a bundle of thread into the end that was next to it, and which was left open for that purpofe.   This expedient, befides occafioning various pleafant phenomena, was obferved to make the force of the conductor much ftronger. *

The ufe of the globe was immediately adopted in the univerfity of Leipfic, where Mr. Winkler, the profeffor of languages, fubftituted a *cufhion* inftead of the hand, which had before been employed to excite the globe.   But the beft rubber for the globe, as well as the tube, was, long after this, ftill thought, by all electricians, to be the human hand, dry, and free from moifture. †

Mr. P. Gordon, a Scotch Benedictine monk; and profeffor of philofophy at Erford, was the firft who ufed a *cylinder* inftead of a globe.   His cylinders were eight inches long, and four inches in diameter.   They were made to turn with a bow, and the whole inftrument was portable.   Inftead of a cake of rofin, he infulated by means of a frame furnifhed with net work of filk. ‡

The apparatus likewife of many of the German electricians was very *various*, and expenfive.   Mr. Winkler, in a paper read at the Royal Society, March 21ft 1745, ‖ defcribes a machine for rubbing tubes, and another for rubbing globes, and compares the effects of them both.   He obferves, that the fparks which are producedfrom glafs veffels drawn to and fro were larger, and more vehemently pungent, provided that thofe veffels were of the

* Phil. Tranf. abridged, Vol. 10. p. 271. † Ib. p. 272. ‡ Hiftoire, p. 31.
‖ Phil. Tranf. abridged, Vol. 10. p. 273.

fame

same magnitude with the globes; but that the flux of effluvia was not so constant as from the globes. Mr. Winkler also invented a machine, which he describes at large
in his works, by means of which he could give his globe
six hundred and eighty turns in a minute. *

THE German electricians generally used more globes than
one at a time, and imagined they found the effects proportionable, though this fact was called in question by Dr. Watson, and
others; and Mr. Nollet preferred globes made blue with
zaffre, which were carefully tried, and rejected by Dr. Watson afterwards. †

SUCH a prodigious power of electricity could they excite
from these globes, whirled by a large wheel, and rubbed with
woollen cloth, or a dry hand (for we find both these methods
were in use among them about this time) that, if we may
credit their own accounts, the blood could be drawn from
the finger by an electric spark; the skin would burst, and a
wound appear, as if made by a caustic. They say, that if
several globes or tubes were used, the motion of the heart and
arteries of the electrified person would be very sensibly increased; and that, if a vein were opened under the operation,
the blood, issuing from it, would appear like lucid phosphorus,
and run out faster than when the man was not electrified.
Analogous to this last experiment, they observed, that water,
running from an artificial fountain electrified, was scattered in
luminous drops, and that a larger quantity of water was
thrown out in a given time than when the fountain was not

* Histoire, p. 32.　　† Phil. Transf. adridged, Vol. 10. p. 277.

electrified.

electrified. *   Part of this account we know might be true, but some part must have been exaggerated. It is certain that Mr. Gordon increased the electric sparks to such a degree, that they were felt from a man's head to his foot, and small birds were killed by them. †

The thing that strikes us most in their experiments, performed by these machines, is their setting fire to inflammable substances. This they were, probably, led to attempt, from observing the vivid appearance of electrical light, the burning pain that was felt by a smart stroke from the conductor, and the many analogies the electrical fluid evidently bore to phosphorus and common fire.

The first person who succeeded in this attempt was Dr. Ludolf of Berlin, towards the beginning of the year 1744; who kindled, with sparks excited by the friction of a glass tube, the etherial spirits of Frobenius. This he did at the opening of the Royal Academy, and in the presence of some hundreds of persons. He performed the experiment by electrical sparks proceeding from an iron conductor. John Henry Winkler, Greek and Latin professor at Leipsick, did the same in the May following, by a spark from his own finger; and kindled, not only the highly rectified spirit abovementioned, but French brandy, corn spirits, and other spirits still weaker, by previously heating them. He also says, that oil, pitch, and sealing wax might be lighted by electric sparks, provided those substances were first heated to a degree next to kindling. ‡

* Phil. Transf. abridged, Vol. 10. p. 277.   † Nollet's Recherches, p. 172.
‡ Phil. Transf. abridged, Vol. 10. p. 271.

THE German electricians likewise constructed a machine, by which they could give friction to a glass cylinder in vacuo. By these means they contrived to electrify a wire which terminated in the open air, and there shewed a considerable electric power. They also electrified that end which was in the open air, and made the other end which was in vacuo exert its electricity. *

THE same Germans also mention an experiment, which, if pursued, would have led them to discover, that the friction of the glass globe did not produce, but only collect the electric matter. But that was a discovery reserved, as we shall find, for Dr. Watson. It seems that both Mr. Boze and Mr. Allamand had suspended the machine, and the man who worked it, upon silk; and observed, that, not only the conductor, but also the man, and the machine gave signs of electricity; though they did not attend accurately to all the circumstances of that curious fact, which did not at all answer their expectations. For, imagining that part of the electric power was continually going off to the ground by the machine, they supposed that the effect of insulating it, would have been a stronger electricity. †

IN this period it was that Ludolf the younger demonstrated, that the luminous barometer is made perfectly electrical by the motion of the quicksilver; first attracting, and then repelling bits of paper, &c. suspended by the side of the tube. Before this experiment, those effects had been ascribed to other causes. ‡

---

* Phil. Transf. abridged, Vol. 10. p. 275. † Wilson's Essay, preface, p. 14. Watson'
Sequel, p. 34. ‡ Histoire, p. 89.

ABOUT

ABOUT the same time also, Mr. Boze took a great deal of pains to determine, whether the weight of bodies would be affected by electricity, but he could not find that it was.

THE electrical *star*, made by turning swiftly round an electrified piece of tin, cut with points equidistant from the center; and also the *electrical* bells, which will be described hereafter among the surprising and diverting experiments performed by the help of electricity, were of German invention. *
Lastly, to these it may be added, that Mr. Winkler contrived a wheel to move by electricity; that Mr. Boze conveyed electricity from one man to another by a jet of water, when they were both placed upon cakes of rosin, at the distance of six paces; and that Mr. Gordon even fired spirits by a jet of water. †

THE firing of the effluvia of bodies, which was first done in Germany, was soon after repeated in England, and among others by Dr. Miles; who, as appears by a paper of his, read at the Royal Society, March the 7th 1745, kindled phosphorus by the application of the excited tube itself to it, without the intervention of any conductor. ‡

THIS gentleman's tube happening to be in excellent order upon this occasion, he observed, and was perhaps the first who observed, *pencils of rays*, which he calls *coruscations*, darting from the tube, without the aid of any conductor approaching it. Of these coruscations he gave a drawing, which answers pretty exactly to the appearance of such pencils as are now very common, particularly since Mr. Canton has

* Nollet's Recherches, p. 187.   † Phil. Transf. abridged, Vol. 10. p. 276.   ‡ Ib. p. 272.

taught

taught us the ufe of the amalgama, by which a tube may be excited much more ftrongly than it could have been before.*

BUT the moft diftinguifhed name in this period of the hiftory of electricity, is that of DR. WATSON. He was one of the firft among the Englifh who took up, and improved upon, the difcoveries made by the Germans; and to his ingenuity, and intenfe application, we owe many curious improvements and difcoveries in electricity. His firft letters to the Royal Society on this fubject are dated between March 28th. and October 24th. 1745.

DR. WATSON's attention to the fubject of electricity feems firft, or principally, to have been engaged by the accounts of the Germans having fired fpirit of wine by it. In this experiment he fucceeded; and, moreover, found that he was able to fire, not only the phlogifton of Frobenius, and rectified fpirit of wine, but even common proof fpirit. He alfo fired air made inflammable by a chymical procefs.† He even fired both fpirit of wine, and inflammable air, by a drop of cold water, thickened with a mucilage made with the feed of flea wort, and even with ice. ‡ He alfo fired thefe fubftances with a hot poker electrified, when it would not fire them in any other ftate. ‖ He fired gun powder, and difcharged a mufket by the power of electricity, when the gunpowder had been ground with a little camphor, or a few drops of fome inflammable chymical oil. § Laftly, it was a difcovery of Dr. Watfon's, that thefe fubftances were capable of

* Phil. Tranf. abridged, Vol. 10. p. 272.   † Ib. p. 286.   ‡ Ib. p. 290.   ‖ Ib. p. 288.
§ Ib. p. 289.

being

being fired by what he calls *the repulsive power of electricity*; which was performed by the electrified person holding the spoon which containeth the substance to be fired, and another person, not electrified, bringing his finger to it. * Before this time, the substance to be fired had always been held by a person not electrified.

In his attempts to fire electrics per se, as turpentine, and balsam of capivi, by this repulsive power, he thought he confuted an opinion which had prevailed amongst many persons, that electricity only floated on the surfaces of bodies, for he found that the fume of these substances could not be fired but by a spark fetched from the spoon which contained them. This spark must therefore pass through the electric, from the surface of the spoon below, which was in contact with the electrified conductor. But it must be observed, that no substance in a state of fluidity is electric, except air.

ELECTRIFYING a number of pieces of fine spun glass, and other pieces of wire, of the same length and thickness; he was agreeably amused by observing, that the threads of glass jumped to the electrified body, and adhered to it without any snapping; whereas the wires jumped up and down very fast, giving a snap, and a small flame, every time.†

In a paper read at the Royal Society, February 6th 1746, he observed, that electric sparks appeared different in colour and form, according to the substances from which they proceeded; that the fire appeared much redder from rough bodies, as rusty iron, &c. than from polished bodies, though

* Phil. Transf. abridged, Vol. 10. p. 281.   † Ib. p. 286.

they

they were ever fo fharp, as from polifhed fciffars, &c. He judged that the different appearance was owing, rather to the different reflection of the electric light from the furface of the bodies from which it was emitted, than to any difference in the fire itfelf. *

HE alfo obferved, that electricity fuffered no refraction in pervading glafs; having found, by exact obfervations, that its direction was always in right lines, even through glaffes of different forms, included one within another, and large fpaces left between each glafs; † that if books or other non electrics were laid upon glafs and interpofed between the excited electric and light bodies, the direction of the virtue was ftill in right lines, and feemed inftantly to pafs through both the books and the glafs. In thefe experiments he con-ftantly obferved, that the electric attraction through glafs was much more powerful, when the glafs was made warm, than when it was cold. ‡ He fometimes found electricity to pervade, though in fmall quantities, electrics of above four inches thick. ‖

HE fays, that in electrifying fubftances of great extent, the attractive power was firft obferved, at that part of it which was moft remote from the excited electric.

HE made fome experimen:s which fhowed, that the fire of electricity was affected, neither by the prefence, nor the abfence of other fire. One of his experiments was made with a chymical mixture, thirty degrees below the freezing

* Phil. Tranf. abridged, Vol. 10. p. 290.   † Ib. p. 291.   ‡ Ib. p. 292.
‖ Ib. p. 295.

point

point of Fahrenheit's thermometer; from which, when electrified, the flashes were as powerful, and the strokes as smart, as from red hot iron. *

In a sequel to the above experiments, read the 30th of October 1746, Dr. Watson mentions his having lined a glass globe to a considerable thickness with a mixture of wax and rosin; but he found no difference between that and the other globes. †

He also made various experiments with a number of globes, whirled at the same time, and having one common conductor; and concluded from them, that the power of electricity was increased by the number and size of the globes, to a certain degree, but by no means in proportion to their number and size. Yet the Doctor allows a very great increase, in an inference he makes from these very experiments. As bodies to be electrified, he says, will only contain a certain quantity of electricity; when that quantity is acquired, *which is soonest done by a number of globes*, the surcharge is dissipated as fast as it is excited. So that, it is plain, more fire was collected by the number of globes, though the form of the conductor he made use of was such as could not retain it. The great power of his four globes united is manifest from his own account of them. For, he says, that when two pewter plates were held, one in the hand of an electrified person, and the other in the hand of one who stood upon the floor; the flashes of pure and bright flame were so large, and succeeded each other so fast, that, when the room was darkened, he

* Phil. Tranf. abridged, Vol. 10. p. 293.   † Ib. p. 295.

could

could diſtinctly ſee the faces of thirteen perſons who ſtood round the room. *

Lastly, the Doctor found, that the ſmoke of original electrics was a conductor of electricity, and alſo that flame would conduct the whole of it undiminiſhed; by obſerving that two perſons, ſtanding upon electrics, could communicate the virtue to each other, with nothing interpoſed but the ſmoke in the one caſe, and flame in the other. †

It was in this period that Mr. Du Tour diſcovered, that flame would deſtroy electricity; as he informed the Abbé Nollet, in a letter dated 21ſt Auguſt 1745. The ſame was alſo diſcovered by Mr. Waitz, and an account of it was publiſhed in a diſſertation, which was honoured with a prize the ſame year at the academy in Berlin.

---

* Phil. Tranſ. abridged, Vol. 10. p. 295.   † Ib. p. 296.

PERIOD

PERIOD VIII.

THE HISTORY OF ELECTRICITY, FROM THE DISCOVERY OF THE LEYDEN PHIAL IN THE YEAR 1746, TILL DR. FRANKLIN's DISCOVERIES.

SECTION I.

THE HISTORY OF THE LEYDEN PHIAL ITSELF, TILL DR. FRANKLIN's DISCOVERIES RELATING TO IT.

THE year 1746 was famous for the moſt ſurpriſing diſcovery that has yet been made in the whole buſineſs of electricity, which was the wonderful accumulation of its power in glaſs, called at firſt the LEYDEN PHIAL; becauſe firſt made by Mr. Cuneus a native of Leyden, as he was repeating ſome experiments which he had ſeen with Meſſrs. Muſchenbroeck, and Allamand, profeſſors in the univerſity

M of

of that city. * But according to other accounts, it was Mr. Muſchenbroeck himſelf who firſt felt the ſhock, as he was uſing an iron cannon, ſuſpended on ſilk lines, for a conductor. †

The views which led to this diſcovery were, as I have been informed, as follows. Profeſſor Muſchenbroeck and his friends, obſerving that electrified bodies, expoſed to the common atmoſphere, which is always replete with conducting particles of various kinds, ſoon loſt their electricity, and were capable of retaining but a ſmall quantity of it, imagined, that, were the electrified bodies terminated on all ſides by original electrics, they might be capable of receiving a ſtronger power, and retaining it a longer time. Glaſs being the moſt convenient electric for this purpoſe, and water the moſt convenient non electric, they firſt made theſe experiments with water, in glaſs bottles : but no conſiderable diſcovery was made, till the profeſſor, or Mr. Cuneus, happening to hold his glaſs veſſel in one hand, containing water, which had a communication with the prime conductor, by means of a wire; and, with the other hand, diſengaging it from the conductor (when he imagined the water had received as much electricity as the machine could give it) was ſurpriſed by a ſudden ſhock in his arms and breaſt, which he had not in the leaſt expected from the experiment.

It is extremely curious to obſerve the deſcriptions which philoſophers, who firſt felt the electrical ſhock, give of it ; eſpecially as we are ſure we can give ourſelves the ſame ſen-

* Dalibard's Hiſtoire abrigée. p. 33. † Hiſtoire de l'electricité, p. 29.

ſation,

fation, and thereby compare their defcriptions with the reality.   Terror and furprife certainly contributed not a little to
the exaggerated accounts they gave of it; and, could we not
have repeated the experiment, we fhould have formed a very
different idea of it from what it really is, even when given in
greater ftrength than thofe who firft felt this electrical fhock
were able to give it.   It will amufe my readers if I give
them an example or two.

Mr. Muschenbroeck, who tried the experiment with a
very thin glafs bowl, fays, in a letter to Mr. Reaumur,
which he wrote foon after the experiment; that he felt
himfelf ftruck in his arms, fhoulders and breaft, fo that he
loft his breath, and was two days before he recovered from
the effects of the blow and the terror.   He adds, that he
would not take a fecond fhock for the kingdom of France. *

The firft time Mr. Allamand made this experiment (which
was only with a common beer glafs) he fays, that he loft the
ufe of his breath for fome moments; and then felt fo intenfe
a pain all along his right arm, that he at firft apprehended
ill confequences from it, though it foon after went off without any inconvenience. †   But the moft remarkable account
is that of Mr. Winkler of Leipfick.   He fays, that the firft
time he tried the Leyden experiment, he found great convulfions by it in his body; and that it put his blood into great
agitation; fo that he was afraid of an ardent fever, and was
obliged to ufe refrigerating medicines.   He alfo felt an heavi-
nefs in his head, as if a ftone lay upon it.   Twice, he fays,

* Hiftoire, de l'electricité, p. 30.   † Phil. Tranf. abridged, Vol. 10. p. 321.

M 2

it

it gave him a bleeding at his nofe, to which he was not in-clined ; and that his wife (whofe curiofity, it feems, was ftronger than her fears) received the fhock only twice, and found herfelf fo weak, that fhe could hardly walk ; and that, a week after, upon recovering courage to receive another fhock, fhe bled at the nofe after taking it only once. *

WE are not, however, to infer from thefe inftances, that all the electricians were ftruck with this panic. Few, I be-lieve, would have joined with the cowardly profeffor, who firft felt this fhock, in faying that he would not take a fecond for the kingdom of France. Far different from thefe were the fentiments of the magnanimous Mr. Boze, who with a truly philofophic heroifm, worthy of the renowned Empe-docles, faid he wifhed he might die by the electric fhock, that the account of his death might furnifh an article for the memoirs of the French academy of fciences. † But it is not given to every electrician to die the death of the juftly envied Richman.

IT was this aftonifhing experiment that gave eclat to elec-tricity. From this time it became the fubject of general con-verfation. Every body was eager to fee, and, notwithftanding the terrible account that was reported of it, to *feel* the expe-riment ; and in the fame year in which it was difcovered, numbers of perfons, in almoft every country in Europe, got a livelihood by going about and fhowing it.

WHILE the vulgar of every age, fex, and rank were viewing this prodigy of nature and philofophy with wonder and amazement ; we are not furprized to find all the electri-cians of Europe immediately employed in repeating this

* Phil. Tranf. abridged, Vol. 10. p. 327.   † Hiftoire, p. 164.

great

great experiment, and attending to the circumstances of it. Mr. Allamand remarked, that, when he first tried it, he stood simply upon the floor, and not upon cakes of rosin. He said, that it did not succeed with all kinds of glass; for that, though he had tried several, he had had perfect success with none but those of Bohemia, and that he had tried English glasses without any effect at all.* Professor Muschenbroeck at that time only observed, that the glass must not be all wet on the outside.

It is no wonder that so few of the properties of glass charged with electrical fire were known at first, notwithstanding the attention that was immediately given to the subject by all the electricians in Europe. The experiment is, to this day, justly viewed with astonishment by the most profound electricians: for, though some remarkable phenomena of it have been excellently accounted for by Dr. Franklin, and others, still much remains to be done; and, in many respects, the circumstances attending it are still inexplicable. What will result from more attention being given to it time only can show.

Dr. Watson, who gives an account of this famous experiment in the Philosophical Transactions, observes, that it succeeded best when the phial, which contained the water, was of the thinnest glass, and the water warmer than the ambient air. He says he tried the effect of increasing the quantity of water in glass vessels of different sizes, as far as four gallons, without in the least increasing the stroke. He

---

* Phil. Transf. abridged, Vol. 10. p. 321.

also

alfo obferved, that the force of the ftroke did not increafe in proportion to the fize of the globe, or the number of globes employed upon the occafion; for that he had been as forcibly ftruck with a phial charged by means of a globe of feven inches in diameter, as from one of fixteen, or from three of ten; and that, at Hamburgh, a fphere had been employed of a Flemifh ell in diameter, without the expected increafe of power. He found, that if mercury was ufed, inftead of water, the ftroke was by no means increafed in proportion to its fpecific gravity. He alfo firft obferved, that feveral men, touching each other, and ftanding upon electrics, were all fhocked, though only one touched the gun barrel; but that no more fire was vifible from them all, than if one only had difcharged it.

SEVERAL of thefe obfervations fhow how imperfectly this great experiment was underftood, for fome time after it was firft made. Dr. Watfon, however, obferved a circumftance attending the charging of the phial, which, if purfued, would have led him to the difcovery, which was afterwards made by Dr. Franklin. He fays, that when the phial is well electrified, and you apply your hand thereto, you fee the fire flafh from the outfide of the glafs, wherever you touch it, and it crackles in your hand. *

HE alfo obferved, that when a fingle wire only was faften-ed round a phial, properly filled with warm water, and charg-ed; upon the inftant of its explofion, the electrical corrufca-tions were feen to dart from the wire, and to illuminate the water contained in the phial.

* Phil. Tranf. abridged, Vol. 10. p. 298.

SEVERAL

Several other very important circumstances, relating to the discharge of the phial, were observed by Dr. Watson. He found that the stroke was, *cæteris paribus*, as the points of contact of the non electrics on the outside of the glass. And upon showing Dr. Bevis the experiments which proved this assertion, the Doctor suggested a more clear and satisfactory method of proving it, and which has been the means of accumulating and increasing the force of charged glass far beyond what was expected from the first discovery of it. This method was coating the outside of the phial, very near to the neck, with sheet lead, or tinfoil. When a bottle was prepared in this manner, and nearly filled with water, they observed, that a person who only held in his hand a small wire communicating with that coating, felt as strong a shock as he would have felt, if his hand had been in actual contact with every part of the phial touched by the coating.*

Dr. Watson also discovered, that the electrical power, in the discharge of the phial, darts *rectissimo cursu*, as he styles it, between the gun barrel and the phial; and, though it is not strictly true, that the shock goes the nearest way, yet it does so *cæteris paribus*, which alone was a considerable discovery for that time. He observed, that, in a company joining hands, a person touching two other persons in the circle, who did themselves touch one another, felt nothing of the shock, his body making no necessary part of the circle; and also, if a man, holding a wire, which communicated with the outside of the phial, as it hung upon the conductor,

* Phil. Transf. abridged, Vol. 10. p. 299.

should

should touch the conductor with it; the explosion was made, but the man felt nothing. *

In a paper read at the Royal Society, January 21st. 1748, Dr. Watson mentions another discovery relating to the Leyden phial, which Dr. Bevis suggested, and he compleated. Having been before fully satisfied, that the shock from the phial was not in proportion to the quantity of matter contained in the glass, but was increased by it, and likewise by the number of points of non electric contact on the outside of the glass; he procured three jars, into which he put round leaden shot, and, joining their wires and coating, discharged them all as one jar. Upon this he observed, that the electrical explosion from two or three of those jars was not double or treble to that from one of them; but that the explosion from three was much louder than that from two, and the explosion from two much louder than that from one. †

This experiment had induced him to imagine, that the explosion from those jars was owing to the great quantity of non electric matter contained in them. And whilst he was considering of some certain method of assuring himself whether the fact was so, Dr. Bevis informed him, that he had found the electric explosion to be as great from covering the sides of a pane of glass, within about an inch of the edge, (which was a curious improvement of Mr. Smeaton's) as it could have been from an half pint phial of water. Upon this Dr. Watson coated large jars with leaf silver, both inside and outside, within an inch of the top, and from the great

* Phil. Transf. abridged, Vol. 10. p. 301.   † Ib. p. 374.

explosion

explosion he produced from them, when so little non electric matter was contained in them, he was of opinion, that the effect of the Leyden bottle was greatly increased by, if it was not principally owing to, not so much the quantity of non electric matter contained in the glass, as the number of points of non electric contact within the glass, and the density of the matter of which those points consisted; provided the matter was, in its own nature, a ready conductor of electricity. He also observed, that the explosion was greater from hot water inclosed in glasses, than from cold; and from his coated jars warmed, than when cold. *

THE Doctor observed, that when the circle for the discharge was not through perfect conductors, the explosion was made slowly, and not all at once. This law, he says, was invariable, but he was not able to account for it. But to prove that the electricity passed with its whole force through the circle of non electrics, he made a circuit consisting of iron bars, and spoons filled with spirits between each bar, (but at some small distance from them) and, upon the explosion, all the spoons were on fire at once. This was the first time, as he observes, that spirits were ever fired without either the spirits, or the non electric on which they were placed, being insulated, or put upon original electrics. And yet, he says, though we know, from its effects, that the electricity goes through the whole circuit of non electrics, with all its vigour; its progress is so quick, as not to affect, by attracting or otherwise, any light bodies disposed very near the non electrics, through which it must necessarily pass. †

* Phil Transf. abridged, Vol. 10. p. 377.   † Ib. p. 378.

N   It

IT is curious to obſerve in what manner Dr. Watſon explained the ſhock of the Leyden phial, about the time that he firſt made the experiment with it. He had then been led (by a courſe of experiments which will be mentioned hereafter) to the notion both of the *afflux*, and *efflux* of electric matter in all electrical experiments. To apply this principle to the caſe in hand, he ſuppoſed, that the man who felt the ſhock parted with as much of the fire from his body, as was accumulated in the water and the gun barrel; and that he felt the effect in both arms, from the fire which was in his body, ruſhing through one arm to the gun barrel, and through the other to the phial. He imagined alſo, that as much fire as the man parted with was inſtantly replaced from the floor of the room, and that with a violence equal to the manner in which he loſt it. It alſo appears from Dr. Watſon's remarks on ſome ſubſequent experiments of Mr. Monnier, that he then imagined, that though a conſiderable quantity of the electric matter pervaded the glaſs (as he thought was ſeen upon preſenting a non electric body to it, when it ſtood upon the glaſs ſtand, and without which it could not be charged at all) yet, that the loſs of the electric matter this way was not equal to what came in by the wire; the thinneſs of the glaſs permitting it not wholly, but partially to ſtop the electricity. *

AFTERWARDS, when (from a courſe of experiments which will alſo, be recited in their proper place) Dr. Watſon changed his opinion about this afflux and efflux of electric

* Phil. Tranſ. abridged, Vol. 10. p. 348.

matter,

matter, with a generofity and franknefs becoming every inquirer after truth, he retracted this hypothefis; and, in refutation of it, he further adds, that the charged phial would
explode with equal violence, if the hoop of the wire was bent,
fo as to come near the coating of the phial, without any other
non electric body being near, from which fuch a quantity
could be fupplied. He had alfo obferved, that if a man ftood
upon glafs, and difcharged the phial, he felt the fame fhock
as if he had ftood upon the floor. I fhall fubjoin a remarkable paragraph of the Doctor himfelf upon this occafion, as
I think it very applicable even to us in this more advanced
ftate of the fcience.

" I TAKE notice of thefe," fays the Doctor, " in as
" much as, notwithftanding the very great progrefs which
" has been made in our improvements in this part of natural
" philofophy, within thefe few years; pofterity will regard
" us as only in our *noviciate*; and therefore it behoves us,
" as far as we can be juftified therein by experiment, to cor-
" rect any conclufions we may have drawn, if others yet more
" probable prefent themfelves." *   The Doctor has lived to
fee, not only that *pofterity* would confider him and his affiftants at that time, as in their noviciate; but he *himfelf*, already
in the courfe of a few years, looks upon both himfelf and them
in the fame light. And, confidering the quick advances ftill
making in this fcience, it is to be hoped he may ftill live to
fee, even the electricians of the prefent year, to have been
only in their noviciate.

* Phil. Tranf. abridged, Vol. 10. p. 373.

N 2   ALL

HAVING ſeen what was done by Dr. Watſon towards ex-plaining the electric ſhock, before it was undertaken by Dr. Franklin; let us ſee what obligation we are under to other Engliſh electricians, and particularly Mr. Wilſon.

MR. WILSON ſays that, as early as the year 1746, he diſcovered a method of giving the ſhock to any particular part of the body without affecting the reſt. * He increaſed the ſtrength of the ſhock by plunging the phial in water, thereby giving it a coating of water on the outſide, as high as it was filled on the inſide. †

IN a letter to Mr. Smeaton, dated Dublin October 6th. 1746, he mentions his having made ſome experiments, in order to diſcover the law of accumulation of the electric matter in the Leyden bottle; and found, that it was always in proportion to the thinneſs of the glaſs, the ſurface of the glaſs, and that of the non electrics in contact with the inſide and outſide thereof. The experiments, he ſays, were made with water a little warmed, which was poured into the bottle, while the outſide was immerged in a veſſel filled with water, but a little colder; leaving three inches, or thereabouts, uncovered, which was preſerved dry and free from duſt. An account of this experiment he wrote to Mr. Folkes, and it was read before the Royal Society, October 23d. 1746, as appears by their minutes of that day, though the original was loſt or miſlaid.

ANOTHER curious experiment Mr. Wilſon made, in order to prove an hypotheſis, which he conceived very early, of the influence of a ſubtle medium ſurrounding all bodies, and

* Wilſon's Eſſay, p. 88. † Ib. p. 71.

reſiſting

refifting the entrance or exit of the electric fluid. To determine this, he made the Leyden experiment with a chain, and
confidered each link of it as having two furfaces, at leaft ; fo
that lengthening or fhortening the chain, in each experiment,
would occafion different refiftances ; and the event, he fays,
proved accordingly. When he made the difcharge with one
wire only, he found the refiftance to be lefs than when
a chain was ufed. But to leave no room for doubt, he
caufed the chain to be ftretched with a weight, that the
links might be brought nearer into contact, and the event
was the fame as when a fingle wire had been ufed. *

Two circuits being made, one confifting of the arms of a
man, and the other of the links of a chain ; he found, that
the fire would take the arms of the man ; but that if the
chain were ftretched, it would take the chain. No perfon,
he fays, who has not made the experiment, would imagine,
with how much force the chain muft be ftretched before the
experiment will anfwer, and the electric fluid pafs through it
without producing a fpark at any of the links ; that is, before the links can be brought into abfolute contact with one
another, their own weight being by no means fufficient. †

MR. WILSON obferved, that if one part of the Leyden
phial was ground very thin, and covered with fealing wax
till it was charged, and then had the fealing wax taken off,
and a conductor communicating with the earth touched the
thin part, the charge would be diffipated in nearly half
the time that it otherwife would have been. ‡

* Letter to Hoadley.   † Wilfon and Hoadley, p. 65.   ‡ Wilfon's Effay, p. 74.

HE

HE obſerved that bodies, placed without the electric circuit, would be affected with the ſhock, if they were only in
contact with any part of it, or very near it. To ſhew this to
the moſt advantage, he ſet a charged phial upon a glaſs ſtand,
and placed ſeveral pieces of braſs upon the ſtand, one of
them in contact with the chain that formed the circuit, and
others a twentieth of an inch from it, or from one another;
and, upon making the diſcharge, there was a ſpark viſible
between each of them. *

ANALOGOUS, in ſome reſpects, to this, was Mr. Wilſon's
obſervation, that if the circuit was not made of metals, or
other very good conductors, the perſon who laid hold of them,
in order to perform the experiment, felt a conſiderable ſhock
in that arm which was in contact with the circuit.

HE alſo obſerved, that when the phial was coated, within
and without, with metals, the firſt exploſion bore the greateſt
proportion to the ſubſequent ones, the whole charge, being
diſſipated almoſt at once; whereas, when water was uſed,
the ſubſequent exploſions were more in number, and more
conſiderable; and that when the phial was charged with nothing but a wire inſerted in it, the firſt exploſion and the
ſubſequent ones were ſtill more nearly equal.

MR. WILSON once happening to break a ſmall wire by the
convulſive ſhock given to his arms by the Leyden phial, he
faſtened to his hands, well guarded with leather, a larger
wire, of the thickneſs of a ſlender knitting needle, and placed
himſelf in ſuch a manner, that it would neceſſarily be ſtretched, if his arm ſhould be convulſed again. He accordingly

* Wilſon's Eſſay, p. 90.

discharged

difcharged the phial, and this wire was broken, like the former. *

MR. GEORGE GRAHAM fhewed how feveral circuits for the difcharge of the Leyden phial might be made at the fame time, and the fire be made to pafs through them all. He made a number of perfons take hold of a plate of metal, communicating with the outfide of the phial; and all together, likewife, lay hold of a brafs rod with which the difcharge was made; when they were all fhocked at the fame time, and in the fame degree. †

LASTLY Mr. Canton found, that if a charged phial was placed upon electrics, the wire and the coating would give a fpark or two alternately; and that, by continuing this operation, the phial would be difcharged. ‡ This difcovery, which is the firft that I find recorded of this excellent philofopher, to whom the fcience of electricity owes fo much, has a near affinity to the great difcovery of Dr. Franklin; but he did not then obferve, that thofe alternate fparks proceeded from the two contrary electricities. This hiftory will furnifh many more inftances of perfons being on the eve of great difcoveries, without actually making them.

WE have feen what obfervations the Englifh philofophers had made upon the Leyden experiment before the time of Dr. Franklin; let us now take a view of what was done by electricians in other parts of the world, within the fame period.

As Mr. Mufchenbroeck's letter to Mr. Reaumur, concerning the experiment of the phial, came at a time when many

* Wilfon's Effay, p. 84.   † Ib. p. 128.   ‡ Ib. p. 64.

learned

learned men were employed about electricity, the Abbé Nollet, and Mr. De Monnier, gentlemen of the academy, zealous to search into so an extraordinary a phenomenon, divesting themselves of the fear with which the professor's letter might justly have inspired them, made the experiment upon themselves, and, in like manner, said they found the commotion very terrible. The report of it instantly spread through the court and the city, from whence all ranks of men crouded to see this new kind of thunder, and to experience the effect of it. *

The Abbé Nollet was the first who made experiments upon the phial in France, and the result of many of them was the same with what Dr. Watson had discovered, for which reason I shall not recite them here. They may all be seen at one view in his *Leçons de physique*, p. 481. The cicumstances which the English philosophers had not attended to, are the following.

The Abbé received a shock from a bottle out of which the air had been exhausted, and into which the end of his conductor had been inserted. This was an accidental discovery; for he received the shock as he was holding one hand to the glass vessel, in order to observe the beautiful irradiations of the electric light towards it in vacuo, and putting his other hand to the conductor, in order to adjust something about it. The blow he received was greater, he says, than he ever felt from the Leyden experiment in any other form. †

* Nollet's Leçons de Physique, p. 452   † Recherches, p. 426.

In

IN the fame place he obferves, that he never confidered the water in the phial as of any ufe, but to convey the electric matter into the infide of the glafs; and that he afcribed the force of the glafs in giving a fhock, to that property of it, whereby it retained it more ftrongly than conductors do, and was not fo eafily divefted of it as they are.

MR. MONNIER is faid, by Mr. Buffon, to have been the firft who difcovered that the Leyden phial would retain its electricity a confiderable time after it was charged, and to have found it to do fo for thirty fix hours, in time of froft. He frequently electrified his phial at home, and brought it in his hand, through many ftreets, from the college of Har-court to his apartments in the King's garden, without any confiderable diminution of its efficacy. *

IT feems that in France experiments were firft made to try how many perfons might feel the fhock of the fame phial. The Abbé Nollet, whofe name is famous in electricity, gave it to one hundred and eighty of the guards, in the King's prefence; and at the grand convent of the Carthufians in Paris, the whole community formed a line of nine hundred toifes, by means of iron wires between every two perfons (which far exceeded the line of one hundred and eighty of the guards) and the whole company, upon the difcharge of the phial, gave a fudden fpring, at the fame inftant of time, and all felt the fhock equally. †

MR. NOLLET alfo tried the effect of the electric fhock upon two birds, one of which was a fparrow, and the other

* Phil. Tranf. abridged, Vol. 10. p. 333.   † Ib. p. 335.

O   a a chaffinch, which, as far as I can find, were the first brute animals of any kind that ever received it. The confequence was, that upon the first fhock, they were both inftantaneoufly ftruck motionlefs, and, as it were, lifelefs, though for a time only; for they recovered fome few minutes after. Upon the fecond fhock, the fparrow was ftruck dead, and, upon exami- nation, was found livid without, as if it had been killed with a flafh of lightning; moft of the blood veffels in the body being burft by the fhock. The chaffinch revived as before. * Fifhes were alfo killed with the electric fhock, by the Abbé, and others.

THE circumftance of the blood veffels of the fparrow being burft is, I imagine, a miftake. I have feen no fuch effect, when fmaller animals have been killed by a fhock fifty times as great as, it is probable, the Abbé ufed upon this occafion.

THE Abbé Nollet was the firft electrician who mentioned the burfting of glafs veffels by the electric explofion. They were pierced, he fays, with round holes, three or four lines in diameter. †

IT feems that the French philofophers, as well as the Englifh, had obferved, that, if the phial ftood upon glafs, it could not be charged, except a perfon's hand, or fome other non electric fubftance were brought near to it. Upon this they imagined the fire ftreamed out of the hand, and paffed through the fubftance of the phial into the water. ‡ This fact fur- prifed them very much, as it well might. They alfo obferv-

* Phil. Tranf. abridged, Vol. 10. p. 336. † Nollet's Lettres, Vol. 1. p. 42. ‡ Phil. Tranf. abridged, Vol. 10. p. 331.

ed,

ed, that a light body would be attracted by a charged phial, as it ftood upon the table, if any perfon touched the wire; but they fay that, if the phial itfelf were touched, the light body would be repelled with a force equal to its attraction in the former cafe. * They likewife found, that when the charged phial ftood upon glafs, it might be handled with all fafety.† Thefe experiments feem not to have been made with proper circumfpection: for by an attention to thefe very circumftances, Dr. Franklin was afterwards led to the great difcovery of the different quality of the electricity, on different fides of the glafs.

* Phil. Tranf. abridged, Vol. 10. p. 335.   † Ib. p. 337.

O 2   SECTION

SECTION II.

The methods used by the French and English philosophers, to measure the distance to which the
electric shock can be carried, and the veolcity
with which it passes.

We are now come to an ampler field of electrical experiments, in which we shall be spectators, not of
what might be exhibited in a private room, and by a few
operators; but where we shall find an amazing apparatus necessary, and a great number of assistants in the management
of it; as well as the greatest judgement, and the most unwearied patience in the conduct of it.

The French philosophers were the first to appear in this
field, but they did little more than excite the English to go
far beyond them in these great undertakings. It has been
said already, that a circuit was made of nine hundred toises,
consisting of men holding iron wires betwixt each two,
through which the electric shock was sensibly felt. At another time, they made the shock pass through a wire two
thousand

thousand toises in length, that is near a Paris league, or a-bout two English miles and a half; though part of the wires dragged upon wet grass, went over charmil hedges, or palisades, and over ground newly ploughed up. Into another chain they took the water of the bason in the Thuilleries, the surface of which was about an acre, and the phial was discharged through it. *

Mr. Monnier the younger, also endeavoured to determine the velocity of the electric matter; and, for this purpose, made the shock pass through an iron wire of nine hundred and fifty toises in length, but he could not observe, that it spent a quarter of a second in passing it. He also found, that when a wire of one thousand three hundred and nineteen feet, with its extremities brought near together, was electrified; that the electricity ceased at one end, the moment it was taken off at the other. This fact refuted the opinion of those who maintained, that it was the force of the electrical *shock*, which threw the electric matter with so great velocity.

But all these attempts of the French would scarce deserve to have been mentioned, but that they preceded the greater, the more accurate, and the more numerous experiments of the English. The names of the English gentlemen, animated with a truly philosophical spirit, and who were indefatigable in this business, deserve to be transmitted to posterity in every work of this nature.

The principal agent in this great scene was Dr. Watson. He planned and directed all the operations, and never failed

* Phil. Transf. abridged, Vol. 10. p. 336.

to be prefent at every experiment. His chief affiftants were Martin Folkes Efq. prefident of the Royal Society, Lord Charles Cavendifh, Dr. Bevis, Mr. Graham, Dr. Birch, Mr. Peter Daval, Mr. Trembley, Mr. Ellicott, Mr. Robins, and Mr. Short. Many other perfons, and fome of diftinction, gave their attendance occafionally.

DR. WATSON, who wrote the hiftory of their proceedings, in order to lay them before the Royal Society, begins with obferving (what was verified in all their experiments) that the electric fhock is not, ftrictly fpeaking, conducted in the fhorteft manner poffible, unlefs the bodies through which it paffes conduct equally well; for that, if they conduct unequally, the circuit is always formed through the beft conductors, though the length of it be ever fo great.

THE firft attempt thefe gentlemen made, was to convey the electric fhock acrofs the river Thames, making ufe of the water of the river for one part of the chain of communication. This they accomplifhed on the 14th. and 18th. of July 1747, by faftening a wire all along Weftminfter bridge, at a confiderable height above the water. One end of this wire communicated with the coating of a charged phial, the other being held by an obferver, who, in his other hand, held an iron rod, which he dipped into the river. On the oppofite fide of the river, ftood a gentleman, who, likewife, dipped an iron rod in the river, with one hand; and in the other, held a wire, the extremity of which might be brought into contact with the wire of the phial.

UPON making the difcharge, the fhock was felt by the obfervers on both fides the river, but more fenfibly by thofe

who

who were stationed on the same side with the machine; part of the electric fire having gone from the wire down the moist stones of the bridge, thereby making several shorter circuits to the phial; but still all passing through the gentlemen who were stationed on the same side with the machine. This was, in a manner, demonstrated by some persons feeling a sensible shock in their arms and feet, who only happened to touch the wire, at the time of one of the discharges, when they were standing upon wet steps which led to the river. In one of the discharges made upon this occasion, spirits were kindled by the fire which had gone through the river. *

UPON this, and the subsequent occasions, the gentlemen made use of wires, in preference to chains, for this, among other reasons, that the electricity which was conducted by chains was not so strong, as that which was conducted by wires. This, as they well observed, was occasioned by the junctures of the links not being sufficiently close, as appeared by the snapping and flashing at every juncture, where there was the least separation. These lesser snappings, being numerous in the whole length of a chain, very sensibly lessened the great discharge at the gun barrel.

THEIR next attempt was to force the electrical shock to make a circuit of two miles, at the new river at Stoke Newington. This they performed on the 24th. of July 1747, at two places; at one of which the distance by land was eight hundred feet, and by water two thousand: in the other, the distance by land was two thousand eight hundred feet, and by

* Phil. Transf. abridged, Vol. 10. p. 349, &c.

water

water eight thousand. The disposition of the apparatus was similar to what they before used at Westminster bridge, and the effect answered their utmost expectations. But, as in both cases, the observers at both extremities of the chain, which terminated in the water, felt the shock, as well when they stood with their rods fixed into the earth twenty feet from the water, as when they were put into the river; it occasioned a doubt, whether the electric circuit was formed through the windings of the river, or a much shorter way, by the ground of the meadow: for the experiment plainly shewed, that the meadow ground, with the grass on it, conducted the electricity very well.

By subsequent experiments, they were fully convinced, that the electricity had not, in this case, been conveyed by the water of the river, which was two miles in length, but by land, where the distance was only one mile; in which space, however, the electric matter must necessarily have passed over the new river twice, have gone through several gravel pits, and a large stubble field. *

July 28th. they repeated the experiment, at the same place, with the following variation of circumstances. The iron wire was, in its whole length, supported by dry sticks, and the observers stood upon original electrics; the effect of which was, that they felt the shock much more sensibly, than when the conducting wire had lain upon the ground, and when the observers had likewise stood upon the ground, as in the former experiment.

* Phil. Transf. abridged, Vol. 10. p. 360.

AFTERWARDS,

AFTERWARDS, every thing elfe remaining as before, the obfervers were directed, inftead of dipping their rods into the water, to put them into the ground, each one hundred and fifty feet from the water. They were both fmartly ftruck, though they were diftant from each other above five hundred feet. *

THE fame gentlemen, pleafed with the fuccefs of their former experiments, undertook another, the object of which was, to determine, whether the electric virtue could be conveyed through dry ground; and, at the fame time, to carry it through water to a greater diftance than they had done before. For this purpofe, they pitched upon Highbury barn beyond Iflington, where they carried it into execution on the 5th. of Auguft 1747. They chofe a ftation for their machine, almoft equally diftant from two other ftations for obfervers upon the new river; which were fomewhat more than a mile afunder by land, and two miles by water. They had found the ftreets of London, when dry, to conduct very ftrongly, for about forty yards; and the dry road at Newington about the fame diftance. The event of this trial anfwered their expectations. The electric fire made the circuit of the water, when both the wires and the obfervers were fupported upon original electrics, and the rods dipped into the river. They alfo both felt the fhock, when one of the obfervers was placed in a dry gravelly pit, about three hundred yards nearer the machine than the former ftation, and one hundred yards diftant from the river: from which the gentlemen were fatif-

* Phil. Tranf. abridged, Vol. 10. p. 357.

P   fied,

fied, that the dry gravelly ground had conducted the electricity as strongly as water.

FROM the shocks which the observers received in their bodies, when the electric power was conducted upon dry sticks, they were of opinion; that, from the difference of distance simply considered, the force of the shock, as far as they had yet experienced, was very little, if at all impaired. When the observers stood upon electrics, and touched the water, or the ground, with the iron rods, the shock was always felt in their arms or wrists; when they stood upon the ground with their iron rods, they felt the shock in their elbows, wrists, and ancles; and when they stood upon the ground without rods, the shock was always felt in the elbow and wrist of that hand which held the conducting wire, and in both ancles. *

THE last attempt of this kind which these gentlemen made, and which required all their sagacity and address in the conduct of it, was to try whether the electric shock was perceptible at twice the distance to which they had before carried it, in ground perfectly dry, and where no water was near; and also to distinguish, if possible, the respective velocity of electricity and sound.

FOR this purpose, they fixed upon Shooter's hill, and made their first experiment on the 14th. of August 1747, a time, when, as it happened, but one shower of rain had fallen during five preceding weeks. The wire communicating with the iron rod, which made the discharge, was six thousand

* Phil Transf. abridged, Vol. 10. p. 360.

seven

seven hundred and thirty two feet in length, and was supported all the way upon baked sticks; as was also the wire which communicated with the coating of the phial, which was three thousand eight hundred and sixty eight feet long, and the observers were distant from each other two miles. The result of the explosion demonstrated, to the satisfaction of the gentlemen present, that the circuit performed by the electric matter was four miles, viz. two miles of wire, and two of dry ground, the space between the extremities of the wires. A distance which, without trial, as they justly observed, was too great to be credited. A gun was discharged at the instant of the explosion, and the observers had stop watches in their hands, to note the moment when they felt the shock: but, as far as they could distinguish, the time in which the electric matter performed that vast circuit might have been instantaneous. *

IN all the explosions where the circuit was made of any considerable length, it was observed, that though the phial was very well charged, yet that the snap at the gun barrel, made by the explosion, was not near so loud as when the circuit was formed in a room; so that a by-stander, says Dr. Watson, though versed in those operations, would not imagine, from seeing the flash, and hearing the report, that the stroke, at the extremity of the conducting wire, could have been considerable; the contrary whereof, when the wires were properly managed, he says, always happened.

STILL the gentlemen, unwearied in these pursuits, were desirous, if possible, to ascertain the absolute velocity of elec-

* Phil. Transf. abridged, Vol. 10. p. 363.

P 2

tricity

tricity at a certain diftance; becaufe though, in the laft experiment, the time of it's progrefs was certainly very fmall,
if any; they were defirous of knowing, fmall as that time
might be, whether it was meafurable, and Dr. Watfon had
contrived an excellent method for that purpofe.

ACCORDINGLY, on the 5th. of Auguft 1748, the gentlemen met once more, and the laft time, at Shooter's hill;
when it was agreed to make an electric circuit of two miles,
by feveral turnings of the wire, in the fame field. The
middle of this circuit, they contrived to be in the fame room
with the machine, where an obferver took in each hand one
of the extremities of the wires, each of which was a mile in
length. In this excellent difpofition of the apparatus, in
which the time between the explofion and the fhock might
have been obferved to the greateft exactnefs, the phial was
difcharged feveral times; but the obferver always felt himfelf fhocked at the very inftant of making the explofion.
Upon this the gentlemen where fully fatisfied, that, through
the whole length of this wire, which was 12276 feet in
length, the velocity of the electric matter was inftantaneous. *

THESE experiments excited the admiration of all foreign
electricians. Profeffor Mufchenbroeck, who was greatly fatisfied with the extent and fuccefs of them, faid, in a letter
to Dr. Watfon, upon the occafion, *Magnificentiffimis tuis experimentis fuperafti conatus omnium.*

IT is faid by fome, that the laft of thefe experiments
go upon a wrong fuppofition, and therefore can be of no ufe;
it being fuppofed that the very fame particles of the electric

* Phil. Tranf. abridged, Vol. 10. p. 363.

fluid

fluid, which were thrown on one fide of the charged glafs, actually made the whole circuit of the intervening conductors, and arrived at the oppofite fide : whereas Dr. Franklin's theory only requires that the deficiency on one fide of the glafs be fupplied from the neighbouring conductors; which may, in return, receive as much as they parted with, from the fide of the glafs that was overcharged. So that, to be a little more particular, the redundancy of electric matter on the charged fide of a pane of glafs, only paffes into the bodies which form that part of the circuit which is contiguous to it, driving forward that part of the fluid which was natural to them ; till, at length, the fluid which refided in thofe conductors which formed the laft part of the circuit, paffes into the exhaufted fide of the glafs.

But fhould this be cafe (though in great difcharges it fuppofes the natural quantity of electricity in bodies to be very confiderable) and fhould Dr. Watfon, and other philofophers at that time, have conceived otherwife; it does not follow, that the experiments could poffibly determine *nothing*: for there ftill remains fomething to be meafured, viz. the time required for the fucceffive diflodging the electric fluid in the whole length of the circuit.

Were the whole mafs of the electric matter contained in all the intervening conductors abfolutely folid, no motion could be made at one extremity, without producing an inftantaneous motion at the other; juft as if one end of a rod be ftruck, the motion is inftantly communicated to the other end. But this cannot be the cafe in an elaftic medium, the parts of which yield to one another. In this cafe, the motion is
communicated

communicated in a real fucceffion, like a vibration, running the whole length of the circuit; which muft therefore take up time, and be meafurable. The motion of found may be meafured, though no particle of the vibrating air be finally difplaced. Thefe great experiments of Dr. Watfon, therefore, had a real object, only it appeared to be too fmall to be afcertained by them.

SECTION

SECTION III.

Miscellaneous discoveries of Dr. WATSON, and others, till the time of Dr. FRANKLIN.

THE first of these discoveries in order of time, and in importance second to none (except that of the shock itself, and Dr. Franklin's discovery of the different electricity of the opposite sides of the charged glass) was that of Dr. Watson, proving that the glass tubes and globes did not contain the electric power in themselves, but only served as first movers, and determiners, as he calls it, of that power.

He was first led to this discovery by observing, that, upon rubbing the glass tube, while he was standing upon cakes of wax (in order, as he expected, to prevent any of the electric power from discharging itself through his body upon the floor) the power was, contrary to his expectation, so much lessened, that no snapping could be observed upon another person's touching any part of his body; but that if a person not electrified held his hand near the tube, while it was rubbed, the snapping was very sensible. *

* Phil. Transf. abridged, Vol. 10. p. 303

THE

THE event was the fame when the globe was whirled in similar circumſtances. For if the man who turned the wheel, and who, together with the machine was ſuſpended upon ſilk, touched the floor with one foot, the electric fire appeared upon the conductor; but if he kept himſelf free from any communication with the floor, no fire was produced.

DR. WATSON by this, and the following experiments in conjunction, diſcovered, what he calls, the complete circulation of the electric matter. He obſerved, that only a ſpark or two would iſſue from his hand to the inſulated machine, unleſs he, at the ſame time, formed a communication between the conductor and the floor; but that then there was a conſtant and copious flux of the electric matter to the machine.

OBSERVING, that while his hand was in contact with the conductor, the man who turned this inſulated machine gave ſparks, which would fire inflammable ſubſtances, and perform other electrical experiments which were uſually performed at the conductor; he naturally imagined, that the fire iſſued from the man, for the very ſame reaſon that all electricians had before imagined that it came from the conductor; and ſeeing that the man gave no fire unleſs there was a communication between the floor and the conductor, he concluded that, in this caſe, the fire was ſupplied by that communication, ſo that the courſe of the electricity was inverted, as he expreſſes it. *

* Phil. Tranſ. abridged, Vol. 10. p. 305.

It was not then suspected, that the eye could not distinguish in what direction an electric spark proceeded. Electricians naturally imagined that all electric powers, and consequently the electric fluid, which they supposed to be the cause of these powers, existed in the excited electric, whatever it was; and that whatever powers were exerted by electrified bodies proceeded from a real communication of electric matter to them. Accordingly, when Dr. Watson found that, by cutting off the communication of the electric with the floor, all electrical operations were stopped, he concluded, that the electric fluid was collected from the floor to the rubber, and thence conveyed to the globe. For the same reason, seeing the rubber, or the man who had a communication with it, give no sparks but when the conductor was connected with the floor, he would as naturally conclude that the globe was supplied from the conductor, as he had before concluded that it was supplied from the rubber.

Comparing both these experiments together Dr. Watson was led to infer, that, in all electrical operations, there was both an *afflux* of electric matter to the globe, and the conductor, and likewise an *efflux* of the same electric matter from them.*

Finding that a piece of leaf silver was suspended between a plate electrified by the conductor, and another communicating with the floor, he reasons from it in the following manner. " No body can be suspended in equilibrio but by the joint " action of two different directions of power : so here, the

* Phil. Transf. abridged, Vol. 10. p. 311.

Q

" blast

" blaſt of electric ether from the excited plate blows the
" ſilver towards the plate unexcited, and this laſt, in its turn,
" by the blaſt of electric ether from the floor ſetting through
" it, drives the ſilver towards the plate electrified. We find
" from hence, likewiſe, that the draught of electric ether
" from the floor is always in proportion to the quantity thrown
" by the globe over the gun barrel, or the equilibrium by
" which the ſilver is ſuſpended could not be maintained." *

DR. WATSON obſerves, that the Abbé Nollet, two years
before he made this communication, had given it as his opinion (though without any experiment which proved it) that
the electric matter did not only proceed from the electrified
bodies, but from all others about them, to a certain diſtance.†

SOMETIME after this, Dr. Watſon obſerves, in a paper
read at the Royal Society, January 21ſt. 1748, that Dr. Bevis
had carried his experiment, to prove that rubbing the tube
or the globe only conveyed, and did not produce the electric
matter, further than he had done. For he had obſerved, a-
bove a year before, that placing one man upon electrics, to
rub the tube or globe, and another alſo upon electrics to
touch them, as the conductor; both the man who rubbed, and the man who touched the excited glaſs would give
a ſpark; and further, that if they touched one another, the
ſnapping was much greater than if either of them touched a
perſon ſtanding upon the floor. Upon this the Doctor ſeems
to have corrected his former opinion of the afflux and efflux
of electric matter: for he accounts for this fact by ſuppoſing,

*Phil. Tranſ. abridged, Vol. 10. p. 310.   † Ib. p. 315.

that

that as much electricity as was taken from the person rubbing was given to him who touched the conductor, being conveyed by the globe. By this means the electricity of the former of these persons, he observes, was more rare than it naturally was, and that of the latter more dense; so that the density of electricity between these two persons differed more than that between either of them, and another person standing upon the floor. In this manner did Dr. Watson discover, what Dr. Franklin observed, about the same time, in America, and called the *plus* and *minus* in electricity. *

Dr. Watson observed that the flame at the end of an electrified wire was sensible to the hand, as a cool blast of wind, and that when light substances were attracted and repelled between an electrified plate and one communicating with the floor, the succession of these alternate attractions and repulsions was extremely quick, so that sometimes the eye could hardly keep pace with it; and that when a glass globe, of about an inch in diameter, very light and finely blown, was put into a plate of metal, and another plate hung on the conductor over it, the strokes from the alternate attractions and repulsions were almost too quick for the ear. From this last experiment he likewise deduced an argument to prove the extreme velocity with which they were attracted and repelled. He says, that if they were let fall from the height of six feet or more upon a wooden floor, or even a plate of metal, they were rarely broken; but that by the attraction and repulsion of them between these plates, though at the distance of no more than one sixth of an inch, they were frequently beaten to pieces. †

* Phil. Transf. abridged, Vol. 10. p. 369.   † Ib. p. 309.

THE

THE Doctor also proved, that the electric matter passed through the substance of the metal of communication, and not over the surface of it, by covering a wire with a mixture of wax and rosin, and discharging a phial through it.

MR. MONNIER the younger discovered, that electricity is not communicated to homogeneous bodies in proportion to their masses or quantity of matter, but rather in proportion to their surfaces; and yet that all equal surfaces do not receive equal quantities of electricity, but that those receive the most which are most extended in length; that a square sheet of lead, for instance, received a much less quantity of electricity than a small strip of the same metal with a surface equal to that of the square sheet.*

MR. WILSON, whose curious observations on the Leyden phial have been mentioned in a former section, claims no small share of honour in this. As early as the latter end of the year 1746, he made the same discovery that Dr. Watson had done, that the electric fluid did not come from the globe, but from the earth itself, and from all other non electric bodies about the apparatus. He suggested a method of proving this in a letter to Mr. Ellicott from Chester; and mentions his having completed the experiment himself soon after, in a letter to Mr. Smeaton, from Dublin.

HAVING conceived that the difference between electric and non electric bodies was owing to the different resistance, which a subtle medium, as he calls it, on the surfaces of all bodies gave to the passage of the electric fluid; and conceiving

* Phil. Transf. abridged, Vol. 10. p. 338.

that

that heat would rarify this medium, and thereby convert electrics into non electrics, he made some experiments which
confirmed him in that supposition. He found that one person might communicate electricity to another, notwithstanding the intervention of a considerable quantity of red hot glass.
He also made other experiments of a similar nature, as discharging phials by means of hot glass, hot amber, and various other heated electrics. These, however, as Mr. Canton
afterwards observed, might be owing to the hot air upon the
surfaces of those bodies, which he found to transmit electricity very well. But another experiment, which Mr. Wilson
made upon melted rosin, does not seem liable to that objection. He poured the melted rosin into a phial, and found
that he could give shocks with it; but he observed, that these
shocks diminished as the rosin grew cold, and that when it
was quite cold, they entirely ceased. *

Mr. Wilson mentions a curious experiment (of which,
however, he does not say that he was the inventor) which
he made with paper vanes stuck in a cork, and suspended by
a magnet. These, he says, if they were brought near the
point of any body proceeding from the prime conductor,
would turn round very swiftly, but would not turn at all in
vacuo. This blast he thought was occasioned by the issuing
of the electric matter out of the point, which caused a current in the air; but he did not try what would be the consequence of presenting the vanes to a point which received the
electric fluid. †

* Wilson's Essay, p. 143.   † Ib. p. 141.

Lastly,

LASTLY, Mr. Wilfon obferved, that if a needle were prefented to a piece of down hanging to the conductor, it would
cling clofe to it; but that, upon prefenting any thing that
was blunt, it would be repelled again; and fays that Mr.
Canton made feveral curious experiments of the fame kind. *

MR. SMEATON, within this period, obferved, that if a
man who was infulated preffed againft the globe with the
flat part of his hand, while another perfon, ftanding on the
floor, did the fame, in order to excite it, the perfon who
was infulated would hardly be electrified at all; but that, if
he only laid his fingers lightly on the globe, he would be
electrified very ftrongly. † The fame ingenious perfon alfo
obferved, that upon heating the middle of a large bar of iron
to a glowing heat, and electrifying it, the electric power of
the part that was heated was as ftrong as that of the cold
part. ‡

FOR feveral curious difcoveries relating to electricity, made
within this period, we are indebted to the ingenious Dr.
Miles. In a paper read at the Royal Society January 25th.
1746, he fays that having excited a ftick of black fealing
wax with white and brown paper, or clean dried flannel,
he was able to kindle common lamp fpirits with it. Comparing the ftick of wax with the glafs tube, he obferved
a remarkable difference between the appearance of fire from
both, though he did not underftand the reafon of it. He fays
he found the luminous effluvia to proceed in a much greater

* Wilfon's Effay, p. 153.   † Ib. p. 24.   ‡ Ib. p. 129.

quantity

quantity from the top of his finger to the stick of wax, than they did to the glass. He several times observed a small globular spot of fire to appear first on his finger, from which issued regular streams towards the wax, in the form of a comet's tail. This is now well known to be the constant appearance of the electric fire between an unelectrified body and an electric excited negatively. *

DR. MILES found a stick of sulphur to perform very well; but not at all, when he had put an iron rammer in the center of it, to strengthen it. It is remarkable, that after setting this stick upright in a cupboard, it lost all its electric virtue, and could never afterwards be excited in the least degree. This effect the Doctor attributed to its being put up without any cover.

DR. MILES also mentions his having got a tube of green glass, which he could never excite but with great difficulty, and then but to a small degree. †

THE same ingenious gentleman, some time after, made an experiment upon pieces of leaf brass in a bottle hermetically sealed. To these he found he could give motion by the approach of the excited tube, in the same manner as if they had been in the open air; but one appearance struck him of which he by no means gives a satisfactory account. He observed that when he removed the tube from the exhausted glass slowly, no commotion was seen in the leaf brass, but a very brisk one upon removing the tube suddenly. Indeed this fact could not have been understood but by comparing it with

* Wilson's Essay, p. 317.   † Phil. Transf. abridged, Vol. 10. p. 320.

other

other facts depending upon the same principle, and which were not discovered till some years after. *

FROM England, to which, as an Englishman, I would give the preference only in matters of absolute indifference, I pass over to France, where, next to those made in England, the most important discoveries, and the greatest number of them were made in the period of which I am treating. And, without all dispute, the greatest name in France, in this or any other period, except that of Mr. Du Faye, his friend and associate, is that of the Abbé Nollet.

THE favourite observation of Mr. Nollet, on which he built his darling theory of affluences and effluences was, that bodies not insulated, plunged in electric atmospheres, shewed signs of electricity. He observed a sensible blast from the hand of a person not electrified, in the abovementioned circumstances, also the attraction and repulsion of light bodies by them, the appearance of flame, the diminution of their weight by increased evaporation and perspiration, and almost every other appearance and effect of electricity. Moreover observing that this globe contracted a foulness while it was whirling, even when rubbed with a clean hand, he had the curiosity to collect a quantity of the matter which formed that foulness; and finding that, when it was put into the fire, it had the smell of burnt hair, he concluded that it was an animal substance; and that it had been carried by the affluent electricity from his own body to the globe. †

THE only mistake of this ingenious philosopher in these

* Phil. Transf. abridged, Vol. 10. p. 326.   † Nollet's Recherches, p. 142.

experiments,

experiments, and which was the fource of many others, which, in the end, greatly bewildered and perplexed him was, that the electricity of the body, which was plunged in the atmofphere of an electrified body was of the fame nature with that of the electrified body. Had he but preferved the diftinction, which Mr. Du Faye had difcovered, between the two electricities, and imagined that the body electrified, and that which was plunged in its atmofphere were poffeffed of thefe two different and oppofite electricities, he might have been led to the great difcoveries made by Mr. Canton, Dr. Franklin, and Mr. Wilke; which, we fhall find, arofe from that fingle obfervation; and he would have avoided a great deal of debate and contention, which has not ended to his advantage.

This partial difcovery of Mr. Nollet is by no means the only one of his, that the hiftory of electricity prefents in this period. He made feveral experiments on pointed bodies, and obferved, that thofe which had the fmalleft points fooneft threw out brufhes of electric light, but did not fhow other figns of electricity fo ftrong as bodies that were not pointed. *

He took a great deal of pains in making experiments, in order to determine the degree in which different fubftances conducted the electric-fluid; and found that the fmoke of gum lac, turpentine, karabé, and fulphur did not carry away the electricity of an excited tube fo foon as the fmoke of linen, wood, and more efpecially the fteam of water, and the effluvia of burning tallow, and of other fatty fubftances. In fhort, he found, that vapours which were not watery did

* Recherches, p. 146.

R   very

very little, or no injury to electrical experiments, provided the tube was not exposed to them near the fire which caused them. A smoky room did not prevent his performing experiments, at least in any great degree; nor were odoriferous effluvia at all prejudicial to them. *

SEVERAL curious observations were made by the Abbé upon heat, and heated bodies. He found, that a piece of iron glowing hot, so as to throw off ignited particles, did not leave the smallest trace of electricity in an excited tube, to which it had been brought within five or six inches, and only held there two or three seconds; but it ceased to affect the tube at the same distance before it ceased to be red, and had no influence at all long before it was cold. The electricity of the tube, in this instance, was probably conveyed through the air heated by the iron; as it can hardly be supposed, that the iron emitted any effluvia capable of producing that effect. †

HE found that the excited tube lost nothing of its electricity in the focus of a burning mirror. That the flame of a candle, or the near approach of it, would destroy electricity had been known before: he observed, that the flame was sensibly disturbed by the approach of the excited tube, and he mentions Mr. Du Tour, and the Abbé Needham's having found, that the interposition of the thinnest piece of glass, or of any other substance, between the candle and the tube prevented the dissipation of the electricity. From this fact it was inferred, that the dissipation was owing to some effluvia proceeding from the candle. ‡

* Recherches, p. 194, &c.   † Ib. p. 216.   ‡ Ib. p. 219.

CONTINUING

CONTINUING his obfervations on what increafed or impeded electrical experiments, he found, that a light body, placed on a non electric ftand, moved more brifkly upon the approach of an electrified body, than when it was placed upon an electric ftand. * Several electrical experiments, he obferved, fucceeded beft when there was a number of fpectators prefent, and when they drew near, and ftood clofe together to fee his experiments; provided they did not occafion fo great a perfpiration as made his glaffes moift. † This obfervation we fhall find accounted for hereafter by Mr. Wilke.

THE Abbé moiftened, with water or fpirit of wine, a flender and pointed bar of iron, and thought that the blaft from the point of it was more fenfible than when it was not moiftened; which he attributed to the electric fluid carrying away with it fome of the particles of the water, and of the fpirit of wine. ‡

SOME few obfervations the Abbé made on the difference between excited and communicated electricity, and between the electricity of glafs and that of fulphur. He obferved, that the electricity of an excited globe or tube, caufed an odd fenfation upon the face, as if a fpider's web were drawn over it; whereas that effect was feldom produced by communicated electricity. Excited electricity, he alfo fays, might be perceived by the fmell, at more than a foot diftance, when communicated electricity could not. ∥

HE melted fulphur in a glafs globe, by turning it over a chafing difh of burning coals; when he obferved, that fmall

* Recherches, p. 122.   † Ib. p. 123.   ‡ Ib. p. 140.   ∥ Ib. p. 136.

pieces

pieces of fulphur, before they were melted, were attracted and repelled by the glafs within, at the fame time that the afhes of the coals were attracted without. * Holding a piece of excited fulphur in one hand, with a piece of down fticking to it, and ready to fly off, the down, he fays, would cling faft to the fulphur, upon prefenting to it an excited glafs tube, which he held in his other hand. †

I SHALL, in the laft place, recite the Abbé Nollet's experiments made in vacuo. He found that glafs, and other electrics, might be excited in vacuo, but not fo ftrongly as in the open air. ‡ He obferved that there was a remarkable difference between the appearance of the electric light in vacuo, and in the open air; being much more diffufe, and unbroken in vacuo. ‖ Inferting the extremity of his conductor into an exhaufted glafs veffel, he obferved the veffel to be full of light, whenever he brought his hand to it; that the light was confiderably increafed when he fpread his hand over it; and that when a fpark was taken from the conductor, the whole veffel feemed to be full of light. He alfo obferved, that fmall pieces of metal, inclofed in the veffel, adhered clofe to the glafs; but detached themfelves from it, on the approach of the finger, or of any conductor on the outfide.

THERE are a few other names of electricians in France, whofe experiments and obfervations, made within this period, deferve to be mentioned. Of thefe is Mr. Boulanger. He took great pains to determine the degree in which different fubftances are capable of being excited. The experiments,

---

* Recherchés, p. 184. † Ib. p. 124. ‡ Ib. p. 236. ‖ Ib. p. 243.

he

he fays, were made with the greatest care : and though the state of the fcience did not admit of this bufinefs being determined with great accuracy, it may not be difagreeable to fee the refult of them ; which he has comprifed in the following table, beginning with thofe that are leaft excitable in every column.

FIRST COLUMN.

Ebony.
Gaillac.
Box wood.
Sandal wood.
Oak.
Elm.
Afh.
Linden tree.
Rofe.
Willow.
Ozier.
Cork.
Dry wood of all kinds.
All dry plants.

SECOND COLUMN.

Shells of all kinds.
Whalebone.

Bones.
Ivory.
Horn.
Scales.
Parchment.
Hair.
Wool.
Feathers.
Cotton.
Silk.

THIRD COLUMN.

Allum.
Sugar Candy.
The Phofphorus of Berne.
Yellow and white wax.
Japan varnifh.
Sandarac.
Maftic.
Amber.

Jet.

Jet.
Pitch.
Gum copal.
Gum lac.
Colophonia.
Sulphur.
Sealing wax.
All salts which have suffici-
ent consistence.
All resins.

FOURTH COLUMN.

Loadstone.
Hand stone.
Marble of all colours.
Slate.
Free stone.
Granite.
Porphyry.
Jasper.
Varnished earth.
Cornelians.
Agates.
All opaque precious stones.

Porcelaine.

FIFTH COLUMN.

Hyacinth.
Opal.
Emerald.
Amethyst.
Topaz.
Ruby.
Sapphire.
Cat's eye.
Peridote.
Granite.
Rock chrystal.
Venice and Muscovy talk.
Coloured diamonds, especi-
ally yellow.
White diamonds, especially
the brilliant.
All transparent precious
stones.
Glass, and all vitrifications,
without excepting those
of metals.

THE inference which this author draws from this catalogue
is, that the most brittle, and the most transparent substances,

are

are always the moſt electric; and he has recourſe to an awk-
ward hypotheſis to account for the marcaſites not being excitable at all, notwithſtanding they are both brittle and tranſparent.   He ſays it is owing to condenſed air contained in
thoſe ſubſtances, which is known to prevent excitation. *

THE ſame author ſays, that mineral waters are much more
ſenſibly affected with electricity than common water; that
black ribbons are much ſooner attracted than thoſe of other
colours; and, next to them, the brown, and deep red. †

MR. LE CAT, a phyſician at Rouen, who has diſtinguiſh-
ed himſelf by ſeveral performances in the learned world, ſuſpended ſeveral pieces of leaf gold at his conductor, and obſerved that they hung at different diſtances, according to their
ſizes, the ſmaller pieces placing themſelves nearer the conductor, and the larger receding farther from it.   This he
compares to the diſtances at which the planets make their revolutions round the ſun, and he ſuppoſed the cauſe to be the
ſame in both.   The ſame author very particularly compares
the electric ſhock, which had juſt been diſcovered, to thunder. ‡

GERMANY affords but few articles for the electrical hiſtory of this period; one of them, however, is curious, and
well deſerves to be tranſmitted to poſterity.   Mr. Gordon of
Erford excited the electricity of a cat ſo ſtrongly, that, when
it was communicated by iron chains, it fired ſpirits of wine. ‖

IT has been mentioned before, that ſeveral gentlemen in
Germany, as well as in England, had found, that if the man

* Boulanger, p. 74.   † Ib. p. 124.   ‡ Hiſtoire, p. 84---85.   ‖ Nollet's Recherches, p. 98.

who

who rubbed the globe ſtood upon electrics, ſparks were per-ceived upon touching him; but Meſſrs. Klingſtierna and Stro-ema, two German profeſſors, were the firſt who properly e-lectrified by the rubber; and their experiments were publiſh-ed in the acts of the Royal Academy of ſciences at Stockholm for the year 1747. *

MR. JALLABERT, profeſſor of philoſophy at Geneva, found that a coating of pitch did not prevent the conductor from being electrified, which proved that the electric fluid enters the ſubſtance of metals. He alſo proved, that ice was a conductor of electricity, by making the Leyden experiment with a bottle in which water was frozen. †

THE amazing and extenſive effects of electricity now began to make philoſophers look for it where it had not been ſuſ-pected before. Mr. Hawkeſbee had been convinced that the glaſs was principally concerned in producing the light, which was viſible upon the ſhaking of mercury in glaſs veſſels, ex-hauſted or not exhauſted. Mr. Allimand, as we find in a pa-per read at the Royal Society February 13th. 1746, repeated ſome of his experiments, and obſerved, that an attractive power attended upon this electric light. He brought ſome down of feathers near a glaſs tube, through which mercury was made to run from one end to the other, and ſaw that the down was attracted as the mercury paſſed by it. ‡ An obſer-vation ſimilar to this had been made by Ludolf the younger, as was mentioned before.

THE firſt account that is given of woollen garments being

* Wilke, p. 112.   †. Hiſtoire, p. 95, 96.
‡ Phil. Tranſ. abridged, Vol. 10. p. 321.

observed

obferved to exhibit figns of electricity, when they were put off, after the flafhes of light they gave were known to be owing to electricity, was fent to the Royal Society by Mr. Coke of the Ifle of Wight, who fays, that a lady of his acquaintance obferved it; and that it was alfo at laft found, that it was only new flannel, and after fome time wearing, which gave that appearance, and that this property was loft when it was wafhed. *

The fame appearance, he obferves, upon another occafion, was moft confpicuous in frofty weather; in which feafon he takes notice, that there is generally, not only a greater purity of the air, and abfence of moifture, but that all hairy and horny fubftances (for hairs, as he fays, are only fmall horns) are more elaftic, and confequently fufceptible of, and more capable of exciting ftrong vibrations. He fays, that the flannel being rendered damp with fea water, and afterwards dry, would heighten the electric appearances. †

But though this was the firft appearance of the kind that was obferved, after it was known to arife from electricity; fimilar appearances had been feveral times noted before. Bartholin, who flourifhed in 1650, wrote a book *De luce animalium*, in which he fuppofes, that unctuous effluvia had a great fhare in thofe appearances. The fame writer fays, that Theodore Beza might be feen by a light proceeding from his eye brows; and that fparks would flafh from the body of Charles Gonzaga, Duke of Mantua, upon being gently rubbed. But he does not fay whether he had any particular hairy, or fcaly fuperficies to his fkin. ‡

* Phil. Tranf. abridged, Vol. 10. p. 343.   † Ib. p. 344.   ‡ Ib. p. 344.

Dr.

DR. SIMPSON, who publifhed a philofophical difcourfe on fermentation, dedicated to the Royal Society in 1675, alfo takes notice of the light proceeding from animals on frication, or pectation as he calls it, and inftances in the combing of a woman's head, the currying of a horfe, and the ftroking of a cat's back. *

MR. CLAYTON alfo, in a letter to Mr. Boyle, dated June 23d. 1684, at James town in Virginia, gives him an account of a ftrange accident, as he calls it, which happened to one Mrs. Sewall, whofe wearing apparel emitted a flaffing of fparks, which were feen by feveral perfons. The like happened to Lady Baltimore her mother in law. †

I SHALL conclude this fection with what I can find, in this period, about increafing the power of electricity, and meafuring its effects.

MR. MONNIER the younger, whofe name has been frequently mentioned in the courfe of this hiftory, ufed glafs fpheroids inftead of globes, and endeavoured to increafe his electrical power by ufing feveral of thefe fpheroids at a time; but he found, upon trial, that they did not anfwer his expectations; and was thence difpofed to conclude, that there might be a *ne plus ultra* in the intenfity of electricity, as well as in the heat communicated to boiling water.‡

THE power of glafs in electrifying being found to be fo great, it is no wonder that philofophers fhould endeavour to find what kind of glafs was capable of being excited to the greateft degree. Among other propofals we find a very me-

* Phil. Tranf. abridged, Vol. 10. p. 279.   † Ib. p. 278.   ‡ Ib. p. 330.

morable

morable one communicated to the Royal Society, April 6th.
1749, by Mr. George Mathias Bofe, of Wittemburgh.   He
fays, that a glafs ball which has often been employed in violent diftillations, and other chymical operations fends forth
electricity incomparably more ftrong, than any glafs which
had never been expofed to fo violent a fire.   This article is
the more curious, as it fhews us how much philofophers at
this time piqued themfelves upon difcoveries in electricity.
He afferts his being the firft perfon who ever mentioned this
notable circumftance, as he calls it, and defires Dr. Watfon,
to whom he communicated it, to let him have the honour of
that improvement in the Philofophical Tranfactions *

IT was within this period that Dr. Watfon contrived to
improve the ftrength of electricity by moiftening the rubber
of his globe, though he was not aware of all the reafons for
it.   He obferved that the man who ftood on the floor, to excite the globe by his hand, did it more ftrongly than a cufhion.   This, he fays, he could not conceive to be owing to
any other difference, than to his hand being more moift, and
confequently more readily conducting the electricity from the
floor ; wherefore he ordered his machine, and even his cufhion, to be made damp ; and then found that the electricity
was as ftrong as when the globe was rubbed by the
hand. †

A GENTLEMAN at Chartres in France, greatly increafed the
effects of electricity by means of moifture, for afferting which
he is very much ridiculed by the author of *Hiftoire de l'Elec-
tricite.*

* Phil. Tranf. abridged, Vol. 10. p. 329.   † Ib. p. 312.

S 2   MR.

MR. WILSON says, that if the cushion (which he made of leather) was gilt with silver, brass, or copper, it would do very well; and that the silk line on which the conductor hung should be red or yellow. * The table, he says, should stand on moist ground, or a wire pass from the machine to the moist ground, +

DR. WATSON also found, that though no electricity could be produced by rubbing the globe with original electrics perfectly dry, yet that they answered very well when they had been made moist; the water imbibed by those substances serving as a canal of communication to the electricity between the hand, or the cushion, and the globe; in the same manner as the air, replete with vapours in damp weather, prevents the accumulation of the electric matter in any considerable degree, by conducting it as fast as excited to the nearest non electrics. He observed, on the contrary, that most vegetable substances, though made as dry as possible, furnished electricity, though in small quantities. He excited electricity not only from linen, cotton, &c. but even from sheet lead, and a deal board. ‡

THE Abbé Nollet says, that he found oil of turpentine upon a piece of woollen cloth excited glass very powerfully, but that the least water mixed with it prevented the excitation.‖

MR. BOULANGER says, that if two cylinders be made of the same kind of glass, and of the same fashion, one of them transparent, and the other tinged with any colour, the transparent cylinder will be excited more easily than the coloured one. § He acknowledges, however, that sometimes the

* Wilson's Essay, p. 5, 6.   + Ib. p. 8.   ‡ Ib. p. 380.   ‖ Recherches, p. 186.
§ Boulanger, p. 64.

most

moſt tranſparent, and the moſt brittle glaſs is capable of acquiring but little electricity.* In another place he ſays,
that a cylinder of three or four lines in thickneſs will acquire
a ſtronger, and a more laſting electricity than a cylinder of one
line thick. † He alſo ſays, that one perſon's two hands, or
one cuſhion is better than more. ‡

About the ſame time that Dr. Watſon made his firſt experiments upon the Leyden phial, Mr. Canton diſcovered a
method by which the quantity of electricity accumulated in
the phial might be meaſured to a good degree of exactneſs.
He took the charged phial in his hand, and made it give a
ſpark to an inſulated conductor, which ſpark he took off with
his other hand. This operation he repeated till the whole
was diſcharged, and he eſtimated the height of the charge by
the number of the ſparks. This is a pretty certain and exact
method of knowing how high a phial *has been* charged : but
what electricians chiefly want is a method of aſcertaining how
high a phial *is* charged, or the exact force of the charge
while it is contained in the glaſs.

Something of this kind was done by Mr. Ellicott, in the
ſame year 1746. He propoſed to eſtimate the ſtrength of
common electrification, by its power to raiſe a weight in one
ſcale of a balance, while the other ſhould be held over the
electrified body, and pulled to it by its attractive power. ‖

The Abbé Nollet applied the threads that Mr. Grey and
and Du Faye had uſed in electrical experiments, to ſhow the

* Boulanger, p. 64.　† Ib. p. 135.　‡ Ib. p. 136.　‖ Ib. p. 324.

degree

degree of electricity. He hung two of them together, and observed the angle of their divergence, by means of the rays of the fun, or the light of a candle, and their fhadow upon a board placed behind them. Mr. Waitz alfo thought of the fame kind of electrometer, with this improvement, that he loaded the ends of the threads with fmall weights. *

* Hiftoire, p. 58.

SECTION

SECTION IV.

EXPERIMENTS ON ANIMAL AND OTHER ORGANIZED BODIES
IN THIS PERIOD, AND OTHER EXPERIMENTS CONNECTED
WITH THEM, MADE CHIEFLY BY THE ABBÉ NOLLET.

HITHERTO the effect of electricity upon human bodies
had not been attended to, farther than the mere shock
of the Leyden phial.   But we shall now see a curious set of
experiments on this subject exhibited by the Abbé Nollet.
The English philosophers, who led the way in almost every
other application of electricity, were among the last to try its
effects upon animals, and other organized bodies.   The only
article that I can find communicated by any Englishman
upon this subject, before the discoveries of the Abbé Nollet,
is one of Mr. Trembley's; who says, that several persons
had observed, that while they were electrified, their pulse
beat a little faster than before.   He says, that he himself, after having been electrified a long time together, had

felt

felt an odd fenfation all over his body, and that fome perfons had felt very fharp pains after being electrified.*

THE ingenious Abbé Nollet begins his experiments with the evaporation of fluids by electricity. They were made with the greateft attention, and the following obfervations were the refult of them.

" 1. ELECTRICITY augments the natural evaporation of " fluids; fince, excepting mercury, which is too heavy; and " the oil of olives, which is too vifcous, all the others which " were tried fuffered a diminution which could not be afcrib- " ed to any other caufe than electricity.

" 2. ELECTRICITY augments the evaporation of thofe " fluids the moft, which are moft fubject to evaporate of " themfelves. For the volatile fpirit of fal ammoniac fuffer- " ed a greater lofs than fpirit of wine, or turpentine; thefe " two more than common water; and water more than vine- " gar, or the folution of nitre.

" 3. ELECTRICITY has a greater effect upon fluids " when the veffels which contain them are non electrics; " the effects always feeming to be a little greater when " the veffels were of metal, than when they were of " glafs.

" 4. THIS increafed evaporation was more confiderable " when the veffel which contained the liquor was more open, " but the effects did not increafe in proportion to their aper- " tures. For when thefe liquors were electrified in veffels " whofe aperture was four inches in diameter, though they

* Phil. Tranf. abridged, Vol. 10. p. 321.

" prefented

" presented to the air a surface sixteen times larger than
" when they were contained in vessels whose aperture was
" one inch in diameter, they were, nevertheless, far from
" suffering a diminution proportioned to that difference.

" 5. ELECTRIFICATION does not make any liquors eva-
" porate through the pores, either of metal, or of glass; since,
" after experiments which were continued ten hours, there
" was found no diminution of their weight, when the ves-
" sels in which they were contained were well stopped." *

AFTER having made experiments on fluids, he began another course on solids of various kinds, the result of which
was, that they lost weight only in proportion to the moisture
they contained, and the openness of their pores. †

THE Abbé also extended his experiments to other sensible
qualities of bodies, as their smell, their taste, and chymical
properties; but found no change in any of them, after a
strong and continued electrification of a variety of substances.
Electrification did not affect the power of the magnet, and
neither retarded nor accelerated the heating or cooling of
bodies. ‡

HE then proceeded to the electrification of capillary tubes,
full of water; it having been observed by Mr. Boze, who
communicated the observation to Mr. Nollet, ‖ that the
water would issue in a constant stream when they were electrified; whereas it would only drop very slowly without that
operation. Every body, at first sight, would judge that the
stream was accelerated, and that the electrified vessel would

---

* Nollet's Recherches, p. 327, &c.　† Ib. p. 335.　‡ Ib. p. 341.　‖ Ib. p. 343.

　　　　　　　　　　　　soon

foon be empty: but this accurate philofopher was unwilling to rely on firft appearances, and therefore refolved to afcertain the fact, by meafuring the time, and the quantity of liquor running out. And, in order to know if the acceleration, fuppofing there were any, was uniform, during the whole time of the running out, he made ufe of veffels of different capacities, terminating in pipes of different bores, from three lines in diameter, to the fmalleft capillaries.

As the Abbé did not find it fo eafy a matter to draw a fafe conclufion in this cafe as might at firft be imagined, he gives us in grofs the following refult of above an hundred experiments. *

" 1. The electrified ftream, though it divides, and car-
" ries the liquid farther, is neither fenfibly accelerated nor
" retarded, when the pipe through which it iffues is not lefs
" than a line in diameter.

" 2. Under this diameter, if the tube is wide enough to
" let the liquid run in a continued ftream, electricity accele-
" rates it a little; but lefs than a perfon would imagine, if he
" judged by the number of jets which are formed, and by
" the diftance to which they go.

" 3. If the tube be a capillary one, from which the wa-
" ter only drops naturally, the electrified jet not only be-
" comes a continued ftream, and even divided into feveral
" ftreams, but is alfo confiderably accelerated; and the
" fmaller the capillary tube is, the greater, in proportion, is
" this acceleration.

* Recherches, p. 327. Phil. Tranf. abridged, Vol. 10. p. 382.

" 4. So

" 4. So great, is the effect of the electric virtue, that it
" drives the water in a constant stream out of a very small
" capillary tube, out of which it had not before been able
" even to drop."

The most unaccountable of these experiments, as the ingenious Abbé acknowledges, are those which suppose a re-
tardation of the electrified current, and he long doubted the
fact; but a great number of experiments, carefully noted in
his journal, obliged him to admit it, though still with hesita-
tion, and to account for it in the best manner he could;
which, indeed, was not very satisfactory. *

The beautiful appearance of these streams of electrified
water, when the experiment was exhibited in the dark, is
particularly described by this author, after Messrs. Boze and
Gordon, who first observed it. †

These last experiments served as a basis to the Abbé's future enquiries. He considered all organized bodies as assem-
blages of capillary tubes, filled with a fluid that tends to run
through them, and often to issue out of them. In consequence of this idea, he imagined, that the electric virtue
might possibly communicate some motion to the sap of vegetables, and also augment the insensible perspiration of animals. He began with the following experiments, the result of which confirmed his supposition. ‡

He electrified, for four or five hours together, fruits, green
plants, and spunges, dipped in water which he had carefully
weighed; and found that, after the experiment, all those bo-

* Recherches, p. 351.   † Ib. p. 354.   ‡ Ib. p. 355.

T 2   dies

dies were remarkably lighter than others of the same kind, weighed with them, both before and after the experiment, and kept in the same place and temper. *

THE electrification of growing vegetables was first begun in Britain. Mr. Maimbray at Edinburgh electrified two myrtle trees, during the whole month of October 1746; when they put forth small branches and blossoms sooner than other shrubs of the same kind, which had not been electrified. Mr. Nollet, hearing of this experiment, was incouraged to try it himself. †

HE took two garden pots, filled with the same earth, and and sowed with the same seeds. He kept them constantly in the same place, and took the same care of them; except that one of the two was electrified fifteen days together, for two or three, and sometimes four hours a day. The consequence was, that the electrified pot always shewed the sprouts of its seeds two or three days sooner than the other. It also threw out a greater number of shoots, and those longer in a given time; which made him believe, that the electric virtue helped to open and display the germs, and thereby to facilitate the growth of plants. This, however, our cautious philosopher only calls a conjecture, which required further confirmation. The season, he says, was then too far advanced to allow him to make as many experiments as he could have wished, but he says the next course of experiments had greater certainty, and they are not less interesting. ‡

* Phil Transf. abridged, Vol. 10. p. 383.   † Recherches, p. 356.   ‡ Ib. p. 358, &c.
Phil. Transf. abridged, Vol. 10. p. 383.

THE

The same experiments were carrying on about the same time by Mr. Jallabert, Mr. Boze, and the Abbé Menon, principal of the college of Bueil at Angers, who all drew the same conclusions from them. *

The Abbé chose several pairs of animals of different kinds, cats, pigeons, chaffinches, sparrows, &c. All these he put into separate wooden cages, and weighed them. One of each pair he electrified for five or six hours together, and then weighed them again. The result was, that the electrified cat was commonly sixty five or seventy grains lighter than the other, the pigeon from thirty five to thirty eight grains, the chaffinch or sparrow six or seven grains. In order to have nothing to charge upon the difference that might arise from the temperament of the individuals he happened to pitch upon, he repeated the same experiments, by electrifying that animal of each pair which had not been electrified before; and, notwithstanding some small varieties which happened, the electrified animal was constantly lighter than the other in proportion. †

After these experiments, he had no doubt but that electricity increased the insensible perspiration of animals, but it was not certain whether this increase was in the ratio of their bulks, or in that of their surfaces. The Abbé's opinion was, that it was neither in the one, nor the other, strictly speaking, but in a ratio much more nearly approaching the latter than the former; so that he imagined, there was no room to apprehend, that a human person electrified would lose

* Recherches, p. 357.   † Ib. p. 366, &c.

near

near a fiftieth part of his weight, as it appeared to him that it had happened to one fort of bird; nor the 140th. part, as to the pigeon, &c. All that he had then obferved upon that head was, that a young man or woman, between the ages of twenty and thirty, from being electrified five hours together, had loft feveral ounces of their weight, more than they were wont to lofe when they were not electrified. *

THE Abbé obferves, that no inconvenience whatever was felt by the perfons who fubmitted to be electrified in this manner. They only found themfelves a little exhaufted, and had got a better appetite. He adds, that none of them found themfelves fenfibly warmer, and that he could not perceive that their pulfe was increafed. ‡

THESE laft experiments on human bodies, he juftly ob-ferves, are difficult to purfue with exactnefs, becaufe the cloathing, which cannot ftrictly be compared to the hairs or feathers of animals, retains a confiderable fhare of the per-fpired matter, and prevents our forming a good judgment of the whole effect of the electric virtue.

THE foregoing experiments, he fays, convinced him of the reality of the *effluent* matter, carrying away with it the perfpirable parts of bodies, and what could be evaporated from their furfaces. And he was convinced of the *affluent* matter, by obferving all thofe effects produced, if, inftead of electrifying bodies themfelves, they were only brought near a large body which was electrified. He moiftened a thick fpunge in water, and cut it into two pieces, and then weighed the

* Phil. Tranf. abridged, Vol. 10. p. 384. Recherches, p. 382.   ‡ Ib. p. 389.

parts

parts feparately, and placing the whole near a large electrified body; he found that, after an electrification of five or fix hours, that part of the fpunge which was nearer to the electrified body had loft more weight than the other. From this fact he concluded, that if any part of an animal body was prefented to a large electrified fubftance, it would perfpire more than the other, and that perhaps obftructions might by this means be removed from the pores of it. *

The experiments above recited of Mr. Nollet by no means fatisfied the Englifh philofophers, and particularly Mr. Ellicott, who made experiments to refute the theory which the author had deduced from them. He obferved that the fyphon, though electrified, would only deliver the water by drops, if the bafon in which the water was contained was electrified too. But this does not invalidate Mr. Nollet's curious experiments upon the fubject of evaporation and perfpiration. For when an animal body is electrified, there is always non electric matter enough in the atmofphere, to anfwer the purpofe of the unelectrified bafon, in the experiment of the capillary tube; thereby to caufe a continual exhalation of the perfpirable matter from the pores of the fkin. Befides, the capillary tube will, in fact, unite the water in a conftant ftream, when it has only the open air to throw it into. In all debates upon fubjects in natural philofophy, facts ought only to be oppofed to facts. The veracity of the Abbé Nollet is not to be called in queftion; though it muft be acknowledged, that, in his later writings, at a time when his

* Phil. Tranf. abridged, Vol. 10. p. 385.

favourite

favourite syftem was in danger, he makes many miftakes with
refpect to the facts that nearly affect it.

To account for the appearance of light, which feems, in
fome cafes, to iffue from a non electric body prefented to an
excited electric, and which Mr. Nollet thought to be the
affluent matter, Mr. Ellicott fuppofes that it was the light
which had come from the electric. In accounting for
the fufpenfion of leaf gold between an electrified and an unelectrified plate, Mr. Ellicott's theory made it neceffary to fuppofe (what Dr. Franklin afterwards found not to be fact) that
the leaf gold will always be fufpended nearer the unelectrified
than the electrified plate.

In his anfwer to Mr. Nollet, Mr. Ellicott alfo endeavours
to account for the electric matter iffuing from a point at the
extremity of the conductor, more fenfibly than if it had terminated round or flat. He fays that the effluvia, in rufhing
from the globe along the conductor, as they approached the
point, were brought nearer together, and therefore were
denfer there than in any other part of the rod. Confequently,
he fays, if the light be owing to the denfity and velocity of
the effluvia, it will be vifible at the point, and no where
elfe. This, as far as I can find, was the firft attempt to account for this phenomenon; but it by no means accounts
for the whole virtue of the conductor being diffipated from
fuch points. Indeed, it is no wonder that the influence of
points, which are but imperfectly underftood even at this day,
furnifhed too difficult a problem fo many years ago. *

* Phil. Tranf. abridged, Vol. 10. p. 393.

It

It will, now, be univerfally acknowledged, that there was very great merit in thefe experiments of the Abbé Nollet, made upon animal and other organized bodies.   He opened a new and noble field of electrical difcoveries, and he purfued them with great attention, perfeverance, and expence.   This laft circumftance, I fuppofe, may have been the reafon why his experiments have not, as far as I can find, been refumed and purfued by any electrician fince his time, though there feems to be great room to improve upon what he began. The only method in which they can be conducted to any purpofe, would be by the help of a machine for perpetual electrification, to go by wind or water; which would, like-wife, ferve for many other capital experiments in electricity. This application of electricity, in particular, may perhaps be of more ufe in medicine, than any other mode in which it has hitherto been adminiftered.

U   SECTION

SECTION V.

THE HISTORY OF THE MEDICATED TUBES, AND OTHER
COMMUNICATIONS OF MEDICINAL VIRTUES BY ELECTRI-
CITY, WITH THEIR VARIOUS REFUTATIONS.

IN the course of this history we have seen frequent instances
of self deception, for want of attending to all the essential circumstances of facts; but nothing we have yet seen
equals what was exhibited in the years 1747 and 1748. Mr.
Grey's deceptions were chiefly owing to his mistaking the
cause of real appearances; but in this case we can hardly help
thinking, that, not only the imagination and judgment, but
even all the external senses of philosophers must have been
imposed upon. It was asserted by Signior Pivati at Venice
(who has all the merit of these extraordinary discoveries) and,
after him, by Mr. Verati at Bologna, Mr. Bianchi at Turin,
and Mr. Winkler at Leipsick, that if odorous substances were
confined in glass vessels, and the vessels excited, the odours
and other medicinal virtues would transpire through the glass,

infect

infect the atmosphere of the conductor, and communicate the
virtue to all persons in contact with it; also that those substances, held in the hands of persons electrified, would communicate their virtues to them; so that medicines might be
made to operate without being taken into the stomach. They
even pretended to have wrought many cures by the help of
electricity applied this way. Some of the more curious of
these pretended experiments deserve to be recorded, for the
entertainment and instruction of posterity.

The forementioned Signior Johannes Francisco Pivati, a
person of eminence at Venice, says, in an Italian epistle,
printed at Venice with all the usual licences, in the year 1747,
that a manifest example of the virtue of electricity was shown
in the balsam of Peru, which was so concealed in a glass cylinder, that, before the excitation of it, not the least smell
could by any means be discovered. A man who, having a
pain in his side, had applied hyssop to it by the advice of a
physician, approached the cylinder thus prepared, and was
electrified by it. The consequence was, that when he went
home, and fell asleep, he sweated, and the power of the balsam
was so dispersed, that even his cloaths, the bed, and chamber, all smelled of it. When he had refreshed himself by
this sleep, he combed his head, and found the balsam to have
penetrated his hair; so that the very comb was perfumed. *

The next day, Signior Pivati says, he electrified a man in
health in the same manner, who knew nothing of what had
been done before. On his going into company half an hour

* Phil. Transf. abridged, Vol. 10. p. 400.

afterwards,

afterwards, he found a gradual warmth diffusing itself through his whole body, and he grew more lively and chearful than usual. His companion was surprized at an odour, and could not imagine whence it proceeded, but he himself perceived that the fume arose from his own body, at which he also was much surprized, not having the least suspicion that it was, owing to the operation which had been performed upon him by Signior Pivati. *

MR. WINKLER of Leipsick, being struck with so extraordinary a relation, says, that he was desirous of trying the power of electricity on certain substances in the same manner, and that he found the event to confirm what had been related.†

HE put some beaten sulphur into a glass sphere, so well covered and stopped, that, on turning it over the fire, there was not the least smell of sulphur perceived. When the sphere was cold he electrified it; when, immediately, a sulphureous vapour issued from it, and, on continuing the electricity, filled the air, so as to be smelled at the distance of more than ten feet. He called in a friend well versed in electricity, professor Haubold, and several others, as witnesses and judges of this fact; but they were presently driven away by the stench of the sulphur. He staid a little longer himself in this sulphureous atmosphere, and was so impregnated thereby, that his body, cloaths, and breath retained the odour, even the next day. On repeating this experiment in the presence of a person who was conversant with the effects

* Phil. Transf. abridged, Vol. 10. p. 401.   † Ib.

of

of fulphur, the figns of an inflamed blood were vifible in his mouth on the third day. *

After this he tried the effect of a more agreeable fmell, and filled the fphere with cinnamon. When he had heated this as before, the fmell of cinnamon was foon perceived by the company, and the whole room was in a fhort time fo perfumed by it, that it immediately affected the nofes of all who came in, and the odour remained the next day.

He tried the balfam of Peru with the like fuccefs, when his abovementioned friend (whofe teftimony, he fays, he did not care to be without) after he had received the power of the balfam, fmelled fo ftrong of it, that, going abroad to fupper, he was often afked by the company what perfume he had about him. The next day, when Mr. Winkler was drinking tea, he fays, he found an unufual fweet tafte, owing to the fumes of the balfam that ftill remained in his mouth. †

In a few days, when the fphere had loft all the fcent of the balfam, they let a chain out of the chamber window, and extended it through the open air, into another room detached from the former. Here they fufpended the chain on filken lines, and gave it into the hand of a man, who alfo ftood on extended filken lines, and knew nothing of their purpofe. When the electricity had been excited for fome time, the man was afked whether he fmelled any thing; and, on fnuffing up his nofe, he faid he did. Being afked again what fmell it was, he faid he did not know. When the electrifi-

* Phil. Tranf. abridged, Vol. 10. p. 401.   † Ib.

cation

cation had been continued for about a quarter of an hour, the room smelled so strong of it, that the man, who knew nothing of the balsam, said his nose was filled with a sweet smell, like that of some sort of balsam. After sleeping in a house, a considerable distance from the room where the experiment was tried, he rose very chearful in the morning, and found a more pleasant taste than ordinary in his tea. *

I SHALL only give an account of two instances of the effect of medicine applied in this manner. The assistance of Signior Pivati, the celebrated inventor of this improvement in electricity, was implored by a young gentleman, who was miserably afflicted by a quantity of corrupted matter collected in his foot, that eluded all the attempts of the physicians. Signior Pivati filled a glass cylinder with proper materials; and, having electrified it, drew sparks from the part affected, and continued the operation for some minutes. When the patient went to bed, he had a good night, and a mitigation of his pain. When he awaked in the morning, he found a small red tubercle on his foot, which only itched, as if a cold humour had flowed through the inner part of his foot. He sweated every night for eight days together, and at the end of that time was perfectly well.

AFTER this, Signior Donadoni, bishop of Sebenico, came to Signior Pivati, attended by his physician and some friends. His lordship was at that time seventy five years old, and had been afflicted with pains in his hands and feet for several years. The gout had so affected his fingers, that he was not

* Phil. Transf. abridged, Vol. 10. p. 401.

able

able to move them; and his legs, fo that he could not bend his knees. In this deplorable fituation, the poor old bifhop intreated Signior Pivati to try the effects of electricity on him. The electrician undertook it, and proceeded after the following manner. He filled a glafs cylinder with difcutient medicines, and managed it fo, that the electric virtue might enter into the patient, who prefently felt fome unufual commotions in his fingers; and the action of electricity had been continued but two minutes, when his lordfhip opened and fhut both his hands, gave a hearty fqueeze to one of his attendants, got up, walked, fmote his hands together, helped himfelf to a chair, and fat down, wondering at his own ftrength, and hardly knowing whether it was not a dream. At length he walked out of the chamber down ftairs, without any affiftance, and with all the alacrity of a young man. *

A variety of facts of this nature being publifhed, and feemingly well attefted, engaged all the electricians of Europe to repeat thefe experiments; but none of them could fucceed but thofe mentioned above. An excellent remark of Mr. Baker's, who advifed trying all thefe experiments, notwithftanding their feeming very improbable, deferves to be quoted here. " Romantic as thefe things may feem, they " fhould not be abfolutely condemned without a fair trial, " fince we all, I believe, remember the time, when thofe " phenomena in electricity, which are now the moft com- " mon and familiar to us, would have been thought deferving

* Phil. Tranf. abridged, Vol. 10. p. 403.

" as

" as little credit as the cafes under confideration may feem to
" do, had accounts of them been fent to us from Rome,
" Venice, or Bologna, and had we never experienced them
" ourfelves. *

To fee thefe wonders, and to be affured of their truth or
fallacy, Mr. Nollet, who was deeply interefted in every thing
that related to his favourite ftudy, and who fet no bounds to
his labour or expences, in the purfuit of truth, even paffed
the Alps, and travelled into Italy, where he vifited all the
gentlemen who had publifhed any account of thefe experiments. But though he engaged them to repeat their experiments in his prefence, and upon himfelf; and though he
made it his bufinefs to get all the beft information he could
concerning them, he returned, convinced that the accounts
of cures had been much exaggerated; that in no one inftance
had odours been found to tranfpire through the pores of excited glafs; and that no drugs had ever communicated their
virtues to perfons who only held them in their hands while
electrified.

He had no doubt, however, but that, by continued e-
lectrification, without drugs, feveral perfons had found confiderable relief in various diforders; particularly, that a paralytic perfon had been cured at Geneva, and that a perfon
deaf of one ear, a footman who had a violent pain in his
head, and a woman who had a diforder in her eyes were
cured at Bologna. †

The English philofophers fhowed no lefs attention to this

* Phil. Tranf. abridged, Vol. 10. p. 406.   † Ib. p. 413, &c.

fubject

ſubject than the Abbé Nollet. The Royal Society had received an account from Mr. Winkler of his experiments, to prove the tranſudation of odoriferous matter through the pores of excited glaſs ; and none of them ſucceeding here, the ſecretary was deſired to write to Mr. Winkler, in the name of the ſociety, deſiring him to tranſmit to them, not only a circumſtantial account of his manner of making the experiments, but likewiſe ſome globes and tubes, fitted up by himſelf, for that purpoſe.

THESE veſſels, and directions how to uſe them, Mr. Winkler actually ſent, and the experiments were made with every poſſible precaution, at the houſe of Dr. Watſon (the moſt intereſted and active perſon in the kingdom in every thing relating to electricity) on the 12th of June 1751. There were preſent Martin Folkes preſident of the Royal Society, Nicholas Mann Eſq. vice preſident, Dr. Mortimer, and Peter Daval Eſq. ſecretaries, Mr. Canton a fellow, and Mr. Shroder, a gentleman of diſtinction, well known to, and correſponding with Mr. Winkler. But, notwithſtanding all the pains theſe gentlemen took, purſuing, with the utmoſt exactneſs, the directions of Mr. Winkler, and alſo uſing methods of their own, which they thought ſtill better adapted to force the effluvia through the glaſs, they were unſucceſsful. They were not able to verify Mr. Winkler's experiments even in one ſingle inſtance. *

BUT perhaps the moſt ſatisfactory refutation, both of this pretended tranſudation of odours, and the medicinal effects of

---

* Phil. Tranſ. abridged, Vol. 47. p. 231.

electricity

electricity abovementioned,was made at Venice, the very place where this medical electricity took its rife. The experiments were made by Dr. Bianchini, profeffor of medicine, in the prefence of a great number of witneffes, many of them prejudiced in favour of the pretended difcoveries; but they were all forced to be convinced of their futility, by the evidence of facts; and by experiments made with the greateft care and accuracy, *

After the publication of thefe accounts properly attefted, every unprejudiced perfon was fatisfied, that the pretended difcoveries from Italy and Leipfick, which had raifed the expectation of all the electricians in Europe, had no foundation in fact; and that no method had yet been difcovered whereby the power of medicine could by electricity be made to infinuate itfelf into the human body. †   Dr. Franklin alfo fhowed by feveral experiments the impoffibility of mixing the effluvia or virtue of medicines with the electric fluid. ‡

In fome refpects fimilar to the experiments with the medicated tubes (as thofe mentioned above were ufually called) was that of profeffor Boze of Wittemburgh, which he termed the *beatification*; and which, for a long time, employed other electricians to repeat after him, but to no purpofe. His defcription of this famous experiment was, that if, in electrifying, large globes were employed, and the electrified perfon were placed upon large cakes of pitch, a lambent flame would by degrees arife from the pitch, and fpread itfelf round his feet; and that from thence it would be propagated to his knees and

* Phil. Tranf. abridged, Vol. 48. p. 399.   † Ib. p. 406.   ‡ Franklin's Letters, p. 82.

body,

body, till, at laft, it afcended to his head; that then, by continuing the electrification, the perfon's head would be furrounded by a glory, fuch a one, in fome meafure, as is reprefented by painters in their ornamenting the heads of faints. *

THIS experiment, as well as that of the medicated tubes, fet all the electricians in Europe to work, and put them to a great deal of expence; but none of them could fucceed, fo as to produce an appearance any thing like that defcribed by Mr. Boze. No perfon took more pains in this bufinefs than Dr. Watfon. He himfelf underwent the operation feveral times, fupported by folid electrics three feet high. Upon being electrified very ftrongly, he found, as he fays, feveral other perfons alfo did, a tingling upon the fkin of his head, and in many parts of his body, or fuch a fenfation as would be felt from a vaft number of infects crawling over him at the fame time; and he conftantly obferved the fenfation to be the greateft in thofe parts of his body which were neareft to any non electric, but ftill no light appeared upon his head, though the experiment was feveral times made in the dark, and with fome continuance. †

AT length the Doctor, wearied with thefe fruitlefs attempts, wrote to the profeffor, and his anfwer fhowed that the whole had been a mere trick. He candidly acknowledged, that he had made ufe of a fuit of armour, which was decked with many bullions of fteel, fome pointed like nails,

* Phil. Tranf. abridged, Vol. 10. p. 411.   † Ib.

X 2

some like wedges, and some pyramidal; and that when the electrization was very vigorous, the edges of the helmet would dart forth rays, something like those which are painted on the heads of saints. And this was all his boasted beatification. *

THIS same Mr. Boze, who seems to have had a singular affectation of something mysterious and marvellous in his experiments, in a letter to the Royal Society at London, said that he had been able, by electricity only, to invert the poles of a natural magnet, to destroy their virtue, and restore it again, but he did not describe his method. † Considering that no person in England could succeed in this attempt, and that we are now able to do it but imperfectly, it is hardly probable that he did it at all.

THERE seems to have been some deception in an experiment which the worthy and excellent Dr. Hales communicated to the Royal Society this year, when he says, that he observed the electric spark from warm iron to be of a bright light colour; from warm copper, green; and from a warm egg, of a yellowish flame colour. These experiments, he said, seemed to argue, that some particles of those different bodies were carried off in the electric flashes, whence those different colours were exhibited. ‡

I SHALL conclude this section, which might justly be intitled the *marvellous*, with mentioning the surprizing effect of an electric spark in setting fire to a fustian frock, on a son

* Phil Transf. abridged, Vol. 10. p. 413. † Wilson's Essay, p. 219.
‡ Phil. Transf. abridged, Vol. 10. p. 406.

of

of Mr. Robert Roche, when he was electrified for some disorder. I do not question the fact. The experiment was repeated, and it answered again as well as at the first time, when it was merely accidental. The paper containing this account was read at the Royal Society on the 29th of May 1748. *

* Phil. Transf. abridged, Vol. 10. p. 406.

PERIOD

PERIOD IX.

THE EXPERIMENTS AND DISCOVERIES OF DR. FRANKLIN.

SECTION I.

DR. FRANKLIN's DISCOVERIES CONCERNING THE LEYDEN PHIAL, AND OTHERS CONNECTED WITH THEM.

WE have hitherto seen what had been done in electricity by the English philosophers, and those on the continent of Europe, till about the year 1750; but our attention is now strongly called to what was doing on the continent of America; where Dr. Franklin and his friends were as assiduous in trying experiments, and as successful in making discoveries, as any of their brethren in Europe. For this purpose, we must look back a few years. As Dr. Franklin's discoveries were made intirely independant of any in Europe, I was unwilling to interrupt the former general account, by

introducing

introducing them in their proper year. For the same reason, I imagine it will be generally more agreeable to see, at one view, what was done in America for some considerable space of time, without interrupting this account with what was doing, in the mean time, in Europe. I shall, therefore, digest, in the best manner I can, the three first publications of Dr. Franklin, entitled *New Experiments and Observations on Electricity, made at Philadelphia in America*, communicated in several letters to Peter Collinson Esq. of London, fellow of the Royal Society; the first of which is dated July 28th 1747, and the last April 18th 1754.

Nothing was ever written upon the subject of electricity which was more generally read, and admired in all parts of Europe than these letters. There is hardly any European language into which they have not been translated; and, as if this were not sufficient to make them properly known, a translation of them has lately been made into Latin. It is not easy to say, whether we are most pleased with the simplicity and perspicuity with which these letters are written, the modesty with which the author proposes every hypothesis of his own, or the noble frankness with which he relates his mistakes, when they were corrected by subsequent experiments.

Though the English have not been backward in acknowledging the great merit of this philosopher, he has had the singular good fortune to be, perhaps, even more celebrated abroad than at home; so that to form a just idea of the great and deserved reputation of Dr. Franklin, we must read the foreign publications on the subject of electricity; in many of
which

which the terms *Franklinism, Franklinist,* and the *Franklinian system* occur in almost every page. In consequence of this, Dr. Franklin's principles bid fair to be handed down to posterity, as equally expressive of the true principles of electricity, as the Newtonian philosophy is of the true system of nature in general.

THE zeal of Dr. Franklin's friends, and his reputation, were considerably increased by the opposition which the Abbé Nollet made to his theory. The Abbé, however, never had any considerable seconds in the controversy, and those he had, I am informed, have all deserted him.

THE rise of Dr. Franklin's fame in France was first occasioned by a bad translation of his letters falling into the hands of Mr. Buffon, intendant of the French king's gardens, and author of the Natural History for which he is famous. This gentleman, having successfully repeated Dr. Franklin's experiments, engaged a friend of his, Mr. Dalibard, to revise the translation ; which was afterwards published, with a short history of electricity prefixed to it, and met with a very favourable reception from all ranks of people. What contributed not a little to the success of this publication, and to bring Dr. Franklin's principles into vogue.in France, was a friend of Mr. Dallibard's exhibiting Dr. Franklin's experiments for money. All the world, in a manner, flocked to see these new experiments, and all returned full of admiration for the inventor of them. *

DR. FRANKLIN had discovered, as well as Dr. Watson,

* Nollet's Letters, Vol. 1. p. 4.

that

that the electric matter was not created, but collected by
friction, from the neighbouring non electric bodies. He had
observed, that it was impossible for a man to electrify himself, though he stood upon glass or wax; for that the tube
could communicate to him no more electricity than it had
received from him in the act of excitation. He had observed,
that if two persons stood upon wax, one of which rubbed the
tube, and the other took the fire from it, they would both
appear to be electrified; that if they touched one another
after that operation, a stronger spark would be perceived between them, than when any other person touched either of
them; and that such a spark would take away the electricity
of both. *

THESE experiments led the Doctor to think, that the electric fluid was conveyed from the person who rubbed the tube
to him who touched it; which introduced some terms in
electricity that had not been used before, but have continued
in use ever since. The person who touched the tube was
said, by Dr. Franklin, to be electrified *positively*, or *plus*;
being supposed to receive an additional quantity of electric
fire: whereas the person who rubbed the tube was said to be
electrified *negatively*, or *minus*; being supposed to have lost a
part of his natural quantity of the electric fluid. †

THIS observation was necessary to explain the capital discovery which Dr. Franklin made with respect to the manner of
charging the Leyden phial; which is, that when one side of
the glass was electrified positively or plus, the other side was
electrified negatively or minus; so that whatever quantity of

* Franklin's Letters, p. 14.   † Ib. p. 15.

Y   fire

fire is thrown upon one side of the glass, the same is thrown out of the other; and that there is really no more electric fire in the phial after it is charged than before; all that can be done by charging, being to take from one side, and convey to the other. Dr. Franklin also observed, that glass was impervious to electricity, and that, therefore, since the equilibrium could not be restored to the charged phial by any internal communication, it must be done by conductors externally, joining the inside and the outside. *

THESE capital discoveries he made by observing, that when a phial was charged, a cork ball suspended on silk would be attracted by the outside coating, when it was repelled by a wire communicating with the inside; and that it would be repelled by the outside, when it was attracted by the inside. †
But the truth of this maxim appeared more evident when he brought the knob of the wire communicating with the outside coating within a few inches of the wire communicating with the inside coating, and suspended a cork ball between them; for, in that case, the ball was attracted by them alternately, till the phial was discharged. ‡

THE European electricians had observed, that a phial could not be charged unless some conductor was in contact with the outside; but Dr. Franklin made the observation more general, and also was able, by the principle abovementioned, to give a better account of it. As no more electric fire, he says, can be thrown into the inside of a phial when all is driven from the outside; so, in a phial not yet charged, none

* Franklin's Letters, p. 3.   † Ib. p. 4.   ‡ Ib. p. 5.

can

can be thrown into the infide when none can be got from the outfide. He alfo fhowed, by a beautiful experiment, that, when the phial was charged, one fide loft exactly as much as the other gained, in reftoring the equilibrium. Hanging a fmall linen thread near the coating of an electrified phial, he obferved, that every time he brought his finger near the wire, the thread was attracted by the coating. For as the fire was taken from the infide by touching the wire, the outfide drew in an equal quantity by the thread. *

He proved that, in difcharging the phial, the giving from one fide was exactly equal to the receiving by the other, by placing a perfon upon electrics, and making him difcharge the phial through his body; when he obferved, that no electricity remained in him after the difcharge. † He alfo hung cork balls upon an infulated conductor at the time of the difcharge of a phial hanging to it; and obferved, that if they did not repel before the explofion, they did not repel at the time, nor after. ‡ But the experiment which moft compleatly proved, that the coating on one fide received juft as much as was emitted from the difcharge of the other, was the following.

He infulated his rubber, then, hanging a phial to his conductor, he found it could not be charged, even though his hand was held conftantly to it; becaufe, though the electric fire might leave the outfide of the phial, there was none collected by the rubber to be conveyed into the infide. He then took away his hand from the phial, and forming a communication, by a wire from the outfide coating to the infulated

* Franklin's Letters, p. 5.   † Ib. p. 8.   ‡ Ib. p. 84.

rubber,

rubber, he found that it was charged with eafe. In this cafe, it was plain, that the very fame fire which left the outfide coating was conveyed by the way of the rubber, the globe, the conductor, and the wire of the phial, into the infide. *

DR. FRANKLIN's new theory of charging the Leyden phial led him to obferve a greater variety of facts, relating both to charging and difcharging it, than other philofophers had attended to. He found that the phial would be electrified as ftrongly if it were held by the hook, and the coating applied to the globe or tube, as if it were held by the coating, and the hook applied; and, confequently, that there would be the fame explofion, and fhock, if the electrified phial were held in one hand by the hook, and the coating touched with the other, as when held by the coating, and touched at the hook. To take the charged phial by the hook with fafety, and not diminifh its force, he obferves, that it muft firft be fet down upon electrics per fe. †

DR. FRANKLIN obferved, that if a man held in his hand two phials, the one fully electrified, and the other not at all, and brought their hooks together, he would have but half a fhock: for the phials would both remain only half electrified, the one being half charged, and the other half difcharged. ‡

IF two phials were charged both through their hooks, a cork ball fufpended on filk, and hanging between them, would firft be attracted, and then repelled by both; but if

* Franklin's Letters, p. 83.   † Ib. p. 19.   ‡ Ib. p. 21.

they

they were electrified, the one through the wire, and the o-
ther through the the coating, the ball would play vigoroufly
between them both, till they were nearly difcharged. * The
Doctor did not, at that time, take notice, that if the phials
were both charged through their coatings (by which both
the hooks would have been electrified minus) the ball would
be repelled by them both, as when they were electrified plus.
And when he, afterwards, obferved that two bodies electrified
minus repelled one another, he feems to have been furprifed
at the appearance, and acknowledged that he could not fatif-
factorily account for it. †

IT was known to every electrician, that a globe or tube
wet on the infide would afford little or no fire ; but no good
reafon was given for it, before Dr. Franklin attempted its
explanation by the help of his general maxim. He fays, that
when a tube lined with any non electric is rubbed, what is
collected from the hand by the downward ftroke enters the
pores of the glafs, driving an equal quantity out of the inner
furface, into the non electric lining ; and that the hand, in
paffing up to take a fecond ftroke, takes out again what had
been thrown into the outward furface, the inner furface at
the fame time receiving back again what it had given to
the non electric lining ; fo that the particles of the electric
fluid went in and out of their pores, upon every ftroke given
to the tube. ‡

IF, in thefe circumftances, a wire was put into the tube,
he obferved, that, if one perfon touched the wire, while a-

* Franklin's Letters, p. 21.   † Ib. p. 34.   ‡ Ib. p. 76.

nother

nother was rubbing the tube, and took care to withdraw his finger as foon as he had taken the fpark, which had been made to fly from the infide, it would be charged. *

If the tube was exhaufted of air, he obferves, that a non electric lining in contact with the wire was not neceffary; for that, in vacuo, the electric fire would fly freely from the inner furface, without a non electric conductor. †

Upon the fame principle he accounts for the effects of an excited electric being perceived through the glafs in the vacuum beyond it.   The tube and its excited atmofphere, being brought near a glafs veffel, repels the electric fluid from the inner furface of the glafs; and this fluid, iffuing from the inner furface, acts upon light bodies in the vacuum, both in its paffage from the glafs, and likewife in its return to it, when the excited electric on the outfide is withdrawn. ‡

This maxim, that whatever the phial takes in at one furface it lofes at the other, led Dr. Franklin to think of charging feveral phials together with the fame trouble, by connecting the outfide of one with the infide of another; whereby the fluid that was driven out by the firft would be received by the fecond, and what was driven out of the fecond would be received by the third, &c. By this means he found, that a great number of bottles might be charged with the fame labour as one only; and that they might be charged equally high, were it not that every bottle receives the new fire, and lofes its old with fome reluctance, or rather gives fome fmall refiftance to the charging. This circumftance, he fays, in a

* Franklin's Letters, p. 77.   † Ib.   ‡ Ib. p. 78.

number

number of bottles, becomes more equal to the charging power, and so repels the fire back again on the globe sooner than
a single bottle would do. *

UPON this principle Dr. Franklin constructed an *electrical
battery*, consisting of eleven panes of large sash glass, coated
on each side, and so connected, that charging one of them
would charge them all.   Then having a contrivance to bring
the giving sides in contact with one wire, and all the receiving sides with another, he united the force of all the plates,
and discharged them all at once. †

WHEN Dr. Franklin first began his experiments upon the
Leyden phial, he imagined that the electric fire was all crowded into the substance of the non electric in contact with the
glass; but he afterwards found, that its power of giving a
shock lay in the glass itself, and not in the coating, by the
following ingenious analysis of the bottle.

IN order to find where the strength of the charged bottle
lay, he placed it upon glass; then first took out the cork and
the wire, and finding the virtue was not in them; he touched
the outside coating with one hand, and put the finger of the
other into the mouth of the bottle; when the shock was felt
quite as strong as if the cork and wire had been in it.   He
then charged the phial again, and pouring out the water into an
empty bottle insulated, expected that if the force resided in the
water it would give the shock, but he found it gave none.   He
then judged that the electric fire must either have been lost in
decanting, or must remain in the bottle; and the latter he

* Franklin's Letters, p. 12.   † Ib. p. 26.

found

found to be true; for, filling the charged bottle with freſh water, he found the ſhock, and was ſatisfied that the power of giving it reſided in the glaſs itſelf. *

THE Doctor made the ſame experiment with panes of glaſs, laying the coating on lightly, and changing it as he had before changed the water in the bottle, and the reſult was the ſame in both. †

THAT the electric fire reſided in the glaſs was alſo further evident from this conſideration, that when glaſs was gilt, the diſcharging of it would make a round hole, tearing off a part of the gilding, which, the Doctor thought, could only have been done by the fire coming out of the glaſs through the gilding. He alſo ſays, that when the gilding was varniſhed even with turpentine, this varniſh, though dry and hard, would be burnt by the ſpark driven through it, yielding a ſtrong ſmell, and a viſible ſmoak. Alſo, that when a ſpark was driven through paper, it would be blackened by the ſmoke, which ſometimes penetrated ſeveral of the leaves, and that part of the gilding which had been torn off was found forcibly driven into the hole made in the paper by the ſtroke. He alſo obſerved, that when a thin bottle was broken by a charge, the glaſs was broken inwards, at the ſame time that the gilding was broken outwards. *

LASTLY, Dr. Franklin diſcovered, that ſeveral ſubſtances which would conduct electricity in general, would not conduct the ſhock of a charged phial. A wet packthread, for inſtance, though it tranſmitted electricity very well, ſometimes

* Franklin's Letters, p. 24.   † Ib. p. 25.   ‡ Ib. p. 32.

failed

failed to conduct a shock; as also did a cake of ice.   Dry earth too rammed into a glass tube intirely failed to conduct a shock, and indeed would convey electricity but very imperfectly. *

* Franklin's Letters, p. 33.

SECTION

SECTION II.

Dr. FRANKLIN's DISCOVERIES CONCERNING THE SIMI-
LARITY OF LIGHTNING AND ELECTRICITY.

THE greateſt diſcovery which Dr. Franklin made concerning electricity, and which has been of the greateſt
practical uſe to mankind, was that of the perfect ſimilarity
between electricity and lightning.   The analogy between
theſe two powers had not been wholly unobſerved by philoſophers, and eſpecially by electricians, before the publication
of Dr. Franklin's diſcovery.   It was ſo obvious, that it had
ſtruck ſeveral perſons, but I ſhall give only one inſtance, in
the ſagacious Abbé Nollet.

THE Abbé ſays, *   " If any one ſhould take upon
" him to prove, from a well connected compariſon of phe-
" nomena, that thunder is, in the hands of nature, what elec-
" tricity is in ours, that the wonders which we now exhibit
" at our pleaſure àre little imitations of thoſe great effects

* Leçons de Phyſique, Vol. 4. p. 34.

" which

" which frighten us, and that the whole depends upon the
" same mechanism ; if it is to be demonstrated, that a
" cloud, prepared by the action of the winds, by heat, by a
" mixture of exhalations, &c. is opposite to a terrestrial ob-
" ject ; that this is the electrized body, and, at a certain
" proximity from that which is not ; I avow that this idea,
" if it was well supported, would give me a great deal of
" pleasure ; and in support of it, how many specious reasons
" present themselves to a man who is well acquainted with
" electricity. The universality of the electric matter, the
" readiness of its action, its inflammability, and its activity
" in giving fire to other bodies, its property of striking bo-
" dies externally and internally, even to their smallest parts,
" the remarkable example we have of this effect in the ex-
" periment of Leyden, the idea which we might truly adopt
" in supposing a greater degree of electric power, &c. all
" these points of anology, which I have been some time
" meditating, begin to make me believe, that one might, by
" taking electricity for the model, form to ones self, in
" relation to thunder and lightning, more perfect and
" more probable ideas than what have been offered hither-
" to, &c."

But though the Abbé, and others, had been struck with
the obvious analogy between lightning and electricity,
they went no farther than these arguments *a priori*. It was
Dr. Franklin who first proposed a method of verifying this
hypothesis, entertaining the bold thought, as the Abbé Nollet expresses it, of bringing lightning from the heavens, of
thinking that pointed iron rods, fixed in the air, when the

Z 2　　　　　　　　　　　　atmosphere

atmosphere was loaded with lightning, might draw from it the matter of the thunderbolt, and discharge it without noise or danger into the immense body of the earth, where it would remain as it were absorbed.

MOREOVER, though Dr. Franklin's directions were first begun to be put in execution in France, he himself compleated the demonstration of his own problem, before he heard of what had been done elsewhere : and he extended his experiments so far as actually to imitate all the known effects of lightning by electricity, and to perform every electrical experiment by lightning.

BUT before I relate any of Dr. Franklin's experiments concerning lightning, I must take notice of what he observed concerning the power of pointed bodies, by means of which he was enabled to carry his great designs into execution. He was properly the first who observed the intire and wonderful effect of pointed bodies, both in drawing, and throwing off the electric fire.

MR. JALLABERT was perhaps the first who observed that a body, pointed at one end, and round at another, produced different appearances upon the same body, according as the pointed, or round end was presented to it. But as Mr. Nollet, in whose presence he made the experiment, says, the effect was not constant, and nothing was inferred from it. *
And the Abbé acknowledges, that Dr. Franklin was the first who showed the property of pointed bodies, in drawing off electricity more effectually, and at greater distances than other bodies could do it. †

* Lettres, Vol. 1. p. 130.   † Recherches, p. 132.

HE

He electrified an iron shot, three or four inches in diameter, and observed, that it would not attract a thread, when the point of a needle was presented to it; but that this was not the case, unless the pointed body had a communication with the earth; for, presenting the same pointed body, stuck on a piece of sealing wax, it had not that effect; though the moment the pointed body was touched with his finger, the electricity of the ball to which it was suspended was discharged. The converse of this he proved, by finding it impossible to electrify the iron shot when a sharp needle lay upon it. *

By observing points of different degrees of acuteness, Dr. Franklin corrected the observation of Mr. Ellicott, and other English electricians, that a pointed body, as a piece of leaf gold, would always be suspended nearer to the plate which was unelectrified than that which was electrified, if it were put between them. For the Dr. observed, that it always removed farthest from that plate to which its sharpest point was presented, whether it was electrified or not; and if one of the points was very blunt, and the other very sharp, it would be suspended in the air by its blunt end, near the electrified body, without any unelectrified plate being held below it at all. †

Dr. Franklin endeavoured to account for this effect of pointed bodies, by supposing that the base on which the electric fluid at the point of an electrified body rested, being small, the attraction by which the fluid was held to the body was slight; and that, for the same reason, the resistance to the en-

* Franklin's Letters, p. 56, &c.　　† Ib. p. 67.

trance

trance of the fluid was proportionably weaker in that place than where the surface was flat. *   But he himself candidly owns,..that he was not quite satisfied with this hypothesis. Whatever we think of Dr. Franklin's theory of the influance of pointed conductors in drawing and throwing off the electric fluid, the world is greatly indebted to him for the practical use he made of this doctrine. †

DR. FRANKLIN begins his account of the similarity of the electric fluid and lightning by cautioning his readers not to be staggered at the great difference of the effects in point of degree; since that was no argument of any disparity in their nature.   It is no wonder, says he, if the effects of the one should be so much greater than those of the other. For if two gun barrels electrified will strike at two inches distance, and make a loud report, at how great a distance will 10,000 acres of electrified cloud strike, and give its. fire, and how loud must be that crack ! ‡

I SHALL digest all Dr. Franklin's observations concerning lightning under the several points of resemblance which he observed between it and electricity, mentioning these points of similarity in the order in which he himself remarked them; only bringing into one place the observations which may happen to lie in different parts of his letters, when they relate to the same subject.

1. FLASHES of lightning, he begins with observing, are generally seen crooked, and waving in the air. The same, says he, is the electric spark always, when it is drawn from

* Franklin's Letters, p. 56.   † Ib. p. 62.   ‡ Ib. p. 44.

an

an irregular body at some distance. * He might have added, when it is drawn by an irregular body, or through a space in which the best conductors are disposed in an irregular manner, which is always the case in the heterogenious atmosphere of our globe.

2. LIGHTNING strikes the highest and most pointed objects in its way preferably to others, as high hills, and trees, towers, spires, masts of ships, points of spears, &c. In like manner, all pointed conductors receive or throw off the electric fluid more readily than those which are terminated by flat surfaces. †

3. LIGHTNING is observed to take the readiest and best conductor. So does electricity in the discharge of the Leyden phial. For this reason, the Doctor supposes that it would be safer, during a thunder storm, to have ones cloaths wet than dry, as the lightning might then, in a great measure, be transmitted to the ground, by the water, on the outside of the body. It is found, says he, that a wet rat cannot be killed by the explosion of the electrical bottle, but that a dry rat may. ‡

4. LIGHTNING burns. So does electricity. Dr. Franklin says, that he could kindle with it hard dry rosin, spirits unwarmed, and even wood. He says, that he fired gunpowder, by only ramming it hard in a cartridge, into each end of which pointed wires were introduced, and brought within half an inch of one another, and discharging a shock through them. ‖

5. LIGHTNING sometimes dissolves metals. So does e-

* Franklin's Letters, p. 46.　　† Ib. p. 47.　　‡ Ib. p. 47.　　‖ Ib. p. 48, 92.

lectricity,

lectricity, though the Doctor was miftaken when he imagined it was by a cold fufion, as will appear in its proper place. The method in which Dr. Franklin made electricity melt metals was by putting thin pieces of them between two panes of glafs bound faft together, and fending an electric fhock through them. Sometimes the pieces of glafs, by which they were confined, would be fhattered to pieces by the dif-charge, and be broken into a kind of coarfe fand, which once happened with pieces of thick looking glafs; but if they re-mained whole, the piece of metal would be miffing in feveral places where it had lain between them, and inftead of it, a metallic ftain would be feen on both the glaffes, the ftains on the under and upper glafs being exactly fimilar in the mi-nuteft ftroke. *

A PIECE of leaf gold ufed in this manner appeared not on-ly to have been melted, but, as the Doctor thought, even vitri-fied, or otherwise fo driven into the pores of the glafs, as to be protected by it from the action of the ftrongeft aqua regis. Sometimes he obferved that the metallic ftains would fpread a little wider than the breadth of the thin pieces of metal. True gold, he obferved, made a darker ftain, fomewhat red-difh, and filver a greenifh ftain. †

MR. WILSON fuppofes that, in this experiment, the gold was not driven into the pores of the glafs, but only into fo near a contact with the furface of the glafs, as to be held there by an exceeding great force; fuch an one, he fays, as is ex-erted at the furface of all bodies whatever. ‡

* Franklin's Letters, p. 48, 65.   † Ib. p. 68.   ‡ Hoadley and Wilfon, p. 68.

6. LIGHT-

6. LIGHTNING rends fome bodies. The fame does electricity. * The Doctor obferves, that the electric fpark would ftrike a hole through a quire of paper. When wood, bricks, ftone, &c. are rent by lightning, he takes notice, that the fplinters will fly off on that fide where there is the leaft refiftance. In like manner, he fays, when a hole is ftruck through a piece of pafteboard by an electrified jar, if the furfaces of the pafteboard are not confined and compreffed, there will be a bur raifed all round the hole on both fides of the pafteboard; but that if one fide be confined, fo that the bur cannot be raifed on that fide, it will all be raifed on the other fide, which way foever the fluid was directed. For the bur round the outfide of the hole is the effect of the explofion, which is made every way from the center of the electric ftream, and not an effect of its direction. †

7. LIGHTNING has often been known to ftrike people blind. And a pigeon, after a violent fhock of electricity, by which the Doctor intended to have killed it, was obferved to have been ftruck blind likewife. ‡

8. IN a thunder ftorm at Stretham, defcribed by Dr. Miles, ‖ the lightning ftripped off fome paint which had covered a gilded molding of a pannel of wainfcot, without hurting the reft of the paint. Dr. Franklin imitated this, by pafting a flip of paper over the filleting of gold, on the cover of a book, and fending an electric flafh through it. The paper was torn off from end to end, with fuch force, that it was broken in feveral places; and in others there was brought

* Franklin's Letters, p. 49.   † Ib. p. 124.   ‡ Ib. p. 63.
‖ Phil. Tranf. Vol. 45. p. 387.

away part of the grain of the Turkey leather in which the book was bound. This convinced the Doctor, that if it had been paint, it would have been ftripped off in the fame manner with that on the wainfcot at Stretham. *

9. LIGHTNING deftroys animal life. Animals have likewife been killed by the fhock of electricity. The largeft animals which Dr. Franklin and his friends had been able to kill were a hen, and a turkey which weighed about ten pounds. †

10. MAGNETS have been obferved to lofe their virtue, or to have their poles reverfed by lightning. The fame did Dr. Franklin by electricity. By electricity he frequently gave polarity to needles, and reverfed them at pleafure. A fhock from four large jars, fent through a fine fewing needle, he fays, gave it polarity, fo that it would traverfe when laid on water. What is moft remarkable in thefe electrical experiments upon magnets is, that if the needle, when it was ftruck, lay Eaft and Weft, the end which was entered by the electric blaft pointed North; but that if it lay North and South, the end which lay towards the North, would continue to point North, whether the fire entered at that end or the contrary; though he imagined, that a ftronger ftroke would have reverfed the poles even in that fituation, an effect which had been known to have been produced by lightning. He alfo obferved, that the polarity was ftrongeft when the needle was ftruck lying North and South, and weakeft when it lay Eaft and Weft. He takes notice that, in thefe experiments, the needle, in fome cafes, would be finely blued, like the

* Phil. Tranf. Vol. 45. p. 64.   † Franklin's Letters, p. 86. 153.

fpring

spring of a watch, by the electric flame; in which case the colour given by a flash from two jars only might be wiped off, but that a flash from four jars fixed it, and frequently melted the needles. The jars which the Doctor used held seven or eight gallons, and were coated and lined with tinfoil. *

To demonstrate, in the completest manner possible, the sameness of the electric fluid with the matter of lightning, Dr. Franklin, astonishing as it must have appeared, contrived actually to bring lightning from the heavens, by means of an electrical kite, which he raised when a storm of thunder was perceived to be coming on. This kite had a pointed wire fixed upon it, by which it drew the lightning from the clouds. This lightning descended by the hempen string, and was received by a key tied to the extremity of it; that part of the string which was held in the hand being of silk, that the electric virtue might stop when it came to the key. He found that the string would conduct electricity even when nearly dry, but that when it was wet, it would conduct it quite freely; so that it would stream out plentifully from the key, at the approach of a person's finger. †

At this key he charged phials, and from electric fire thus obtained, he kindled spirits, and performed all other electrical experiments which are usually exhibited by an excited globe or tube.

As every circumstance relating to so capital a discovery as this (the greatest, perhaps, that has been made in the whole

* Franklin's Letters, p. 90, &c.   † Ib. p. 106.

compass

compaſs of philoſophy, ſince the time of Sir Iſaac Newton)
cannot but give pleaſure to all my readers, I ſhall endeavour
to gratify them with the communication of a few particulars
which I have from the beſt authority.

The Doctor, after having publiſhed his method of verifying his hypotheſis concerning the ſameneſs of electricity with
the matter of lightning, was waiting for the erection of a
ſpire in Philadelphia to carry his views into execution; not
imagining that a pointed rod, of a moderate height, could
anſwer the purpoſe; when it occurred to him, that, by means
of a common kite, he could have a readier and better acceſs
to the regions of thunder than by any ſpire whatever. Preparing, therefore, a large ſilk handkerchief, and two croſs
ſticks, of a proper length, on which to extend it; he took
the opportunity of the firſt approaching thunder ſtorm to take
a walk into a field, in which there was a ſhed convenient for
his purpoſe. But dreading the ridicule which too commonly attends unſucceſsful attempts in ſcience, he communicated
his intended experiment to no body but his ſon, who aſſiſted
him in raiſing the kite.

The kite being raiſed, a conſiderable time elapſed before
there was any appearance of its being electrified. One very
promiſing cloud had paſſed over it without any effect; when,
at length, juſt as he was beginning to deſpair of his contrivance, he obſerved ſome looſe threads of the hempen ſtring
to ſtand erect, and to avoid one another, juſt as if they had
been ſuſpended on a common conductor. Struck with this
promiſing appearance, he immediately preſented his knucle

to

to the key, and (let the reader judge of the exquifite pleafure he muft have felt at that moment) the difcovery was complete. He perceived a very evident electric fpark. Others fucceeded, even before the ftring was wet, fo as to put the matter paft all difpute, and when the rain had wet the ftring, he collected electric fire very copioufly. This happened in June 1752, a month after the electricians in France had verified the fame theory, but before he heard of any thing they had done.

Besides this kite, Dr. Franklin had afterwards an infulated iron rod to draw the lightning into his houfe, in order to make experiments whenever there fhould be a confiderable quantity of it in the atmofphere; and that he might not lofe any opportunity of that nature, he connected two bells with this apparatus, which gave him notice, by their ringing, whenever his rod was electrified. *

The Doctor being able, in this manner, to draw the lightning into his houfe, and make experiments with it at his leifure; and being certain that it was in all refpects of the fame nature with electricity, he was defirous to know if it was of the pofitive or negative kind. The firft time he fucceeded in making an experiment for this purpofe was the 12th of April 1753, when it appeared that the lightning was negative. Having found that the clouds electrified negatively in eight fucceffive thunder gufts, he concluded they were always electrified negatively, and formed a theory to account for it. But he afterwards found he had concluded too foon. For, on the

* Franklin's Letters, p. 112.

fixth of June following, he met with one cloud which was electrified pofitively ; upon which he corrected his former theory, but did not feem able perfectly to fatisfy himfelf with any other. The Doctor fometimes found the clouds would change from pofitive to negative electricity feveral times in the courfe of one thunder guft, and he once obferved the air to be ftrongly electrified during a fall of fnow, when there was no thunder at all. *

BUT the grand practical ufe which Dr. Franklin made of his difcovery of the famenefs of electricity and lightning, was to fecure buildings from being damaged by lightning, a thing of vaft confequence in all parts of the world, but more efpecially in feveral parts of North America, where thunder ftorms are more frequent, and their effects, in that dry air, more dreadful, than they are ever known to be with us.

THIS great end Dr. Franklin accomplifhed by fo eafy a method, and by fo cheap, and feemingly trifling apparatus, as fixing a pointed metalline rod higher than any part of the building, and communicating with the ground, or rather the neareft water. This wire the lightning was fure to feize upon, preferable to any other part of the building ; whereby this dangerous power would be fafely conducted to the earth, and diffipated, without doing any harm to the building. †

DR. FRANKLIN was of opinion, that a wire of a quarter of an inch of thicknefs would be fufficient to conduct a greater quantity of lightning than was ever actually difcharged from the clouds in one ftroke. He found, that the gilding

* Franklin's Letters, p. 112, &c.   † Ib. p. 62, 124.

of

of a book was fufficient to conduct the charge of five large jars, and thought that it would probably have conducted the charge of many more. He alfo found by experiment, that if a wire was deftroyed by an explofion, it was yet fufficient to conduct that particular ftroke, though it was thereby rendered incapable of conducting another. *

THE Doctor alfo fuppofed, that pointed rods erected on edifices might likewife often prevent a ftroke of lightning in the following manner. He fays, that an eye fo fituated as to view horizontally the underfide of a thunder cloud, will fee it very ragged, with a number of feparate fragments, or petty clouds, one under another, the loweft fometimes not far from the earth. Thefe, as fo many ftepping ftones, affift in conducting a ftroke between a cloud and a building. To reprefent thefe by an experiment, he directs us to take two or three locks of fine loofe cotton and connect one of them with the prime conductor, by a fine thread of two inches (which may be fpun out of the fame lock) a- nother to that, and a third to the fecond, by like threads. He then bids us to turn the globe, and fays we fhall fee thefe locks extending themfelves towards the table (as the lower fmall clouds do towards the earth) but, that, on prefenting a fharp point, erect under the loweft, it will fhrink up to the fecond, the fecond to the firft, and all together to the prime conductor, where they will continue as long as the point con- tinues under them. A moft ingenious and beautiful experi- ment! May not, he adds, in like manner, the fmall electri-

* Franklin's Letters, p. 124, 125.

fied

fied clouds, whofe equilibrium with the earth is foon reftored by the point, rife up to the main body, and by that means occafion fo large a vacancy, as that the grand cloud cannot ftrike in that place. *

* Franklin's Letters, p. 121, &c.

SECTION

S E C T I O N   III.

MISCELLANEOUS DISCOVERIES OF DR. FRANKLIN AND HIS FRIENDS IN AMERICA DURING THE SAME PERIOD.

DR. FRANKLIN, retaining the common opinion, that electrified bodies have real atmospheres of the electric fluid (confisting of particles at some diftance from the furface of the body, but always going along with it) obferved that thefe atmofpheres and the air did not feem to exclude one a-nother; though, he fays, this be difficult to conceive, confidering that they are generally fuppofed to repel one another.

An electric atmofphere, he fays, raifed round a thick wire, inferted into a phial, drives out none of the air it contained; nor, on withdrawing that atmofphere, will any air rufh in, as he found by a very curious experiment, accurately made; whence he alfo concluded, that the elafticity of the air was not affected by it. *

THE experiment, as the Doctor informs me, was made

* Franklin's Letters, p. 98.

B b   with

with a small glass syphon, one leg passing through the cork into the bottle. The other leg had in it a drop of red ink, which readily moved on the least change of heat or cold in the air contained in the phial; but not at all on the air's being electrified.

He also made an experiment which would seem to prove the immobility, as we may say, of these atmospheres by any external force, if they have any existence at all; but others may think it is rather an argument against their existence. He electrified a large cork ball fastened to the end of a silk string three feet long; and, taking the other end in his hand, he whirled it round, like a sling, a hundred times in the open air, with the swiftest motion he could possibly give it; and observed, that it still retained its electric atmosphere, though it must have passed through eight hundred yards of air. *

To show that a body, in different circumstances of dilatation and contraction, is capable of receiving, or retaining more or less of the electric fluid on its surface, he made the following curious experiment. He electrified a silver can, in which was about three yards of brass chain, one end of which he could raise to what height he pleased, by means of a pully and a silk cord. He suspended a lock of cotton by a silk string from the ceiling of the room, making it to hang near the cup; and observed, that every time he drew up the chain, the cotton approached nearer to the cup, and as constantly receded from it when the chain

* Franklin's Letters, p. 97.

was

was let down.   From this experiment it was evident, he fays, that the atmofphere about the cup was diminifhed by raifing the chain, and increafed by lowering it; and that the atmofphere of the chain muft have been drawn from that of the cup when it was raifed, and have returned to it again when it was let down. *

To make electric atmofpheres in fome meafure vifible, the Doctor ufed to drop rofin on hot iron plates held under bodies electrified; and, in a ftill room, the fmoke would afcend, and form vifible atmofpheres round the bodies, making them look very beautiful.   In trying in what circumftances, the repellency between an electrified iron ball, and a fmall cork ball would be altered, he obferved, that this fmoke of rofin did not deftroy their repellency, but was attracted both by the iron and the cork. †

THE Doctor obferved, that filver expofed to the electric fpark would acquire a blue ftain, that iron would feem corroded by it; but he could never perceive any impreffion made on gold, brafs, or tin.   The fpots on the filver or iron were always the fame, whether they received the fpark from lead, brafs, gold, or filver; and the fmell of the electric fire was the fame, through whatever bodies it was conveyed. ‡

WHILE we are attending to what was done by Dr. Franklin at Philadelphia, we muft by no means overlook what was done by Mr. Kinnerfley, the Doctor's friend while at Bofton in New England.   Some of his obfervations, of which an account is given in the Doctor's letters, are very curious; and

* Franklin's Letters, p. 121.   † Ib. p. 55.   ‡ Ib. p. 81, 98.

B b 2   some

some later accounts, which he himself has transmitted to England, seem to promise, that, if he continue his electrical inquiries, his name, after that of his friend, will be second to few in the history of electricity.

He first distinguished himself by re-discovering Mr. Du Faye's two contrary electricities of glass and sulphur, with which both he and Dr. Franklin were at that time wholly unacquainted. But Mr. Kinnersley had a great advantage over Mr. Du Faye; for, making his experiments in a more advanced state of the science, he saw immediately, that the two contrary electricities of glass and sulphur were the very same positive and negative electricities, which had just been discovered by Dr. Watson and Dr. Franklin.

He observed, that a cork ball, electrified by a conductor from excited glass, would be attracted by excited amber and sulphur, and repelled by excited glass and china; that electrifying the ball with the wire of a charged phial, it would be repelled by excited glass, but attracted by excited sulphur; and that when he electrified it by sulphur or amber, till it became repelled by them, it would be attracted by the wire of the phial, and repelled by its coating. These experiments surprized him very much, but by analogy he was led to infer, *a priori*, the following paradoxes, as he calls them, which were afterwards verified by Dr. Franklin, at his request. *

1. " If a glass globe be placed at one end of a prime
" conductor, and a sulphur one at the other, both being e-

* Franklin's Letters, p. 99.

" qually

" qually in good order, and in equal motion, not a fpark of
" fire can be obtained from the conductor, but one globe
" will draw out as faft as the other gives in.

2. " If a phial be fufpended on the conductor with a
" chain from its coating to the table, and only one of the
" globes be made ufe of at a time, twenty turns of the
" wheel, for inftance, will charge it; after which, as many
" turns of the other wheel will difcharge it; and as many
" more will charge it again.

3. " The globes being both in motion, each having a
" feparate conductor, with a phial fufpended on one of them,
" and the chain faftened to the other; the phial will become
" charged, one globe charging pofitively, and the other
" negatively.

4. " The phial being thus charged, hang it in like
" manner, on the other conductor. Set both wheels a going
" again, and the fame number of turns that charged it be-
" fore will now difcharge it, and the fame number repeated
" will charge it again.

5. " When each globe communicates with the fame
" prime conductor, having a chain hanging from it to the
" table, one of them, when in motion (but which I cannot
" fay) will draw fire up through the cuſhion, and difcharge
" it through the chain; and the other will draw it up
" through the chain, and difcharge it through the cuſh-
" ion." *

When Mr. Kinnerſley was adviſing his friend to try the ex-

* Franklin's Letters, p. 100.

periments

periments with the fulphur globe, he cautions him not to make ufe of chalk on the cufhion, telling him that fome fine powdered fulphur would do better. And he expreffes his hope that if the Doctor fhould find the two globes to charge the prime conductor differently, he would be able to difcover fome method of determining which it was that charged pofitively.

DR. FRANKLIN, when thefe experiments and conjectures were propofed to him, had no idea of their having any real foundation; but imagined, that the different attractions and repulfions obferved by Mr. Kinnerfley proceeded rather from the greater or fmaller quantities of the electric fire, obtained from different bodies, than from its being either of a different kind, or having a different direction. But finding, upon trial, that the principal of Mr. Kinnerfley's fuppofitions were verified by fact, he had no doubt of the reft. *

IN anfwer to the doubt of Mr. Kinnerfley, whether the glafs, or the fulphur electrified pofitively, the Doctor gave it as his opinion, that the glafs globe charged pofitively, and the fulphur negatively, for the following reafons.

1. BECAUSE, though the fulphur globe feemed to work equally well with the glafs one, yet it could never occafion fo large, and fo diftant a fpark between his finger and conductor as when the glafs globe was ufed. But what he adds to confirm this proof does not feem to be fatisfactory. He fuppofes that bodies of a certain bignefs cannot fo eafily part with the quantity of electric fluid which they have, and hold at-

* Franklin's Letters, p. 102, 103.

tracted

tracted within their fubftance, as they can receive an additional quantity upon their furface, by way of atmofphere ; and
that therefore fo much could not be drawn out of the conductor, as might be thrown on it. *

2. HE obferved that the ftream or brufh of fire, appearing
at the end of the wire connected with the conductor, was
long, large, and much diverging when the glafs globe was
ufed, and made a fnapping or rattling noife ; but that when
the fulphur globe was ufed, it was fhort, fmall, and made a
hiffing noife. He alfo obferved, that juft the reverfe of both
thefe cafes happened when he held the fame wire in his hand,
and the globes were worked alternately.   The brufh was
large, long, diverging, and fnapping or rattling, when the
fulphur globe was turned ; but fhort, fmall, and hiffing,
when the glafs globe was turned.   When the brufh was long,
large, and much diverging, it feemed to the Doctor, that
the body to which it joined was throwing the fire out, and
when the contrary appeared,   it feemed to be drinking
in. †

3. HE obferved, that when he held his knuckle before
the fulphur globe, while it was turning, the ftream of fire
between his knuckle and the globe feemed to fpread on its
furface, as if it flowed from the finger, but before the glafs
globe it was otherwife.

4. HE obferved that the cool wind (or what was called fo)
which is felt as coming from an electrified point, was much
more fenfible when the glafs globe, than when the fulphur

* Franklin's Letters, p. 104.   † Ib.

one

one was ufed.   But thefe, though the beft arguments which the fenfes can furnifh, of the courfe of the electric fluid, the Doctor acknowledges were but hafty thoughts.   Indeed, confidering that the velocity of the electric fluid has been found, by experiment, to be, nearly, inftantaneous in a circuit of many miles, it cannot be fuppofed that the eye fhould be able to diftinguifh which way it goes in the fpace of one or two inches. *

I SHALL conclude this article with obferving, that the experiments, which the Doctor made with globes of glafs and fulphur, are much more eafily exhibited by the conductor and infulated rubber of either of them, all the effects being the reverfe of each other.

I MUST now, for the prefent, take leave of this ingenious writer and his friends, after having brought the hiftory of their labours to the year 1754, and muft return to fee what was doing on the continent of Europe for two or three years preceding this date, while we left it to go over to America.

* Franklin's Letters, p. 105.

PERIOD

PERIOD X.

THE HISTORY OF ELECTRICITY, FROM THE TIME THAT
DR. FRANKLIN MADE HIS EXPERIMENTS IN AMERICA,
TILL THE YEAR 1766.

WE are now entering upon the laſt period into which
the hiſtory of electricity divides itſelf, in which the
great variety of matter preſented to our view muſt oblige an
hiſtorian to have recourſe to the ſtricteſt method; for, otherwiſe, the narration would be extremely perplexed and diſguſting.   As this period contains the events of a larger ſpace of
time than moſt of the others, yet without any convenient
reſting place; as the buſineſs of electricity has been conſiderably multiplied in it, and a greater number of labourers have
been employed in gathering in the harveſt of diſcoveries, the
ſeeds of which were ſown by Dr. Watſon, Dr. Franklin, and
others, in the preceding periods; I am obliged to ſubdivide
this into more diſtinct parts, but I hope they will not be
found to be more than were neceſſary, in order to prevent
confuſion.

C c   HOWEVER

HOWEVER, this circumstance, of the great quantity and variety of materials furnished in this period, in proportion as it tends to embarrass an historian, and exercise his talent for proper distribution and arrangement, is a striking demonstration of a truth, which must give the greatest pleasure to all the lovers of electricity and Natural Philosophy. If the progress continue the same in another period, of equal length, if the harvest of discoveries continue to be more plentiful, and the labourers proportionably more numerous; what a glorious scene shall we see unfolded, what a fund of entertainment is there in store for us, and what important benefits may be derived to mankind!

SECTION

SECTION I.

IMPROVEMENTS IN THE ELECTRICAL APPARATUS, WITH
EXPERIMENTS AND OBSERVATIONS RELATING TO IT.

A S our electrical apparatus has been much improved
within this period, I shall first recite what has occurred
to me upon this subject; particularly the methods which
have, from time to time, been communicated of increasing
the power of electricity, by different circumstances of excitation.

So early as the year 1751, upon occasion of trying Mr.
Winkler's experiments, notice is taken of Mr. Canton's method of rubbing tubes with silk prepared with linseed oil.
These he had found, by the experience of some considerable
time, to produce the greatest effect upon tubes, but he had
not found that they were proportionably useful in rubbing
globes. *

UPON another occasion, Mr. Canton observes, that by
means of this rubber, a solid cylinder of glass, which had

* Phil. Transf. Vol. 47. p. 239.

C c 2   been

been set before the fire till it was quite dry, might be excited as easily as a glass tube, so as to act like one in every respect; that even the first stroke would make it strongly electrical. *

BUT the greatest improvement which Mr. Canton discovered for increasing the power of electricity, was by rubbing on the cushion of the globe, or on the oiled silk rubber of the tube, a small quantity of an amalgam of mercury and tin, with a very little chalk or whiting. By this means, a globe or tube may be excited to a very great degree with very little friction, especially if the rubber be made more damp or dry as occasion may require. †

MR. WILKE says, that a glass tube excited with a woollen cloth, on which some white wax or oil has been put, will throw out flames with a great noise in the dark. ‡ These flames, he says, he never knew to be thrown from a globe, except sometimes when they were first used. ||

OUR electrical apparatus has been much augmented within this period by the discovery of father Windelinus Ammersin of Switzerland, who, in a Latin treatise, published at Lucern, in the year 1754, has shewn us, that wood properly dried, till it becomes very brown, is a non conductor of electricity. He recommends boiling the wood in linseed oil, or covering it over with varnish, after being dried, to prevent any return of moisture into its pores; and adds, that wood, so treated, seems to afford stronger appearances of electricity than even glass. He himself made use of common wooden measures, such as are usually found in granaries,

* Phil. Transf. Vol. 48. pt. 2. p. 784.　† Ib. Vol. 52. pt. 2. p. 461.　‡ Wilke, p. 124.
|| Ib. p. 126.

first

first boiled in oil, and afterwards mounted, so as to be turned by a wheel. *

It appears from the Philosophical Transactions, says, Mr. Wilson, so early as the year 1747, that Dr. Watson, having occasion to support a long wire, in an experiment made near Shooter's hill, with a view to determine the velocity of the electric fluid, used stakes of dry wood, which he told him, were baked, to prevent the electric fluid from escaping into the ground. †

A more extraordinary method of procuring electricity than by baked wood, was one that Signior Beccaria made use of. He put a dry and warm cat's skin upon his glass globe, and rubbing it with his hand, excited a very powerful electricity. ‡

These wooden cylinders electrify positively or negatively as the rubber is silk or flannel, but much more powerfully when negative than when positive, owing to the roughness which there generally is upon their surfaces, and therefore make an agreeable variety in an electrical apparatus. But the oldest and most usual method of procuring negative electricity was by globes of sulphur. These Mr. Le Roi made by putting a coating of sulphur upon a globe of glass, and then smoothing it with an hot iron; but Mr. Nollet preferred melting the sulphur in the inside of the glass globe, and then breaking the glass from off it, because this method made a much finer polish. ||

One globe he made of a mixture of sulphur and pounded glass, but he found that it had the same effect as if it had

* Phil. Transf. Vol. 52. pt. 1. p. 342.   † Ib. Vol. 51. pt. 2. p. 896.
‡ Lettere dell' Elettricismo, p. 58. || Nollet's Letters, Vol. 2. p. 121.

been

been all of fulphur. He fays that, when one part of this globe was excited, the whole furface became electrical. *

BUT fince Mr. Canton's difcovery of the negative power of rough glafs, fome philofophers have made ufe of glafs globes made rough by emery; and the ufual method of taking off their polifh was by rubbing them as they turned upon their axis; but Mr. Speedler, a mathematical inftrument maker at Copenhagen, obferves, in his letters upon the fubject of electricity, that glafs globes, made rough by drawing the ftone, or emery, from pole to pole, have a much greater virtue; this method of taking off the polifh giving them a greater roughnefs with refpect to the rubber †

BUT a better, and a readier method than all thefe of producing negative electricity, is by infulating the rubber of a fmooth globe, and connecting it with an infulated prime conductor, while the common conductor hath a communication with the ground. The rubber, if well infulated, is fure to produce a negative electricity, equal in power to the pofitive of the fame globe. Mr. Dalibard directs a great number of precautions, in order to electrify well at the rubber, and to prevent it from receiving any electric fire in its ftate of infulation. ‡

MR. BERGMAN of Upfal, fays, that very often, when his glafs globes could not be excited to a fufficient degree of ftrength, he lined them with a thin coating of fulphur, and that then they gave a much ftronger pofitive electricity than before. ||

---

* Nollet's Letters, Vol. 2. p. 125, 127. † Wilke, p. 57. ‡ Dalibard's Franklin, p. 110.
|| Phil. Tranf. Vol. 52. pt. 2. p. 485.

In Italy, and other places, Mr. Nollet informs us, it is the custom of electricians to put a coating of pitch, or other resinous matter on the inside of their globes, which they pretend, makes them always work well. †

We are obliged to the Abbé Nollet for some observations on the electrical powers of different kinds of glass, in the sixth volume of his *Leçons de physique* printed in the year 1764.

It is not every sort of glass, says he, that is equally electrizable. There are some sorts which are not so at all, or hardly at all; such, for example, is that of which they make plates of glass at St. Gobin in Picardy. I have tried it, says he, an hundred times, in the form of plates, tubes, and globes, and in all kinds of weather, but have scarce been ever able to draw from it the least sensible signs of electricity.

The glass of which panes for windows are made, and which is also used for drinking glasses, when it is newly manufactured, is excited with great difficulty. I have often, says he, repeatedly rubbed tubes, and other pieces, even in the glass house where they were made, but without success; and it has not been till after some months, and sometimes years, that I could bring them to act.

It is certain, and he says he has constantly observed, that glass becomes more fit for electrical experiments by force of rubbing, and that sometimes it has required some months to bring globes and tubes to act well.

He did not think that these facts could be accounted for

* Lettres, Vol. 2. p. 122.

either

either by the different degrees of tranſparency, or the different colours of glaſs. This, indeed, was evident from ſome globes acquiring electricity from uſe which had it not originally. The glaſs of which bottles are made at Severs ſerved him very well, whereas globes of white glaſs did not become tolerable till after having been uſed a certain time.

He could not tell poſitively why certain kinds of glaſs were electrizable or not by rubbing, but he ſuſpected, that it was principally owing to the degree of its hardneſs and vitrification. He was induced to think ſo, becauſe he found the glaſs at the French manufactory at St. Gobin, and at Cherbourg (the hardeſt, the moſt compact, and the beſt vitrified of all the kinds of glaſs in France) was the moſt difficult to be electrized; whereas the chryſtal glaſs of England, that of Bohemia, &c. which are much ſofter, were the beſt of all for experiments in electricity. He ſays, moreover, that he had procured imperfect glaſſes, which had not been long enough in the furnace to be clear; and that, though they were of the ſame compoſition as plates of glaſs, which, he obſerved before, was not eaſily electrized, yet that theſe were excited very ſenſibly.

He ſays that a globe of ten or twelve inches diameter, and which makes about four revolutions in a ſecond of time, will receive a convenient rubbing; but that we muſt not expect that if the globe be one half, or one fourth part greater or leſs, that the effects will be increaſed or diminiſhed in proportion. *

Upon the ſubject of inſulating bodies, he obſerves, that

* Leçons de Phyſique, Vol. 6. p. 273.---276.

when

when cakes of fulphur, rofin, fealing wax, and bees wax are made ufe of for this purpofe, they ought to be well cooled before they are ufed: for, he fays, he has conftantly obferved, that when they are newly made, they are not fo proper to infulate bodies, as they generally are at the end of fome months. *

It will be proper, under this head, to acquaint young electricians, that globes have been feveral times known to burft during the act of excitation, and that the fragments have been thrown with great violence in every direction, fo as to be very dangerous for the by ftanders. This accident happened to Mr. Sabatelli in Italy, Mr. Nollet in France, Mr. Beraud at Lyons, Mr. Boze at Wittemburgh, Mr. Le Cat at Rouen, and Mr. Robein at Rennes.

The air in the infide of Mr. Sabatelli's globe had no communication with the external air, but that of the Abbé Nollet had. This laft, which was of Englifh flint, which had been ufed two years, and which was more than a line thick, burft like a bomb in the hands of a fervant who was rubbing it; and the fragments (the largeft of which were not more than an inch in diameter) were difperfed on all fides, to a confiderable diftance. The Abbé fays, that all the globes which were burft in that manner exploded after five or fix turns of the wheel; and he afcribes this effect to the action of the electric matter, making the particles of the glafs vibrate in a manner he could not conceive. †

* Leçons de Phyfique, Vol. 6. p. 299.   † Nollet's Letters, Vol. 1. p. 19.

D d   When

WHEN Mr. Beraud's globe burſt (and he was the firſt to whom this accident was ever known to happen) he was making ſome experiments in the dark, on the 8th of February 1750; when a noiſe was firſt heard, as of ſomething rending to pieces; then followed the exploſion, and when the lights were brought in, it was obſerved, that thoſe places of the floor which were oppoſite to the equatorial diameter of the globe were ſtrewed with ſmaller pieces, and in greater numbers than thoſe which were oppoſite to other parts of it. This globe had been cracked, but it had been in conſtant uſe in that ſtate above a year, and the crack had extended itſelf from the pole to the equator. The proprietor aſcribed the accident to the vibration of the particles of glaſs, and thought that the crack had ſome way impeded thoſe vibrations. *

WHEN Mr. Boze's globe broke, he ſays that the whole of it appeared, in the act of breaking, like a flaming coal; a circumſtance which we ſhall ſee accounted for hereafter by Mr. Wilke. †

MR. BOULANGER ſays, that glaſs globes have ſometimes burſt like bombs, and have wounded many perſons, and that their fragments have even penetrated ſeveral inches into a wall. ‡ He alſo ſays, that if globes burſt in whirling by the gun barrel's touching them, they burſt with the ſame violence, the ſplinters often entering into the wall. ‡

THE Abbé Nollet had a globe of ſulphur which alſo burſt, as he was rubbing it with his naked hands, after two or

* Hiſtoire, p. 87.　　† Wilke, p. 124.　　‡ Boulanger, p. 23.　　‖ Ib. p. 144.

three

three turns of the wheel, having firſt cracked inwardly. It broke into very ſmall pieces, which flew to a great diſtance; and into a fine duſt, of which part flew againſt his naked breaſt; where it entered the ſkin ſo deep, that it could not be got off without the edge of a knife. *

* Nollet's Letters, Vol. 2. p. 220.

D d 2   SECTION

SECTION II.

OBSERVATIONS ON THE CONDUCTING POWER OF VARIOUS SUBSTANCES, AND PARTICULARLY MR. CANTON's EXPERIMENTS ON AIR; AND SIGNIOR BECCARIA's ON AIR, AND WATER.

ONE of the principal *desiderata* in the science of electricity, is to ascertain wherein consists the distinction between those bodies which are conductors, and those which are non-conductors of the electric fluid. All that has been done relating to this question, till the present time, amounts to nothing more than observations, how near these two classes of bodies approach one another; and before the period of which I am now treating, these observations were few, general, and superficial. But I shall now present my reader with several very curious and accurate experiments, which, though they do not give us satisfaction with respect to the great *desideratum* abovementioned; yet throw some light upon the subject. They show that substances which had been considered as perfect conductors, or non-conductors, are so only to a certain

de gree

degree; and that, probably, all the known parts of nature have, in some measure, the properties of both.

These experiments were made by two persons, whom, in the style of history, I may justly call two of the greatest heroes of this part of my work; viz. Mr. Canton, whose discoveries in electricity are far more numerous, and more considerable than those of any other person, within this period, in England; and Signior Beccaria, one of the most eminent of all the electricians abroad.

That air was capable of receiving electricity by communication, and of retaining it when received, had not been discovered by any person before Mr. Canton; but, by the help of one of his exquisite contrivances, he was able to ascertain that delicate circumstance, and even measure the degree of it, if it was in the least considerable.

He got a pair of balls, turned in a lathe, out of the dry pith of elder. These he put into a narrow box, with a sliding cover, so disposed that the threads (which were of the finest linen) were kept straight in the box. Holding this box by the extremity of the cover, the balls would hang freely from a pin in the inside. These balls, hung at a sufficient distance from buildings, trees, &c. easily show the electricity of the atmosphere. They also determine whether the electricity of the clouds and the air be positive, by the decrease; or negative, by the increase of their repulsion, at the approach of excited amber or sealing wax.

By the help of this instrument, he observed, that it was possible to electrify the air of a room near the apparatus; and even the air of the whole room in which it was, to a considerable

derable degree, and he was able to do it both positively and negatively.

In a paper read at the Royal Society, December the 6th. 1753, he observes, that the common air of a room might be electrified to a considerable degree, so as not to part with its electricity for some time. Having rendered the air of his room very dry, by means of a fire, he electrified a tin tube (with a pair of balls suspended at one of its extremities) to a great degree; when it appeared, that the neighbouring air was likewise electrified. For, having touched the tube with his finger, or other conductor, the balls, notwithstanding, continued to repel one another, though not at so great a distance as before.* But he observes that their repulsion would decrease as they were moved towards the floor, wainscot, or any of the furniture; and that they would touch each other when brought within a small distance of any conductor. Some degree of this electric power, he has known to continue in the air above an hour after the rubbing of the tube, when the weather had been very dry.

To electrify the air, or the moisture contained in it, negatively, Mr. Canton supported, by silk stretched between two chairs (placed back to back, at the distance of about three feet) a tin tube with a fine sewing needle at one end of it; and rubbed sulphur, sealing wax, or a rough glass tube as near as he could to the other end, for three or four minutes; after which he found the air to be negatively electrical, and that it would would continue so a considerable time after the apparatus was removed into another room. †

* Phil. Transf. Vol. 49. pt. 1. p. 300.   † Ib. Vol. 48. pt. 2. p. 784.

In

IN a paper dated November the 11th. 1754, he says, that dry air, at a great diftance from the earth, if in an electric ftate, will continue fo till it meets with fome conductor, is probable from the following experiment. An excited glafs tube, with its natural polifh, being placed upright in the middle of a room (by putting one end of it in an hole, made for that purpofe, in a block of wood) would, generally, lofe its electricity in lefs than five minutes, by attracting to it a fufficient quantity of moifture, to conduct the electric fluid from all parts of its furface to the floor ; but if, immediately after it was excited, it was placed, in the fame manner, before a good fire, at the diftance of about two feet, where no moifture would adhere to its furface, it would continue electrical a whole day, and how much longer he knew not. *

SIGNIOR BECCARIA, who had no knowledge of what Mr. Canton had done, made the fame difcovery of the communication of electricity to the air, and diverfified the experiment in a much more pleafing and fatisfactory manner. He proves, that the air, which is contiguous to an electrified body, acquires, by degrees, the fame electricity ; that this electricity of the air counteracts that of the body, and leffens its effects; and that as the air acquires, fo it alfo parts with this electricity very flowly.

HE began his experiments by hanging linen threads upon an electrified chain, and obferving, that they diverged the moft after a few turns of his globe. After that, they came nearer together, notwithftanding he kept turning the globe and the excitation was as powerful as ever. †

* Phil. Tranf. Vol. 48. pt. 2. p. 784.   † Lettere dell' Elettricifmo, p. 87.

when

WHEN he had kept the chain electrified a confiderable time, and then difcontinued the friction, the threads collapf-ed by degrees, till they hung parallel. After this, they began to diverge again, without any frefh electrification ; and, if the air was ftill, this fecond divergence would continue an hour, or more.

THIS divergence was leffened by the electrification of the chain. For if the globe was turned again, the threads would firft become parallel, and then begin to diverge again as be-fore. Thus the fecond divergence of the threads took place, when the chain was deprived of its electricity, and when that which the air had acquired began to fhow itfelf.

WHILE the threads were beginning to diverge with the electricity of the air, if he touched the chain, and thereby took off what remained of its electricity, the threads would feparate farther. Thus the more the electricity of the chain was leffened, the more did the electricity of the air appear.

WHILE the threads were in their fecond divergence, he hung two other threads, fhorter than the former, by another filk thread to the chain; and when all the electricity of the chain was taken quite away, they would feparate, like the former threads.

IF he prefented other threads to the former, in their fe-cond divergence, they would all avoid one another. *

IN this complete and elegant manner did Signior Beccaria demonftrate, that air actually receives electricity by commu-nication, and lofes it by degrees; and that the electricity of

* Lettere dell' Elettricifino, p. 90.

the

the air counteracts that of the body which conveys electricity to it.

SIGNIOR BECCARIA also made a variety of other experiments, which demonstrate other mutual affections of the air and the electric fluid; particularly some that prove their mutual repulsion; and that the electric fluid, in passing through any portion of air, makes a temporary vacuum.

HE brought the ends of two wires within a small distance of one another, in a glass tube, one end of which was closed, and the other immerged in water; and observed, that the water sunk in the tube, every time that a spark passed from the one to the other, the electric fluid having repelled the air.*

HE made the electric explosion a great number of times, in the same air, inclosed in a glass tube, in order to ascertain whether the elasticity of the air was affected by it; but he could not find any alteration. After the operation, he broke the tube under water, but neither did any air make its escape, nor any water force its way into the tube. The experiment was made with all the precaution, with respect to heat and cold, that the nature of the case required. †

SIGNIOR BECCARIA's experiments on *water*, showing its imperfection as a conductor, are more surprizing than those he made upon air, showing its imperfection in the contrary respect. They prove that water conducts electricity according to its quantity, and that a small quantity of water makes a very great resistance to the passage of the electric fluid.

HE made tubes, full of water, part of the electric circuit,

* Elettricismo artificiale e naturale, p. 110.   † Ib. p. 81

and obferved, that when they were very fmall, they would not tranfmit a fhock, but that the fhock increafed as wider tubes were ufed. *

BUT what aftonifhes us moft in Signior Beccaria's experiments with water, is his making the electric fpark vifible in it, notwithftanding its being a real conductor of electricity. Nothing, however, can prove more clearly how imperfect a conductor it is.

HE inferted wires, fo as nearly to meet, in fmall tubes filled with water; and, difcharging fhocks through them, the electric fpark was vifible between their points, as if no water had been in the place. The tubes were generally broken to pieces, and the fragments driven to a confiderable diftance. This was evidently occafioned by the repulfion of the water, and its incompreffibility, it not being able to give way far enough within itfelf, and the force with which it was repelled being very great. †

THE force with which fmall quantities of water are thus repelled by the electric fluid, he fays, is prodigious. By means of a charge of four hundred fquare inches, he broke a glafs tube two lines thick, when the pieces were driven to the diftance of twenty feet. Nay he fometimes broke tubes eight, or ten lines thick, and the fragments were driven to greater diftances in proportion. ‡

HE found the effect of the electric fpark upon water greater than the effect of a fpark of common fire on gunpowder; and fays he does not doubt, but that, if a method could be found

Elettricifmo artificiale e naturale, p. 113.   † Ib. p. 114.   ‡ Lettere dell' elettricifmo, p. 74.

of

of managing them equally well, a cannon charged with water would be more dreadful than one charged with gunpowder. He actually charged a glafs tube with water, and put a
fmall ball into it, when it was difcharged with great force, fo
as to bury itfelf in fome clay he placed to receive it. *

THIS refiftance which fmall quantities of water make to
the electric matter, he imagined, was greater than the refiftance made to it by air.† And yet he thought it was poffible,
that, in this cafe, the electric matter might not act upon the
water immediately, but upon the fixed air that was in it.
For when the tubes were not broken, he obferved that a
great number of air bubbles were fet loofe, through the whole
mafs of the water, rofe to the top, and mixed with the common atmofphere. ‡

HE alfo imagined that the electric fluid acted upon the fixed
air in all bodies, though no experiment could make it fenfible. ‖

ON the contrary, he fuppofed that the action of the electric matter tended to fix elaftic air, by exciting a fulphureous
matter, which Dr. Hales fhows to have that property. § But
the experiment abovementioned, of the electric fpark taken
in a clofed tube, doth not favour this fuppofition.

WHEN a fmall drop of water was put between the points
of two wires, and a large fhock paffed through them, the
water was equally difperfed on the infide of a glafs fphere, in
which they were all inclofed. In the fame manner, he con-

---

* Lettere dell' elettricifmo, p. 75, 76.   † Elettricifmo artificiale, &c. p. 115.
              ‡ Ib. p. 116.   ‖ Ib. p. 83.   § Ib.

jectures,

jectures, that the action of the electric matter promotes the evaporation of water. *

DISCHARGING a fhock through a quantity of water, poured on a flat furface, where fome parts of the circuit were purpofely left almoft dry; thofe parts became quite dry fooner than they would have been, if no fhock had paffed through them. †

UPON this principle he accounts for the fuppofed burfting of the blood veffels in fmall birds killed by the electric fhock. ‡   And when a mufcle contracts by the fhock, he fuppofes it is owing to the dilatation of the fluids their fibres contain, as the electric matter paffes through them.

So imperfect a conductor of electricity is mere water, that, he thought, a green leaf conducted a fhock better than an equal thicknefs of water. ‖   If this be true, and vegetable fluids conduct electricity better than water, it will confirm a conjecture which Dr. Franklin told me, he had drawn from fome experiments that he had not properly purfued, viz. that animal fluids conducted electricity better than water. He faid the finew of a deer, which did not feem very moift, conducted a fhock, when a wet thread would not do it.

SIGNIOR BECCARIA alfo found, that even *metal* was not a perfect conductor of electricity, but made fome refiftance to the paffage of the electric fluid. This he even afcertained, by meafuring the time that it was retarded, in its paffing through long and fmall wires, notwithftanding the

---

* Elettricifmo artificiale, &c. p. 117.   † Ib. p. 121.   ‡ Ib. p. 128.
‖ Ib. p. 135.

experiments

experiments which had been made before, that feemed to prove the contrary.

He fufpended a wire of 500 Paris feet, in a large building, and, by means of a pendulum which vibrated half feconds, obferved, that light bodies placed under a ball of gilt paper, at one end, did not move, till, at leaft, one vibration of this pendulum, after he had applied the wire of a charged phial to the other.

Trying the fame with a hempen cord, he could count fix, or more vibrations before they would ftir; but when he had wetted the cord, they were moved after two or three vibrations. * He does not, however, abfolutely fay that, the electric fluid muft have taken up all this time in its progrefs, as it might require a certain quantity of the fluid, before it could raife the light bodies. But he did imagine, that it moved with more velocity, in proportion as the bodies into which it paffed had more or lefs of the fluid before. † And he was confirmed in this opinion by feveral phenomena of the atmofphere, which will be related in their proper place; particularly by feeing, very evidently, the progrefs of a quantity of electric matter in the air, as it advanced to ftrike his kite.

To thefe experiments of Signior Beccaria on the conducting powers of air and water, I fhall fubjoin another curious fet of the fame author, fhowing the manner in which the fmoke of rofin and of colophonia is affected by the approach of an electrified body, as they have a very near affinity to this fubject.

* Elettricifmo artificiale, p. 51.   † Ib.

REPEATING

REPEATING Dr. Franklin's experiments to make electric atmospheres visible with the fume of colophonia, which he preferred, for this purpose, to rosin; he observed several curious circumstances, which had escaped the notice of that ingenious philosopher.

HE heated the colophonia on a coal, which he held in a spoon under an electrified cube of metal; and observed, that when part of the smoke ascended to the cube, another part covered the handle of the spoon, and spread to his hand. *

THE smoke lay higher on the flat parts of the cube than on the edges, and corners.

IF a spark was taken from the conductor, the smoke was thrown into an agitation, but presently resumed its former situation.

THE cube with its atmosphere gave larger, and longer sparks, than a cube not surrounded with one.

A LARGER spark might be taken from it by the spoon, than by any other body.

HAVING insulated the spoon, he observed, that hardly any of the smoke went to the cube; and that what happened to go near it was not affected by it, any more than it would have been by any other body. He put his finger to the spoon, and the former phenomena returned. Taking it off again, the smoke that had settled on the cube soon dispersed. †

UNDER this head of the electricity of various substances, it will not be improper to mention an experiment made by Mr.

Elettricismo artificiale, &c. p. 72.　　† Ib. p. 73, 74.

Henry

Henry Eeles of Lifmore in Ireland, which, he thought, proved that fteam, and exhalations of all kinds, are electrical. The paper containing this account was read at the Royal Society, April the 23d. 1755.

He electrified a piece of down, fufpended on the middle of a long filk ftring, and made fteam and fmoke of feveral kinds pafs under it, and through it; and obferved, that its electricity was not in the leaft diminifhed, as he thought it would have been, if the vapour had been non electric, and confequently had taken away with it part of the electric matter with which the down was loaded. He obferved that the effect was the fame, whether the down was electrified with glafs or wax, which he thought was not eafy to be accounted for. *

To this experiment Dr. Darwin of Litchfield, in a letter addreffed to the Royal Society, and read May the 5th. 1757, anfwers; that many electrified bodies, and particularly all light, dry, animal and vegetable fubftances, will not eafily part with their electricity, though they be touched, for a confiderable time, with conductors. He touched a feather, electrified like that of Mr. Eeles, nine times with his finger, and ftill found it electrified. A cork ball was touched feven times in ten feconds before it was exhaufted. †

Mr. Kinnersley of Philadelphia, in a letter dated March 1761, informs his friend and correfpondent Dr. Franklin, then in England, that he could not electrify any thing by means of fteam from electrified boiling water; from

* Phil. Tranf. Vol. 49. pt. 1. p. 153.   † Ib. Vol. 50. pt. 1. p. 252.

whence

whence he concluded, that, contrary to what had been before fuppofed by himfelf and his friend, fteam. was fo far
from rifing electrified, that it left its fhare of common electricity behind. *

To try the effects of electricity upon air, Mr. Kinnerfley
contrived an excellent inftrument, which he calls *an electrical
air thermometer*. It confifted of a glafs tube, about eleven
inches long, and one inch in diameter, made air tight, clofed
with brafs caps at each end, and a fmall tube, open at both ends,
let down through the upper plate, into fome water at the bottom
of the wider tube. Within this veffel he placed two wires, one
defcending from the brafs cap at the upper end, and the o-
ther afcending from the brafs cap at the lower end; through
which he could difcharge a jar, or tranfmit a fpark, &c. and
at the fame time fee the expanfion of the air in the veffel, by
the rife. of the water in the fmall tube. With this inftrument
he made the following experiments, related in a letter to Dr.
Franklin, dated March the 12th. 1761.

He fet the thermometer on an electric ftand, with the
chain fixed to the prime conductor, and kept it well electrized a confiderable time; but this produced no confiderable
effect : from whence he inferred, that the electric fire, when
in a ftate of reft, had no more heat than the air, and other
matter wherein it refides.

When the two wires within the veffel were in contact, a
large charge of electricity, from above thirty -fquare feet of
coated glafs, produced no rarefaction in the air; which fhow-

* Phil. Tranf. Vol. 53. pt. 1. p. 84.

ed,

ed, that the wires were not heated by the fire paffing through them.

When the wires were about two inches afunder, the charge of a three pint bottle, darting from one to the other, rarefied the air very evidently; which fhowed, that the electric fire produced heat in it felf, as Mr. Kinnerfley fays, as well as in the air, by its rapid motion.

The charge of a jar which contained about five gallons and a half, darting from wire to wire, would caufe a prodigious expanfion in the air; and the charge of his battery of thirty fquare feet of coated glafs would raife the water in the fmall tube quite to the top. Upon the coalefcing of the air, the column of water, by its gravity, inftantly fubfided, till it was in equilibrio with the rarefied air. It then gradually defcended, as the air cooled, and fettled where it ftood before. By carefully obferving at what height the defcending water firft ftopped, the degree of rarefaction, he fays, might be difcovered, which, in great explofions, was very confiderable.

It is obvious to remark, that the firft fudden rife of the water of Mr. Kinnerfley's thermometer, upon an explofion being made in the veffel which contained it, is not to be afcribed to the rarefaction of the air by heat, but to the quantity of air actually difplaced by the electrical flafh. It is only when that firft fudden rife is fubfided, as Mr. Kinnerfley himfelf obferves, that the degree of its rarefaction by the heat can be eftimated, viz. by the height at which the water then ftands above the common level.

Dr. Franklin had faid, that *ice* failed to conduct a fhock of electricity; and Mr. Bergman, in a letter to Mr.

F f   Wilfon,

Wilson, read at the Royal Society November the 20th. 1760, shows (what Signior Beccaria had done before) that a small quantity of water failed as much as the ice had done with Dr. Franklin, who seems to have made use of an icicle which, Mr. Bergman thought, was not large enough for the purpose. From hence he suspected that large quantities of ice would transmit a shock of electricity as perfectly as a large quantity of water. *

However, he seems, afterwards, to have changed his sentiments with respect to ice : for, in a subsequent paper, read at the Royal Society March the 18th. 1762, when he had remarked that snow would not conduct the electric shock ; he says, he believes, if he could procure plates of ice of a proper thickness, he could charge them in the same manner as glass. †

Johannes Franciscus Cigna was so fully persuaded of the non conducting power of ice, that he made use of it in an experiment, designed to ascertain whether electric substances did, according to Dr. Franklin's hypothesis, contain more of the electric matter than other bodies. He inclosed a quantity of ice in a glass vessel, and when he thought he had converted it from an electric to a non electric by melting; he tried whether it was electrified : but, though he found no appearance of its having acquired any more of the fluid than it ought to have in its new state, he does not seem to have given up his opinion. ‡

* Phil. Transf. Vol. 51. pt. 2. p. 908. † Ib. Vol. 52. pt. 2. p. 485.
‡ Memoirs of the Academy at Turin, for the year 1765. p. 47.

In

In the laſt part of this work the reader will find ſome experiments, which, it is imagined, will aſcertain the claſs of bodies in which ice ought to be ranked, by proving its conducting power to be, at leaſt, nearly equal to that of water.

SECTION III.

MR. CANTON's EXPERIMENTS AND DISCOVERIES RELATING TO THE SURFACES OF ELECTRIC BODIES, AND OTHERS MADE IN PURSUANCE OF THEM, OR RELATING TO THE SAME SUBJECT; ALL TENDING TO ASCERTAIN THE DISTINCTION BETWEEN THE TWO ELECTRICITIES.

TILL this last period of the history, the same electricity had always been produced by the same electric. The friction of glass had always produced a positive, and the friction of sealing wax, &c. had always produced a negative electricity. These were thought to be essential, and unchangeable properties of those substances; and hence the one was by many called the vitreous, and the other the resinous electricity; and to electrify negatively, or produce a resinous electricity by means of glass; or to electrify positively, that is, produce a vitreous electricity, by means of sealing wax, &c. would have been thought as great a paradox, as to electrify at all by the friction of brass or iron. For though it was not known why the electric matter should flow from the rubber

to

to the excited glafs, or to the rubber from excited fealing wax, the fact had been invariable ; and nothing is even mentioned to have happened, in the courfe of any experiments, that could lead a perfon to fufpect the poffibility of the contrary.

What then muft have been the furprife of electricians, to find that thefe different powers of glafs and fulphur were fo far from being invariable, that they were even interchangeable ; and that the fame glafs tube could be made to affume the powers of both! And what muft have been their fatisfaction to find the circumftance on which the convertibility of thofe oppofite powers depended, completely afcertained. This furprife and pleafure was given them by Mr. Canton, who fhowed that it depended only on the rubber, and the furface of the glafs, whether it electrified pofitively or negatively.

In what manner, by what train of thought, or by what accident he was led to this difcovery, this excellent philofopher has not been pleafed to inform us ; but it is certainly a difcovery that, in an eminent manner, diftinguifhes this period of my hiftory. It throws great light upon the doctrine of pofitive and negative electricity, and led the way to other difcoveries which throw ftill more light upon it.

This fubject of the two electricities feems to have engaged the attention of electricians in a more particular manner, in the whole courfe of this period, and ever fince the difcovery of Dr. Franklin, that the electricity of the two furfaces of charged glafs are always contrary to one another. Accordingly, the reader will find feveral fections in this period of

the

the history relating to it; but he will find that though much has been done, much yet remains to be done; and that we are still far from thoroughly understanding the nature of the two electricities, with their dependence upon, and relation to one another.

PREVIOUS to the communication of the discovery itself, Mr. Canton observes, that sealing wax might have positive electricity superinduced upon it. He excited a stick of sealing wax about two feet and a half in length, and an inch diameter; and, holding it by the middle, he drew an excited glass tube several times over one part of it, without touching the other. The consequence was, that half which had been exposed to the action of the excited glass was positive, and the other half negative: for the former half destroyed the repelling power of balls electrified by glass, while the other half increased it. †

THE experiments, which prove that the appearances of positive and negative electricity depend upon the surface of the electrics, and that of the rubber, were made in the latter end of December 1753.

HAVING rubbed a glass tube with a piece of thin sheet lead, and flour of emery mixed with water, till its transparency was intirely destroyed, he excited it (when it was made perfectly clean and dry) with new flannel, and found it act in all respects like excited sulphur or sealing wax. The electric fire seemed to issue from the knuckle, or end of the finger, and to spread itself on the surface of the tube, in a very beautiful manner.

* Phil. Tians. Vol. 48. pt. 1. p. 356.

IF

If this rough or unpolished tube was excited by a piece of dry oiled silk, especially when rubbed over with a little chalk or whiting, it would act like a glass tube with its natural polish. In this case the electric fire appeared only at the knuckle, or the end of the finger, where it seemed to be very much condensed before it entered.

But if the rough tube was greased all over with tallow from a candle, and as much as possible of it wiped off with a napkin, then the oiled silk would receive a kind of polish by rubbing it; and, after a few strokes, would make the tube act in the same manner as when excited at first by flannel.

The oiled silk, when covered with chalk or whiting, would make the greased rough tube act again like a polished one; but if the friction was continued till the rubber became smooth, the electric power would be changed to that of sulphur, sealing wax, &c.

Thus, says he, may the positive and negative powers of electricity be produced at pleasure, by altering the surfaces of the tube and rubber, according as the one or the other is most affected by the friction between them. For if the polish be taken off one half of a tube, the different powers may be excited with the same rubber at a single stroke; and, he adds, the rubber is found to move much easier over the rough, than over the polished part of it.

That polished glass electrified positively, and rough glass rubbed with flannel negatively, seemed plain from the appearance of the light between the knuckle, or end of the finger, and the respective tubes. But this, Mr. Canton thought, was farther confirmed by observing, that a polished

glass

glafs tube, when excited by fmooth oiled filk, if the hand
was kept three inches, at leaft, from the top of the rubber,
would, at every ftroke, appear to throw out a great number
of diverging pencils of electric fire; but that none were ever
feen to accompany the rubbing of fulphur, fealing wax, &c.
nor was he ever able to make any fenfible alteration in the
air of a room merely by the friction of thofe bodies; whereas
the glafs tube, when excited fo as to emit pencils, would, in
a few minutes, electrify the air, to fuch a degree, that, after
the tube was carried away, a pair of balls, about the fize of
the fmalleft peas, turned out of cork, or the pith of elder,
and hung to a wire by linen threads of fix inches long, would
repel each other to the diftance of an inch and an half, when
held at arms length in the middle of the room. *

AFTER thefe experiments of Mr. Canton, Mr. Wilfon
made feveral, which throw a little more light upon this curious fubject; but it is difficult to draw any general conclufion from them, and his own is not fufficiently determinate.
It is, that two electrics being rubbed together, the body
whofe fubftance is hardeft, and electric power ftrongeft, will
always be plus, and the fofteft and weakeft minus. † Rubbing
the tourmalin and amber together, he produced a *plus* electricity
on both fides the ftone, and a *minus* on the amber; but
rubbing the tourmalin and diamond together, both fides of the
tourmalin were electrified *minus*, and the diamond *plus*.

THESE experiments, which, he thought, proved this pro-
pofition, encouraged him to try what would be the effect of

* Phil. Tranf. Vol. 48. pt. 2. p. 782. † Ib. Vol. 51. pt. 1. p. 331.

rubbing

rubbing or forcing air against different electrics, and the effects were very considerable. In these experiments he only made use of a common pair of bellows, and his first experiment was upon the tourmalin. This substance he brought near the end of the pipe, and found, that after it had received about twenty blasts, it was electrified *plus* on both sides. Air, therefore, seemed to be less electric than the tourmalin.

INTO the place of the tourmalin, he brought a pane of glass, and blew against it the same number of times as in the former experiment; and when he had examined both sides, he found that they were electrified *plus* also, but less than the tourmalin.

AMBER, treated in the same manner, was electrified less than the glass.

HE next had recourse to a smith's bellows. The difference which these occasioned was only a much stronger electricity in the tourmalin. Amber was still weaker than the glass; and the glass weaker than the tourmalin.

HAVING in view the medium (which, I have observed, he laid great stress upon, as constituting the difference between electrics and non electrics) he considered that heat would rarify it on the surfaces of the particles of air; by which means, air, having its resistance lessened, would more readily part with the electric fluid, and, of consequence, electrify more powerfully.

THE pipe of the bellows being made red hot, he blew against the tourmalin twelve times only, which was eight times less than in the former experiment with cold air. In this experiment the tourmalin was electrified *plus* on both

G g   sides,

fides, but to a confiderable degree more than was done in the former experiments. The hot air had the fame effect upon glafs, but electrified it lefs than the tourmalin: and amber; though, like the other bodies, it received an increafe of power by the fame treatment, was electrified the leaft of all.

From the air electrifying more powerfully when it was hot than when it was cold, and the tourmalin being electrified more than glafs, and glafs more than amber, as appeared by the laft experiments; we feem, fays Mr. Wilfon, to have obtained a proof, that the whole atmofphere is conftantly promoting a flow of the electric fluid, by the alternate changes of heat and cold; and further, that air is not only lefs electric than the tourmalin, but lefs than glafs, or even amber.*

In another paper, read at the Royal Society November the 13th. 1760, Mr. Wilfon recites fome curious experiments, which, he fays, fhew that a *plus* electricity may be produced by means of a *minus* electricity.

Having electrified the infide of a large Leyden bottle *plus*, by means of a conducting wire from an excited glafs globe; he fet it on a ftand of prepared wood, and took away the conducting wire, after which the mouth of the bottle was clofed with a ftopple of glafs. Then the pointed end of an ivory conductor was brought oppofite to the middle of the bottle, and about two inches from it. Upon doing this, the balls were electrified *minus*; and the more fo as the ivory was moved nearer the bottle, in an horizontal direction.

* Phil. Tranf. Vol. 51. pt. 1. p. 332, &c.

But,

But, on removing the ivory to a greater diſtance, the *minus* electricity decreaſed; and, at a certain diſtance, there was not any ſign of it remaining; but when the diſtance was increaſed to about eighteen inches from the bottle, a *plus* electricity appeared, which, continued even after the ivory was removed entirely away. *

With a cylinder of baked wood he electrified the balls hanging to the ivory *minus*, at the diſtance of four feet or more, by holding the cylinder over the middle of the ivory, and continuing it there; and, on moving it nearer, they were more ſtrongly electrified *minus*, but the ſame cylinder, on removing it back again to the diſtance of two or three feet, or more, electrified the balls *plus*.

When another conductor of metal, without edges or points, was uſed, inſtead of the ivory, and without any thing hanging from it, the ſame cylinder held over the metal (as was done in the laſt experiment over the ivory, at the diſtance of two feet) produced a *plus* electricity; and this was rendered weaker as the cylinder was moved nearer; but by leſſening the diſtance to about one foot, the *minus* electricity took place.   In theſe caſes Mr. Wilſon thought, that the *plus* appearance aroſe from the earth, air, or other neighbouring bodies.

When the preceding experiments were firſt made, he was a little embaraſſed, by the uncertain appearances of a *plus* electricity at one time, and a *minus* at another, in the ſame experiment; but, by repeated trials and obſervations, he

---

* Phil. Tranſ. Vol. 51. pt. 2. p. 899, &c.

found,

found, that a *plus* or *minus* electricity may be produced at pleasure, by carefully attending to the three following circumstances; viz. the form of the bodies, their sudden or gradual removal, and the degrees of electrifying.

Mr. Wilson, after this, proceeds to mention some other circumstances of a very nice nature, where, the slightest and almost imperceptible differences in the position or in the course of the friction of two bodies produce, in either of them, the *plus* electricity at one time, and the *minus* at another. Such, says he, are the effects of this subtle and active fluid, when the experiments are carefully made; and therefore they require the most scrupulous attention to trace out the causes which occasion them.

Sealing wax and silver were the bodies used in the two first experiments, but many other substances seemed to perform as well. The sealing wax was clean, and undisturbed by any friction whatever, but that of the air surrounding it, and had been so for some hours. The silver was fixed to a piece of prepared wood, which was also preserved from friction for the same length of time. Then, taking one of those substances in each hand, the silver being at the end of the wood the farthest from the hand, he laid the smoothest part of the silver upon the sealing wax, and moved it along the surface gently, once only, and with a very slight pressure, after which the silver was electrified *plus*, and the wax *minus*.

On repeating the experiment with equal care, and in the same manner, except that the smooth side of the silver was a little inclined, so that the edge of it pressed against the wax; the silver, after moving it as before, was electrified *minus*,

and

and the wax *plus*, contrary to what was obſerved in the laſt experiment.

THESE oppoſite effects, occaſioned by the different applications of the *flat* or *edge* of the ſilver, ſeemed to ariſe from an alteration made in the ſurface of the wax, by deſtroying the poliſh in one caſe, and not in the other; and in this reſpect reſembled the poliſhed and rough glaſs mentioned before.

UPON making uſe of prepared wood inſtead of wax, and employing different degrees of preſſure in the friction, with the ſame edge of the ſilver, he produced the like appearances; the leaſt preſſure cauſing a *plus*, and the greateſt preſſure a *minus* appearance in the ſilver.

A FLAT piece of ſteel well poliſhed, and the edges rounded off, afforded the ſame appearances, by only applying the flat ſurface to the wood, but it required more preſſing to produce the *minus* effect in this caſe than it did in the former, where the edge was concerned.

WHETHER the reaſon offered above for explaining theſe laſt curious appearances be true or not, Mr. Wilſon did not venture to affirm, for want of farther experiments; but thus much he thought might be ſafely advanced, that we have learned to produce at pleaſure a *plus* or *minus* electricity from the ſame bodies, by attending to the manner of their application and friction. *

MR. BERGMAN, in a letter to Mr. Wilſon, read at the Royal Society February the 23d. 1764, gives an account of

* Phil. Tranſ. Vol. 51. pt. 2. p. 899, &c.

ſome

ſome curious experiments of his, which, in conjunction with thoſe of Mr. Canton abovementioned, concerning ſurfaces, may throw conſiderable light upon the doctrine of poſitive and negative electricity.

THE experiments were made with two pieces or ſkains of ſilk, one of which was extended in a frame, while Mr. Bergman held the other in his hand. He obſerved, that if the two ſkains of ſilk were the ſame with reſpect to texture, colour, ſuperficies, and in every thing elſe, as far as could be judged; and if he drew the whole length of the piece which he held in his hand over one part of that which was extended in the frame, the piece in his hand contracted the poſitive electricity, and the piece in the frame a negative. If he drew one part of that which he held in his hand over the whole length of the other the effects were reverſed.

IF the piece in his hand was of a different colour from that in the frame (provided it was not black) the event was the ſame.

IF the piece in his hand was black, it was always negative, which ever way it was rubbed, except the piece in the frame was black too; for then, if the whole length of it was rubbed, it was electrified poſitively.

IN endeavouring to account for theſe effects, he obſerves, that the piece which was moſt rubbed was made *ſmoother*, and *warmer* than the other; and was of opinion, that though ſmoothneſs did diſpoſe bodies to be excited poſitively, yet that other circumſtances were alſo to be taken into conſideration; having found that when he held in his hand a ſkain of ſilk, which, by much friction, was made very ſmooth, and

drew

drew it over one part of another skain, which was rough, and had never been used before, that the rough skain was, nevertheless, positive. From this experiment he concluded, that this effect was, in some measure, owing to the colour; and, in pursuing this thought farther, he was led to the following experiments.

If the piece in his hand was well warmed, though it was drawn over one part of the piece in the frame, it became electrified negatively, and the piece in the frame positively. He made these experiments with the same success upon skains of silk of various colours, blue, green, red, white, &c.

If the piece in the frame was black, it never contracted a positive electricity, although the piece in his hand had been much heated, except this were black too. From these experiments, he thought he might safely conclude, that heat did dispose some substances, at least, to a negative state; and thought that the want of attention to this circumstance might have occasioned mistakes in the event of some experiments, especially those concerning island chrystal.

From the whole he concludes, that there is a certain fixed order with respect to negative and positive electricity, in which all bodies may be placed, while other circumstances remain the same. Let A, B, C, D, E be certain substances, each of which, when rubbed with one which is antecedent to it, is negative, but with a subsequent positive. In this case, the less distance there is between the bodies that are rubbed, the weaker, *cæt. par.* will be the electricity; wherefore it will be stronger between A and E, than it will be between A and B. Heat, he says, disposes bodies to a negative electricity,

tricity, but if the distance abovementioned be considerable, it may not be able to *overcome*, though it may *weaken* that electricity, as is evident from the skains of black silk. When a glass globe grows warm in whirling, we are sensible that its electric power is diminished. Is it not owing, says he, to this circumstance, that by heat it is more disposed to negative electricity, by which means the distance above-mentioned between the glass and the rubber is lessened? *

UPON the subject of this section, I must introduce to the acquaintance of my reader two eminent electricians, whose discoveries will give him the greatest satisfaction; I mean Mr. WILKE, and ÆPINUS, the former of Rostock in Lower Saxony, and the latter of Petersburgh: a circumstance which gives me an occasion of congratulating all the lovers of the sciences, and particularly of electricity, on the extensive spread of their favourite studies. What joy would it have given Mr. Hawkesbee, or Mr. Grey, to have foreseen that two such admirable treatises on the subject of electricity, as those of the persons above-mentioned, would come from countries so remote from the place of its rise!

MR. WILKE relates many curious experiments concerning the generation of what he calls *spontaneous electricity*, produced by the liquifaction of electric substances, which, compared with those of Mr. Canton, throw great light upon the doctrine of positive and negative electricity.

HE melted sulphur in an earthen vessel, which he placed upon conductors; then, letting them cool, he took out the

* Phil. Transf. Vol. 54. p. 86.

sulphur

sulphur, and found it strongly electrical; but it was not so when it had stood to cool upon electric substances.

He melted sulphur in glass vessels, whereby they both acquired a strong electricity in the circumstances above-mentioned, whether they were placed upon electrics or not; but a stronger in the former case than in the latter; and they acquired a stronger virtue still, if the glass vessel into which they were poured was coated with metal. In these cases, the glass was always positive, and the sulphur negative. It was particularly remarkable, that the sulphur acquired no electricity till it began to cool, and contract, and was the strongest when in the state of greatest contraction; whereas the electricity of the glass was, at the same time, the weakest; and was the strongest of all when the sulphur was shaken out, before it began to contract, and acquire any negative electricity.

Pursuing these experiments, he found that melted sealing wax, poured into glass, acquired a negative electricity, but poured into sulphur it acquired a positive electricity, and left the sulphur negative. Sulphur poured into baked wood became negative. Sealing wax also poured into wood was negative, and the wood consequently positive; but sulphur poured into sulphur, or into rough glass, acquired no electricity at all. *

Experiments similar to these were also made by Æpinus. He poured melted sulphur into metal cups, and observed that when the sulphur was cool, the cup and the sulphur together showed no signs of electricity, but showed very strong signs of

* Wilke, p. 44, &c.

H h

it

it the moment they were separated.   The electricity always disappeared when the sulphur was replaced in the cup, and revived upon being taken out again.   The cup had acquired a negative, and the sulphur a positive electricity; but if the electricity of either of them had been taken off while they were separate, they would both, when united, show signs of that electricity which had not been taken off.   This electricity, he observes, was only on the surface of the sulphur. *

MR. WILKE has, likewise, recited several curious experiments, which he made on the friction of various substances, which likewise throw considerable light on the same subject.

SULPHUR and glass rubbed together produced a strong electricity, positive in the glass, and negative in the sulphur.

SULPHUR and sealing wax being rubbed together, the wax became positive, and the sulphur negative.

WOOD rubbed with cloth was always negative.

WOOD rubbed against smooth glass became negative, but against rough glass positive.

SULPHUR rubbed against metals was always positive, and this was the only case in which it was so; but being rubbed against lead it became negative, and the metal positive; lead appearing, thereby, to be not so good a conductor as 'the other metals.

AFTER these experiments, Mr. Wilke gives the following catalogue of the principal substances with which electrical experiments are made, in the order in which they are disposed to acquire positive or negative electricity; any of the substan-

* Æpini Tentamen, p. 66, 70.

ces

ces becoming positively electric when rubbed with any that follow it in the list, and negative when rubbed with any that precede it.

Smooth glass.	White wax.
Woollen cloth.	Rough glass.
Quills.	Lead.
Wood.	Sulphur.
Paper.	Metals. *
Sealing wax.	

In all experiments made to determine the order of these substances, Mr. Wilke says, that great care is necessary, to distinguish original electricity from that which is communicated, or the consequence of friction. †

Mr. Wilke says that smooth glass is in all cases positive, and thence infers that it attracts the electric fluid the most of all known substances; but Mr. Canton tells me he has found, that the smoothest glass will acquire a negative electricity by being drawn over the back of a cat.

Of the same nature with these experiments of Mr. Wilke are the following of Æpinus. He pressed close together two pieces of looking glass, each containing some square inches; and observed, that when they were separated, and not suffered to communicate with any conductor, they each acquired a strong electricity, the one positive, and the other negative. When they were put together again, the electricity of both

* Wilke, p. 54, &c.   † Ib. p. 69.

disappeared,

disappeared, but not if either of them had been deprived of their electricity when they were asunder; for, in that case, the two, when united, had the electricity of the other. The same experiment, he says, may be made with glass, and sulphur, or with any other electrics, or with any electric and a piece of metal. *

* Æpini Tentamen, p. 65.

SECTION

SECTION IV.

MR. DELAVAL's EXPERIMENTS RELATING TO THE TWO ELECTRICITIES, AND HIS CONTROVERSY WITH MR. CANTON UPON THAT SUBJECT.

MR. CANTON, in the course of experiments related in the preceding section, clearly proved, that the production of either of the two electricities depends intirely upon the surface of the excited electric with respect to the rubber, and showed, that the very same glass tube would produce either of them at pleasure; yet, notwithstanding this demonstration, Mr. Delaval, several years afterwards, proposed another theory of the two electricities, which seems to be more ingenious than solid; as it goes upon the supposition of the different powers depending intirely upon the different substances themselves. The account of this theory was read at the Royal Society March the 22d. 1759. It necessarily occasioned some controversy with Mr. Canton, in the course of which some new experiments were made, and some new facts discovered; on account of which I shall, with the utmost

moſt impartiality, report all that was advanced on both ſides.

Mr. Delaval obſerved, that there were two of the pure chymical principles of bodies, viz. *earth* and *ſulphur*, which were each poſſeſſed of a different kind of electricity; one of which we might call a *plus* electricity, the other a *minus*; and thought that it might be expected, that, in a body compounded of both, the oppoſite powers of thoſe ingredients would counter-balance, and deſtroy the effect of each other; and therefore, that bodies in which the negative and poſitive powers were equal, would be neutral, or non electrics. Such a ſubſtance he took metal to be, conſiſting of calx and ſulphur; metals not being calcinable without a degree of heat ſufficient to diſſipate all their ſulphur; as is evident from their not being reducible again to their metallic form, without the admixture of ſome unctuous matter. The ſame diſſipation of ſulphur, he ſays, muſt take place in animal and vegetable ſubſtances, before they become white aſhes. Tranſparent ſtones he conſidered as little more than pure earth, free from the leaſt mixture of oil; judging of others by the chymical reſolution of chryſtal.

To confirm this theory, Mr. Delaval made experiments with dry powders of calcined metals, viz. ceruſs, lead aſhes, minium, calx of antimony, &c. incloſing them in long glaſs tubes, and endeavouring to tranſmit the electric virtue through them, and always finding it impoſſible. Animal and vegetable ſubſtances, when reduced to aſhes, were alike impermeable to electricity, as alſo the ruſt of metals.

He

HE was firſt led to theſe experiments, and to this hypotheſis, by finding that dry mould would not conduct electricity.
This he alſo tried with dry Portland ſtone, ſome of which he
had cut into plates nearly as thin as window glaſs. Theſe
he heated to a proper degree, and coated them on both ſides
with metal, in order to make the Leyden experiment. When
the ſtone was hot enough to ſinge paper, it conducted as perfectly as when cold; but on cooling a little, it began not to
conduct, and afforded ſmall ſhocks; which gradually increaſed in ſtrength for about ten minutes, at which time it was
about its moſt perfect ſtate, and remained ſo near a quarter
of an hour. After that time, the ſhocks gradually decreaſed,
as the ſtone grew cooler; till, at laſt, they ceaſed, and the
ſtone returned to its conducting ſtate again, but this ſtate
appeared before the ſtone was quite cold.

EXPERIMENTS of this kind ſucceeded with all bodies a-
bounding with calx or earth, as ſtones, earth, dry clay, wood
when rotten, or burned in the fire, till the ſurface becomes
black. Among other ſubſtances he tried a common tobacco
pipe, part of which, near the middle, he heated to a proper
degree, and then applied one end of it to an electrical bar,
while the other was held in the hand; and he obſerved, that
the electric fluid paſſed no farther along the pipe than to the
heated part. *

FROM theſe experiments Mr. Delaval inferred, that ſtones
and other earthy ſubſtances were convertible, by ſeveral methods, and particularly by different degrees of heat, from

* Phil. Tranſ. Vol. 51. pt. 1. p. 83.

non

non electrics to electrics. But finding, afterwards, that it was the opinion of some persons (Mr. Canton was the person chiefly hinted at) that this change did not immediately, but only consequentially depend on heat, by evaporating the moisture, which would return again when the substance cooled; he observes, in a paper read at the Royal Society December the 17th. 1761, that the tobacco pipe lost its electricity before it was cold, and therefore before it could have imbibed moisture sufficient to destroy its electricity; and besides, that the substance employed in the experiment was not of that kind of bodies which is apt suddenly to draw moisture from the air.

To these arguments Mr. Canton replies, in a paper read at the Royal Society February the 4th. 1762, *hot air* air may easily be proved to be a conductor of electricity, by bringing a hot iron poker, but for a moment, within three or four inches of a small electrified body; when it would be perceived, that its electric power would be almost, if not intirely destroyed; and by bringing excited amber within an inch of the flame of a candle, when it would lose its electricity before it had acquired any sensible degree of heat.

To confirm this, he mentions his having observed, that the tourmalin, Brasil topaz, and Brasil emerald, would give much stronger signs of electricity when cooling, after they had been held about a minute within two inches of an almost surrounding fire, where the air is a conductor, than they ever will after heating them in boiling water. He adds, that if both sides of those stones be equally heated, in a less degree, than will make the surrounding air a conductor, the electri-

city

city of each fide, whether *plus* or *minus*, would continue fo all the time the ftone was both heating and cooling, but would increafe while it was heating, and decreafe while it was cooling; whereas, if the heat was fufficient to make the furrounding air conduct the electric fluid from the pofitive fide of the ftone to the negative fide of it, while it was heating, the electricity of each fide would increafe while the ftone was cooling, and be contrary to what it was while the ftone was heating.

As to the tobacco pipe, Mr. Canton fays, that it not only attracts the moifture of the air, but abforbs it. Hence a tobacco pipe, after it begins to cool, will become a conductor again fooner than wood. And that it imbibes moifture fafter than wood, he fays, is evident, becaufe when wetted, it will not continue wet fo long as wood, imbibing the moifture prefently.

THAT tobacco pipe does not become a conductor by a particular degree of heat, without evaporating its moifture, is evident, he fays, from the following experiments.

IF three or four inches of one end of a tobacco pipe, of more than a foot in length, be made red hot, without fenfibly heating the other end, this pipe will prove a ready conductor; through the hot air furrounding one part of it, and the moifture contained in the other, although fome part of it muft have the degree of heat of a non conductor. But if the whole pipe be made red hot, and fuffered to cool till it has only fuperficial moifture enough to make it a good conductor, and then three or four inches of one end be again made hot, it will become a non conductor.

I i

IF

IF a nail be placed at, or near each end of a longifh folid piece of any of the abforbent bodies above-mentioned, fo as the point of each nail may be about half the thicknefs of the body within its furface; this body, by heat, may be made a non conductor externally or fuperficially, while it remains a good conductor internally. For the electric fluid will readily pafs from one nail to another, through the middle of the body, when it will not pafs on its furface, and even when the internal parts of the body are in an equal degree of heat with the external, as they muft foon be after it begins to cool. But if the fame body be expofed, for a fhort time, to a greater degree of heat than before, or if it be kept longer in the fame heat, it will become a non conductor intirely. *

MR. DELAVAL, in confirmation of particular bodies requiring particular degrees of heat to render them electric or non electric, independant of moifture, mentions a fubftance which, he fays, is affected by heat in an oppofite manner to the former inftances; fince the degrees of heat, neceffary to render the other fubftances electric, makes this non electric,

THE fubftance was *ifland chryftal* (which is well known for its fingular property of a double refraction) on a piece of which he made the following obfervations. 1. After this piece of chryftal had been rubbed, when the heat of the air was moderate, it fhowed figns of electricity, though not very ftrong ones. 2. If the heat was increafed, fo as to be a little greater than that of the hand, it deftroyed its electric power intirely. 3. By cooling the ftone again, the electric power was reftored.

* Phil. Tranf. Vol. 52. pt. 2. p. 459

HE

He immersed this piece of chrystal into a vessel filled with quicksilver, and surrounded with ice, where it remained near two hours, when the weather was very cold; and observed, that, upon taking it out with a pair of tongs (that it might not be altered by the heat of his hands) and rubbing it again, it was more strongly electric then he had at any other time experienced; but that, on placing it a few minutes upon the hearth, at some distance from the fire, its electric property was again destroyed, for that rubbing would not occasion any signs of it.

Thus, says he, we see two different kinds of fixed bodies, the one of which acquires an electric property with the same heat with which another loses it; while a third set of substances, as glass, &c. retain their electricity through both the degrees of heat necessary to the other two.

Some pieces of island chrystal, which he had procured from different places, had not the property of losing their electricity by a moderate heat. He had, in particular, a piece of that chrystal, - one part of which, when greatly heated, became non electric, while the other part, with the same heat, or even with a much greater one, remained perfectly electric.

He found several other earthy substances, whose electricity was destroyed by different degrees of heat.

From considering that the degree of heat, at which the island chrystal first mentioned was in its most perfect electric state, was less than the usual heat of the air, and that a small increase of that heat rendered it non electric; he did not

think

think it improbable, that many fubftances, which are not known to be electric, might prove fo, if expofed to a greater degree of cold then they have been hitherto examined in. *

To thefe obfervations Mr. Canton replies, that having formerly obferved that the friction between mercury and glafs in vacuo would not only produce the light of electricity, as in the luminous barometer, or within an evacuated glafs ball, but would alfo electrify the glafs on the outfide, he immerged a piece of dry glafs in a bafon of mercury; and found, that by taking it out, the mercury was electrified *minus*, and the glafs electrified *plus*, to a confiderable degree. He alfo found, that amber, fealing wax, and ifland chryftal, when taken out of mercury, were all electrified pofitively. How then, fays he, does it appear, that the electricity which was obferved in rubbing the laft mentioned fubftance, after it was taken out of mercury furrounded by ice, was owing to cold, and not to the friction between it and the mercury in taking it out. Ifland chryftal when warm is a non conductor, and all non conductors may be excited with proper rubbers. †

MR. BERGMAN of Upfal, in a letter to Mr. Wilfon, read at the Royal Society April the 14th. 1761, fays, that he had tried the experiments of Mr. Delaval with ifland chryftal, but that the event had always been contrary to what Mr. Delaval had reported. Trying different pieces of chryftal, he found one which, inftead of having its virtue increafed by cooling, was fenfibly increafed by heating. Afterwards trying all the reft

* Phil. Tranf. Vol. 52. pt. 1. p. 354, &c.   † Ib. Vol. 52. pt. 2. p. 461.

which

which he had by him, whether Swedish chrystal, or island, he found the effect to be the same. From this he inferred, that the chrystals which he had were of a quite different kind from that of Mr. Delaval. *

* Phil. Tranf. Vol. 53. pt. 1. p. 98.

SECTION

SECTION V.

MR. CANTON's EXPERIMENTS AND DISCOVERIES RELAT-
ING TO BODIES IMMERGED IN ELECTRIC ATMOSPHERES,
WITH THE DISCOVERIES OF OTHERS, MADE BY PURSUING
THEM.

IN this section I shall present my reader with the finest
series of experiments that the whole history of electricity
can exhibit, and in which we shall see displayed the genius
and address of four of the most eminent electricians in this
whole period; viz. Mr. Canton, and Dr. Franklin, Englishmen; and Mr. Wilke, and Æpinus, foreigners. Mr. Canton
had the honour to take the lead, and he made all the essential experiments. Doctor Franklin professedly pursued
them; and though *all his strength he put not forth* on this occasion, he diversified the experiments, and made some improvement in the method of accounting for them. But Mr.
Wilke and Æpinus in conjunction carried the experiments
vastly farther, and completed the discovery; which is, certainly, one of the greatest that has been made since the time
of Dr. Franklin. I say the time of Dr. Franklin, though he
himself be one of the persons concerned; for by *the time of*
Dr.

Dr. Franklin will always be underftood the time in which he made his capital difcoveries in America.   This will always be a diftinguifhed epocha in the hiftory of electricity, from which all his own future difcoveries will be dated.

The original experiments in this fection, when Mr. Canton firft publifhed them, in his ufual concife, though perfpicuous manner, without any preamble, to inform us how he was led to them, exhibit fuch a variety of attractions and repulfions of electrified bodies in different circumftances, as looked like the power of magic; and were they conducted with a little art, I do not know any electrical experiments (made without light, or noife) more proper for a deception of this kind.   But when they are attentively confidered, they demonftrate a remarkable property of all electrified bodies, which has often been referred to in the courfe of this hiftory, but which had not been attended to before; nor indeed do I apprehend it was fully underftood, till it was explained in all its extent by Mr. Wilke and Æpinus.   It is, that the electric fluid, when there is a redundancy of it in any body, repels the electric fluid in any other body, when they are brought within the fphere of each other's influence, and drives it into the remote parts of the body; or quite out of the body, if there be any outlet for that purpofe.   In other words, bodies immerged in electric atmofpheres always become poffeffed of the electricity, contrary to that of the body, in whofe atmofphere they are immerged.   And this principle purfued led them to the method of charging a plate of air, like a plate of glafs, and to make the moft perfect imitation of the phenomena of thunder and lightning.

The

THE paper containing an account of Mr. Canton's experiments was read at the Royal Society December the 6th. 1753.

MR. CANTON ſuſpended cork balls, one pair by linen threads, and another pair by ſilk; then holding the excited tube at a conſiderable diſtance from the balls with the linen thread, they ſeparated; and, upon drawing it away, they immediately came together: but he was obliged to bring the excited tube much nearer to the balls hanging by ſilk threads, before they would ſeparate; though when the tube was withdrawn, they continued ſeparate for ſome time.

As the balls in the former of theſe experiments were not inſulated, Mr. Canton obſerves, that they could not properly be ſaid to be electrified; but that when they hung within the atmoſphere of the excited tube, they might attract and condenſe the electric fluid round about them, and be ſeparated by the repulſion of its particles. He conjectures alſo, that the balls, at this time, contain leſs than their common ſhare of the electric fluid, on account of the repelling power of that which ſurrounds them, though ſome may be continually entering and paſſing through the threads. And if that be the caſe, he ſays, the reaſon is plain why the balls hung by ſilk in the ſecond experiment muſt be in a much more denſe part of the atmoſphere of the tube before they will repel each other. He adds, that at the approach of an excited ſtick of wax to the balls, in the firſt experiment, the electric fire is ſuppoſed to come through the threads into the balls, and to be condenſed there, in its paſſage towards the wax; ſince, according to Dr. Franklin, excited glaſs emits the electric fluid, and excited wax receives it.

WHEN two balls, ſuſpended by linen threads upon an in-

ſulated

fulated tin tube, were electrified positively, and had separated; he observed, that the approach of the excited tube would make them come near together; if brought to a certain distance, they would touch; and if brought nearer, they would separate again.

In the return of the tube, they would approach each other, till they touched, and then repel as at first. If the tin tube was electrified by wax, or the wire of a charged phial; the balls would be affected in the same manner at the approach of excited wax, or the wire of the phial. If the cork balls were electrified by glass, their repulsion would be increased at the approach of an excited stick of wax. And the effect would be the same, if the excited glass was brought towards them, when they had been electrified by wax.

The bringing the excited glass to the end, or edge of the tin tube, in the former of these experiments, is by Mr. Canton supposed to electrify it positively, or to add to the electric fire it before contained; and therefore some will be running off through the balls, and they will repel each other. But at the approach of excited glass, which likewise emits the electric fluid, the discharge of it from the balls will be diminished, or part will be driven back, by a force acting in a contrary direction, and they will come nearer together. If the tube be held at such a distance from the balls, that the excess of the density of the fluid round about them above the common quantity in air, be equal to the excess of the density of that within them above the common quantity contained in cork, their repulsion will be quite destroyed. But if the tube be brought nearer, the fluid without being more dense than that

K k

within.

within the balls, it will be attracted by them, and they will recede from each other again.

MR. CANTON farther obferves, that when the apparatus has loft part of its natural ftore of this fluid, by the approach of excited wax to one end of it, or is electrified negatively, the electric fire is attracted and imbibed by the balls, to fupply the deficiency; and that more plentifully at the approach of excited glafs, or a body pofitively electrified, than before; whence the diftance between the balls will be increafed, as the fluid furrounding them is augmented. And, in general, whether by the approach or recefs of any body, if the difference between the denfity of the internal and external fluid be increafed, or diminifhed; the repulfion of the balls will be increafed, or diminifhed accordingly.

HE obferved, that when the infulated tin tube was not electrified, if the excited glafs was brought towards the middle of it, the balls hanging at the end would repel each other, and the more fo as the excited tube was brought nearer. When it had been held a few feconds, at the diftance of about fix inches, and withdrawn, the balls would approach each other till they touched; and, feparating again, as the tube was removed farther, would continue to repel when the tube was taken quite away. This laft repulfion would be increafed by the approach of excited glafs, and diminifhed by that of excited wax; juft as if the apparatus had been electrified by wax, after the manner defcribed in the laft experiment.

HE infulated two tin tubes, which may be diftinguifhed by calling them A and B, fo as to be in a line with each other, and half an inch afunder, and at the remote end of each fufpended

pended

pended a pair of cork balls. Then, upon bringing the excited glafs tube towards the middle of A, and holding it a fhort time at the diftance of a few inches, he obferved each pair of balls to feparate. Upon withdrawing the tube, the balls of A would come together, and then repel each other again, but thofe of B would hardly be affected. By the approach of excited glafs the repulfion of the balls of A would be increafed, and thofe of B diminifhed. *

In the former of thefe experiments, Mr. Canton fuppofes the common ftock of electric matter in the tin tube to be attenuated about the middle, and to be condenfed at the ends, by the repelling power of the atmofphere of the excited glafs tube, when held near it. And perhaps, he fays, the tin tube may lofe fome of its natural quantity of the electric fluid before it receives any from the glafs, as that fluid will more readily run off from the ends or edges of it than enter at the middle; and accordingly, when the glafs tube is withdrawn, and the fluid is again equally diffufed through the apparatus, it is found to be electrified negatively; fince excited glafs brought under the balls will increafe their repulfion.

In the latter of the experiments, Mr. Canton fuppofes that part of the fluid driven out of one tin tube enters the other, which is found to be electrified pofitively, by the decreafing of the repulfion of its balls at the approach of excited glafs.

It will readily be feen that, at the time thefe experiments were made, Mr. Canton retained the common idea of electric atmofpheres; whereas it will appear by the experi-

* Phil. Tranf. Vol. 48.pt. 1. p. 350.

ments

ments of Mr. Wilke and Æpinus (which in fact contain nothing more than those of Mr. Canton) that they tend to refute the common opinion, and are much easier explained upon the supposition, that the portion of fluid belonging to any
electrified body is constantly held in contact, or very nearly
in contact, with the body; but acts upon the electricity of
other bodies at a certain distance.

DR. FRANKLIN pursued, or rather diversified the experiments of Mr. Canton, but retaining, likewise, the common
opinion of electric atmospheres, he thought that the phenomena
were more easily explained upon the supposition, that these
atmospheres, being brought near each other, did not easily
mix, and unite into one atmosphere, but remained separate,
and repelled each other; and moreover, that an electric atmosphere, would not only repel another electric atmosphere,
but also the electric fluid contained in the substance of a body
approaching it, and, without joining or mixing with it, force
it into the other parts of the body that contained it.

THOUGH it must be difficult to assign a reason why the
particles of one atmosphere should repel the particles of another atmosphere, or of the fluid contained in another body
with more force than they repel one another, or the particles
of the fluid contained in the body to which they belong;
since the matter is the same in both: yet this idea of the
mutual repulsion of electric atmospheres, could it once be supposed, will certainly and clearly account for all the facts;
and the theory pleases on account of its simplicity. But
the same appearances will be accounted for, in a manner as
simple and intelligible, upon the supposition, that the portion

of

of electric fluid belonging to each body, being strongly attracted by the body, is held in close contact with it; but that
it acts by repulsion upon the electric fluid belonging to other
bodies, at a distance from them; and that the electric fluid
doth not actually pass out of one body into another, till it
have first repelled the fluid out of the other body, and then
be more strongly attracted by the other body than by its own;
which has already got more than its natural share.

THE paper containing an account of these experiments of
Dr. Franklin was read at the Royal Society December the
18th. 1755. His apparatus, was different from that of Mr.
Canton, but still he exhibited the same effects proceeding
from the same cause. He fixed a tassel of fifteen or twenty
threads, each three inches long, at one end of his prime conductor, which was five feet long and four inches in diameter,
supported by silk lines. The threads were a little damp, but
not wet.

IN these circumstances, an excited tube brought near the
end of the prime conductor, opposite to the threads, so as to
give it some sparks, made the threads diverge, each thread
having thereby acquired its separate electric atmosphere.

IN this state, the approach of the excited tube, without
giving any sparks, made the threads diverge more; but, being withdrawn, they closed as much; the atmosphere of the
conductor being driven by that of the tube into the threads,
and returning again upon withdrawing the tube, which had
then left no part of its atmosphere behind it.

THE excited tube brought under the diverging threads
made them close a little, having driven part of their atmos-
                                                    pheres

pheres into the conductor. Upon being withdrawn, they diverged as much; that portion of their atmospheres which they had lost returning again from the conductor, and the tube having left no part of its own.

THE excited tube, held at the distance of five or six inches from the end of the conductor opposite to the threads, made them separate, and, upon being withdrawn, they came together again : but if, in their state of separation, a spark was taken from the conductor near them, they would close; and, upon removing the tube, would separate. The tube, in both cases, left no part of its atmosphere behind it. It only drove the natural quantity of electricity contained in the conductor towards the threads; and part of that being taken away by the spark, the tube would leave the conductor and threads negative, in which case, they would repel one another, as if they had been electrified positively.

IN this situation, if the excited tube was brought near the conductor, they would close again; the atmosphere of the tube forcing that of the conductor into the threads, to supply the place of what they had lost : but, upon withdrawing the tube, they would open again; the tube, as before, taking its whole atmosphere away with it. When the excited tube was brought under the threads, diverging with negative electricity, they diverged more; the atmosphere of the tube driving away more of the atmospheres of the threads, and giving them none in its place.

LASTLY the Doctor brought the excited tube near the prime conductor, when it was not electrified; and when the threads were, thereby, made to diverge, he brought his fing-

er

Signior Beccaria was of opinion, that the direction of the electric fluid may be determined from the phenomena of pointed bodies. The *pencil* (by which he means the electric fire at a point electrified positively) he says, contracts as it approaches a flat piece of metal not electrified; whereas the *star* (by which he means the electric fire at a point electrified negatively) expands in the same circumstances, and has a small cavity near the point towards the large superficies. The pencil is attended with a snapping noise, the star makes little or no noise. He hardly gives any reason for the first of these phenomena; he only says, that such is the necessary consequence of a fluid issuing out of, or entering into a point. But the greater noise made by the pencil, he thought was made by the impulse given by the electric matter to the air, causing it to vibrate : and this must be greater when the fluid is thrown from the point into the air, than when it comes through different portions of the air, and meets in one point. *

When two points are opposed to one another, he says, the phenomena are much the same in both. †

Signior Beccaria observed that hollow glass vessels, of a certain thinness, exhausted of air, gave a light when they were broken in the dark. By a beautiful train of experiments, he found, at length, that the luminous appearance was not occasioned by the breaking of the glass, but by the dashing of the external air against the inside, when it was broke. He covered one of these exhausted vessels with a receiver, and let-

* Elettricismo artificiale, p. 63.   † Ib.

ting

tricity oppofite to that of the electrified body. From this he concludes, that parts of non electric bodies, plunged in electric atmofpheres, acquire an electricity oppofite to that of the atmofphere in which they were plunged. *

HE placed two large infulated conductors with their ends oppofite to one another, and a cork ball fufpended on filk between them; and obferved, that, upon the application of the excited glafs tube to one end of either of them, the cork ball would play between them very faft; and if the tube were held a while at the fame diftance, would be at reft. Upon withdrawing the tube, the motion of the cork ball began again, and, at length, ceafed gradually as before. If the conductors were removed from one another, while they were within the atmofphere of the tube, they would, upon being brought together again, give a fpark. This experiment confirmed the demonftration, that the part of a body which is immerged in the atmofphere of an electrified body acquires the contrary electricity. †

BUT the moft complete demonftration of this general maxim is an experiment of Æpinus. He placed a fmall weight upon one end of a large metallic conductor, and, by means of a filk ftring, removed it from the conductor, while the end on which it refted was immerged in the atmofphere of an electrified body; and found that it had actually acquired a different electricity from that of the atmofphere. If the end of the conductor, oppofite to that on which the moveable weight was placed, was made to communicate with the

* Wilke, p. 77.   † Ib. p. 78.

earth,

earth, ftill that part of it which was near the excited electric was affected with the oppofite electricity. Placing the moveable weight on the oppofite end of the conductor, when it was infulated, he found that it had fometimes acquired an electricity contrary to that of the excited electric, fometimes the fame electricity, though weak, and fometimes no electricity at all. *

The fame ingenious philofopher confidered that the fame principle muft extend to glafs, and all other electrics; fince they, as well as conductors, contain a certain quantity of the electric fluid, in their natural ftate. To verify this, he took a glafs tube, and electrified one end of it pofitively. The confequence was, that four or five inches of that end were pofitive; but beyond that there were two inches negative, and beyond that the tube was again pofitive, though weakly fo. This experiment he repeated very often with the fame fuccefs; as alfo when, inftead of glafs, he ufed a folid ftick of fulphur. To account for this fact, he fuppofed, that the electricity communicated to the end of the tube repelled the natural quantity of the fluid in the glafs to fome diftance. This natural quantity retiring from its former fituation, he fuppofes to become condenfed, and confequently to repel another quantity of the fluid natural to the glafs from its place; and that thus the whole rod would be alternately pofitive and negative. The author afferts, that it was from theory only that he was led to this curious experiment, the fact exactly correfponding to what he had before deduced, as the neceffary

* Æpini Tentamen, p. 129.

confequence

confequence of Dr. Franklin's principles of negative and pofitive electricity. *

THE hint of thefe experiments Æpinus received from thofe
above-mentioned of Mr. Wilke; and thefe gentlemen, re-
fiding at the fame time at Berlin, purfued thefe curious experiments jointly, till they were led by them to difcover a
method of charging a plate of air in the fame manner as
plates of glafs had ufually been charged, and to throw ftill
more light upon the theory of the famous Leyden experiment.

IN the above-mentioned experiments, thefe gentlemen obferved, that the negative ftate of one of the bodies depended
on the oppofite ftate of the other, which was known to be
exactly the cafe of the two fides of a charged pane of glafs;
and the reafon of the non-communication of the fame electricity was evidently the impermeability of the glafs to the
electric fluid in the one cafe, and the impermeability of
the air in the other.   Upon this hint, they made feveral attempts to give the electric fhock by means of air; and at
length fucceeded, by fufpending large boards of wood covered with tin, with the flat fides parallel to one another, and
at fome inches afunder.   For they found, that upon electrifying one of the boards pofitively, the other was always
negative, agreeable to the former experiment: but the difcovery was made complete and indifputable by a perfon's
touching one of the plates with one hand, and bringing his
other hand to the other plate; for he then received a fhock

* Æpini Tentamen, p. 192.

through

through his body, exactly like that of the Leyden experiment. *

With this plate of air, as we may call it, they made a variety of curious experiments. The two metal plates, being in oppofite ftates, ftrongly attracted one another, and would have rufhed together, if they had not been kept afunder by ftrings. Sometimes the electricity of both would be difcharged by a ftrong fpark between them, as when a pane of glafs burfts with too great a charge. A finger put between them promoted the difcharge, and felt the fhock. If an eminence was made on either of the plates, the felf difcharge would always be made through it, and a pointed body fixed upon either of them prevented their being charged at all.

The ftate of thefe two plates, they excellently obferve, juftly reprefents the ftate of the clouds and the earth during a thunder ftorm ; the clouds being always in one ftate, and the earth in the oppofite; while the body of air between them anfwers the fame purpofe as the fmall plate of air between the boards, or the plate of glafs between the two metal coatings in the Leyden experiment. The phenomenon of lightning is the burfting of the plate of air by a fpontaneous difcharge, which is always made through eminences, and the bodies through which the difcharge is made are violently fhocked. †

This principle, they likewife thought, would explain an obfervation of the Abbé Nollet, that electricity was often obferved to be peculiarly ftrong, when the room was full of

* Wilke, p. 97.   † Ib. p. 101.

company,

company, and more particularly, when numbers of them drew near together, to fee the experiments. The conductor was then in one ftate, and the company in another; fo that, conftituting a large furface, when any of them took a fpark, as he thereby difcharged the electricity of all the company, he would feel it more fenfibly than if he had ftood fingle. *

THIS difcovery, of the methòd of giving the electric fhock by means of a plate of air, may be reckoned one of the greateft difcoveries in the fcience of electricity fince thofe of Dr. Franklin. It is beautiful to obferve how this fine difcovery took its rife from the experiments of Mr. Canton. Mr. Canton's experiments were purfued by Dr. Franklin, and thofe of Dr. Franklin purfued by thefe gentlemen produced the difcovery. It is one and the fame principle that, in different circumftances, accounts for this beautiful feries of experiments.

THIS experiment of charging a plate of air is likewife related by Æpinus, who fays that he was led to the difcovery, by reafoning from the confequences of Dr. Franklin's theory.

FROM thefe experiments he was alfo led to form a more diftinct idea of the impermeability of glafs to the electric fluid. For fince a plate of air might be charged as well as a plate of glafs, that property, whatever it be, muft be common to them both; and could not, as Dr. Franklin once fuppofed, be any thing peculiar to the internal ftructure of glafs. Impermeability, he, therefore, infers, muft be com-

* Wilke, p. 96 &c.

mon

mon to all electrics; and since they can all receive electricity by communication to a certain degree, it must consist in the difficulty and slowness with which the electric fluid moves in their pores; whereas, in perfect conductors, it meets with no obstruction at all. *

IT was chiefly this course of experiments, also, that led Mr. Æpinus to deny the existence of electric atmospheres, consisting of effluvia from electrified bodies.

HE seems, however, to consider this as a bold opinion; since he herein differs, as he says, from all the electricians who had written before him, and even from Dr. Franklin himself. Though the common opinion, he says, is by no means countenanced by the general principles of his theory, which suppose the electric fluid to move with difficulty through every electric substance like the air.

To those who might say, that an electric atmosphere is a thing obvious to the senses, and no matter of theory; since it may be felt like a spider's web upon the hands or face; he replies, that this feeling, together with the sulphureous smell of electrified bodies, are only sensations excited by the action of the fluid in the electrified bodies upon the electric fluid in the nostrils, or the hand; or upon those parts of the body themselves in an unelectrified state; and that they are not felt by a person who is not possessed of the same kind and degree of electricity.

He, therefore, thinks there never was any sufficient reason to admit those atmospheres; and declares, that whenever he

* Æpini Tentamen, p. 82.

uses

ufes the word, he means no more by it than the fphere of action of the electricity belonging to any body. Or, he fays, the neighbouring air, which is electrified by it, may be fo called.

But that thefe atmofpheres have little effect in electrical experiments, he fays, is evident from this circumftance; that if it be blown upon with a pair of bellows, the electricity of the body which it furrounds is not fenfibly diminifhed. The electric fluid, he fuppofes, to refide wholly in the electrified body, and from thence to exert its attraction or repulfion to a certain diftance. *

The fubject of electric atmofpheres had not efcaped the attention of the accurate Signior Beccaria, who was probably prior to Æpinus in fuppofing, that electrified bodies have no other atmofphere than the electricity communicated to the neighbouring air, and which goes with the air, and not with the electrified bodies, agreeable to that curious difcovery of his mentioned above.

He alfo mentions an experiment, which, he thinks, directly proves, that all the electricity communicated to any body adheres to its furface, and does not fpread into the air. He electrified a large conductor of gilt paper, in which the gilding was, in feveral places, taken off quite round; and obferved that whenever he difcharged it, by taking a fpark at the end, other fparks were vifible at all the interruptions; the charge of the more remote parts having come off through the fub-

* Æpini Tentamen, p. 257.

ftance

Plate I.

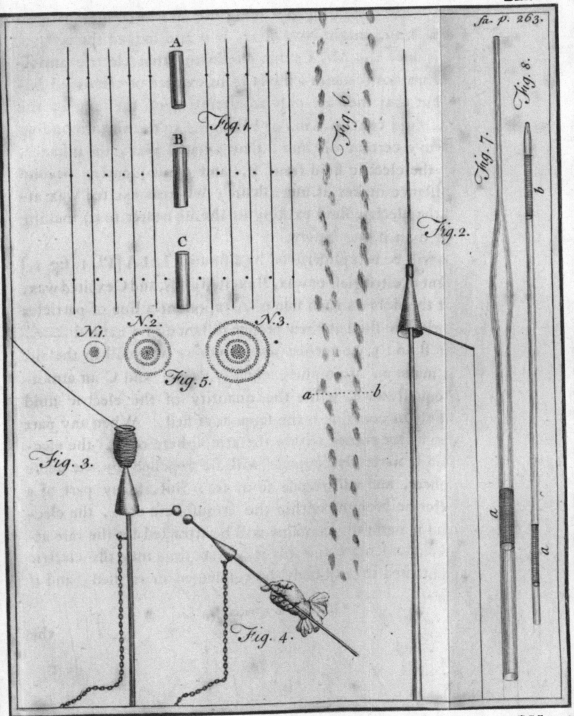

ftance of the metal, and not along the air; as the greateft part of it, at leaft, might have done, if it had lodged there. *

It is now alfo Mr. Canton's opinion, that electric atmof-pheres are not made of effluvia from excited or electrified bo-dies, but that they are only an alteration of the ftate of the electric fluid contained in, or belonging to the air furrounding them, to a certain diftance; that excited glafs, for inftance, repels the electric fluid from it, and confequently, beyond that diftance makes it more denfe; whereas excited wax at-tracts the electric fluid exifting in the air nearer to it, making it rarer than it was before.

This will be beft underftood by a figure. Let A [Pl. 1. fig. 1.] reprefent excited glafs or wax, B excited glafs, and C excited wax; and let the dots on each fide of A reprefent a line of particles of the electric fluid at their proper diftance in a natural ftate.

Let B and C be carried about where you will in the air, B will make an atmofphere equally denfe, and C an atmof-phere equally rare, while the quantity of the electric fluid each of them contains is the fame as at firft. When any part of a conductor comes within the atmofphere of B, the elec-tric fluid it naturally contains will be repelled by the denfe atmofphere, and will recede from it. But if any part of a conductor be brought within the atmofphere of C, the elec-tric fluid it naturally contains will be attracted by the rare at-mofphere, and move towards it. And thus may the electric fluid contained in any body be condenfed or rarified; and if

* Elettricifmo artificiale, p. 54.

the

the body be a conductor, it may be condenfed or rarified in any part of it, and fome may be eafily drawn out of, or an additional quantity put into it.

It was obferved before, that an experiment of Dr. Franklin which he thought proves that electric atmofpheres did not exclude the air, might juftly make us fufpect the exiftence of thofe atmofpheres, fince the electric matter is known to repel the air. Another experiment of the fame nature was made by Dr. Darwin of Litchfield, who fent an account of it to the Royal Society, which was read May the 5th. 1757. He got a glafs tube, open at one end, and having a ball at the other. This ball and half of the tube he coated; and when he had inverted it, and dipped a confiderable part of it in a veffel containing oil of turpentine, he introduced a wire into it, and charged it; and obferved, that the oil did not at all appear to fubfide. From this he concluded, that the electric atmofphere, flowing round the wire and the coating of the tube, above the oil, did not difplace the air, but exifted in its pores. *

An experiment fimilar to that of Dr. Franklin and that of Dr. Darwin was made by Signior Beccaria. He took a coated phial, and when he had inferted into it a fmall glafs tube, bent horizontally when it came out of the phial; he clofed it with cement, and prefented light afhes to the extremity of the tube, the orifice of which was very fine; and always found, that the afhes were blown off, when a fpark was taken into the phial, but they returned towards the end of the tube

* Phil. Tranf. Vol. 50. pt. 1. p. 351.

after-

afterwards. * It is probable, that the metal not being fufficiently in contact with the infide coating, a fpark was made
in the infide, which expelled the air, and caufed the motion
in the afhes. The faireft method of trying it would be with
a phial, in which the metal that received the fire from the
conductor, fhould be a production of the inward coating.

* Lettere dell' elettricifmo, p. 79.

Mm   SECTION

SECTION VI.

MR. SYMMER's EXPERIMENTS RELATING TO THE TWO ELECTRICITIES AND THOSE MADE BY JOHANNES FRANCISCUS ÇIGNA IN PURSUANCE OF THEM.

IT had hitherto been univerſally ſuppoſed, that all the phenomena of electricity were produced by the action of one electric fluid.   Even Mr. Du Faye, at the time that he imagined he had diſcovered another electric fluid, diſtinct from that of glaſs, and peculiar to roſin, &c. thought, however, that it was quite independent of the other, and that their operations were never combined.   Dr. Watſon, and Dr. Franklin thought it was very evident, that the difference between the two electricities conſiſted in the one being a redundancy, and the other a deficiency of the ſame matter. And all the experiments that had been made concerning the two electricities ſeemed to confirm this hypotheſis.   At length, however, Mr. Symmer produces a great number of curious experiments, relating to the ſame ſubject ; and infers from them the probable exiſtence of *two electric fluids*, not independent,

pendent, but always coexiftent, and counteracting one another.

The firft fet of his experiments are very remarkable, but
he does little more than relate naked facts. They were diverfified, and purfued much farther by Johannes Francifcus
Cigna, who has alfo explained them upon the principles of
Dr. Franklin's theory; though he was of opinion, that no
experiments that had yet been made were decifive, in favour
of either of the two hypothefes. Few hiftories of experiments are more entertaining than the firft of thefe of Mr.
Symmer: the fubfequent experiments are lefs fatisfactory.
The papers relating to them all were read at the Royal Society in the year 1759.*

This gentleman had for fome time obferved, that upon
putting off his ftockings, in an evening, they made a crackling or fnapping noife, and that, in the dark, he could perceive them to emit fparks of fire. He had no doubt but that
this proceeded from the principle of electricity, and, after a
great number of obfervations, to determine on what circumftances thofe ftrong electrical appearances depended; he
found, at length, that it was the combination of white and
black that produced the electricity; and that the appearances
were the ftrongeft when he wore a white and black filk
ftocking upon the fame leg. Thefe, however, difcovered no
fign of electricity while they were upon the leg, or hand
(for he found that his hand was fufficient) though they were
drawn backwards and forwards upon it feveral times. Nor

* Phil. Tranf. Vol. 51. pt. 1. p. 340.

M m 2

when

when taken from the hand, and prefented to an electrometer (i. e. Mr. Canton's balls) did they appear to have acquired any more than a very fmall degree of electricity; but the moment they were feparated, they were found, both of them, to be highly electrified, the white pofitively, and the black negatively.

Both the ftockings, when held at a diftance from one another, appeared inflated to fuch a degree, that, when highly electrified, they exhibited the intire fhape of the leg; and when two black, or two white ftockings were held together, they would repel one another, fo as to form an angle, feemingly, of thirty or thirty five degrees.

When a white and black ftocking were prefented to each other; they would be mutually attracted; and, if permitted, would rufh together with furprifing violence. In their approach their inflation gradually fubfided, and their attraction of foreign objects diminifhed, but their attraction of one another increafed. When they actually met, they grew flat, and joined as clofe together, as if they had been fo many folds of filk. When they were feparated, their electricity did not feem to have been in the leaft impaired by the fhock of meeting; for they would be again inflated, attract, repel, and rufh together as before.

When this experiment was performed with two black ftockings in one hand, and two white ones in the other, it exhibited a curious fpectacle. The repulfion of thofe of the fame colour and the attraction of thofe of different colours, threw them into an agitation which was not unentertaining, and made them catch each at the oppofite colour, at a greater diftance than could have been expected.

When

When the ftockings were feparated from one another, they would lofe their power very foon, much like the excited tube; but when they were together, they would retain it an hour or two, or longer, if the air was favourable to electricity. The fharpeft metallic point could not deprive them of it; and when they were one within the other, no means he could think of could procure the leaft perceivable difcharge of the electricity. In this refpect, Mr. Symmer thought there was a confiderable refemblance between the black and the white ftocking, when put within one another, and the Leyden phial.

What was ftill more remarkable in thefe experiments with the white and black ftockings, was the power of elec-trical cohefion which they exhibited. Mr. Symmer perceiv-ed that the white and black ftockings, when electrified, and allowed to come together, not only joined extremely clofe, but actually ftuck to each other. By means of a balance, he found, that in order to feparate them, it required from one to twelve ounces. Another time they raifed feventeen ounces which, was twenty times the weight of that ftocking which fupported them, and this in a direction parallel to its furface.

When one of the ftockings was turned infide out, and put within the other, it required twenty ounces to feparate them, though when they were applied to each other externally, ten ounces were fufficient.

Getting the black ftockings new dyed, and the white ones wafhed, and whitened in the fumes of fulphur; and then putting them one within the other, with their rough
fides

fides together, it required three pounds three ounces to feparate them.   And he had reafon to think that the fulphur
contributed nothing to the experiment.

TRYING this experiment with ftockings of a more fub-
ftantial make, he found the effects more confiderable.   When
the white ftocking was put within the black one, fo that the
outfide of the white was contiguous to the infide of the black,
they raifed nine pounds wanting a few ounces, which was
fifty times the weight of the ftocking.   When the white
ftocking was turned infide out, and put within the black one,
fo that their rough furfaces were contiguous, they raifed fifteen
pounds one pennyweight and a half, which was ninety two
times the weight of the ftocking.

HAVING cut off the ends of the thread, and the tufts of
filk, which had been left in the infide of the ftockings, the
cohefion was confiderably diminifhed.   Preffing them together between his hands contributed much to ftrengthen it. *

WHEN the white and black ftocking were in cohefion, and
another pair, more highly electrified, were feparated from
one another, and prefented to the former, their cohefion
would be diffolved; and each ftocking of the fecond pair
would catch hold of, and carry away with it, that of its oppofite colour.   If the degree of electricity of both pairs were
equal, the cohefion of the former pair would be weakened,
but not diffolved; and all the four would cohere, forming
one mafs.   If the fecond pair were but weakly electrified, the
cohefion of the firft pair would be but little impaired, and
the cohefion of the whole mafs would be fmall in proportion.

* Phil. Tranf. Vol. 51. pt. 1. p. 393.

MR.

Mr. Symmer alſo obſerved, that white and black ſilk, when electrified, not only cohered with each other, but would alſo adhere to bodies with broad, and even with poliſhed ſurfaces, though thoſe bodies were not electrified. This he diſcovered accidentally, having, without deſign, thrown a ſtocking out of his hand, which ſtuck to the paper hangings of the room. He repeated the experiment, and found it would continue hanging near an hour.

Having ſtuck up the black and white ſtockings in this manner, he came with another pair of ſtockings highly electrified; and applying the white to the black, and the black to the white, he carried them off from the wall, each of them hanging to that which had been brought to it.

The ſame experiments held with the painted boards of the room, and likewiſe with the looking glaſs, to the ſmooth ſurface of which both the white and the black ſilk appeared to adhere more tenaciouſly than to either of the former. *

A few obſervations ſimilar to ſome of theſe of Mr. Symmer were made by Signior *Aleſſandro Amadeo Vaudonia*, a friend of Signior Beccaria. He put a beaver ſhirt between two others, which he wore in extreme cold weather; and whenever he put off the uppermoſt ſhirt, which he did every day, he found it adhered to the the beaver ſhirt, and, on the ſeparation, electric ſparks were viſible between them. Whenever he put off the beaver ſhirt, it adhered ſtill more to the under ſhirt, and when held at a conſiderable diſtance from it, would ruſh to it. Theſe attractions would be repeated many times,

* Phil. Tranſ. Vol. 51, pt. 1. p. 366.

but

but they grew more languid by degrees, till they intirely ceafed. Signior Beccaria, upon hearing of this experiment, repeated it with fome variation, and found it to anfwer on himfelf *

THE cohefion of the two ftockings induced Mr. Symmer to try the force of electrical cohefion in electrified panes of glafs. For this purpofe, he got two panes of common window glafs, the thinneft and the fmootheft that he could find, and coated one of the fides of each with tinfoil, leaving a fpace uncovered near the edges. He then put the uncovered fides together, and charging them both as one pane, he found, as he expected, that their cohefion was confiderably ftrong: but he had no apparatus to meafure the ftrength of it. He then turned the plates upfide down, and found that the fame operation which had before charged them, did now uncharge them, according to the anology of the Leyden phial.

PLACING two panes of glafs, each of them coated on both fides, one upon the other, he found that they were both charged feparately, and that there was no cohefion between them.

MR. SYMMER concludes his account of thefe experiments with declaring it to be his opinion, that there were two electric fluids, or emanations of two diftinct electric powers, effentially different from each other; that electricity does not confift in the afflux and efflux of thefe fluids, but in the accumulation of the one or the other of them in bodies electrified; or, in other words, it confifts in the poffeffion of a

* Dell' elettricifmo artificiale, p. 197.

larger

larger portion of one or the other power, than is requifite to maintain an even balance within the body; and laftly, that, according as the one or the other power prevails, the body is electrified in the one or the other manner. Nor will this principle, fays he, of two diftinct electrical powers be found, upon due confideration, to difagree with the general fyftem of nature. It is one of the fundamental laws of nature, that action and reaction are infeparable and equal; and, when we look round, we find that every power which is exerted in the material world meets with a counteracting power, which controuls and regulates its effects, fo as to anfwer the wife purpofes of providence. *

Mr. Symmer also alledges, in proof of his two diftinct powers of electricity, the experiment which Dr. Franklin has related, of piercing a quire of paper with an electric fhock. He thought that the bur which was raifed on both fides of the paper was produced by two fluids, moving in two different directions. To fhow the manner in which this ftroke was made more evidently, he mentions two other fimilar experiments, in which the circumftances of the ftroke were a little varied.

A piece of paper, covered on one fide with Dutch gilding, and which had been left accidentally between two leaves, in a quire of paper in which the former experiment had been made, was found to have the impreffion of two ftrokes upon it, about a quarter of an inch from each other; the gilding being ftripped off, and the paper left bare for a little fpace in

* Phil. Tranf. Vol. 51. pt. 1. p. 389.

N n

both

both places. In the center of one of thefe places was a little round hole, in the other only an indenture or impreffion, fuch as might have been made with the point of a bodkin.

THESE obfervations Mr. Symmer communicated to Dr. Franklin, who, notwithftanding Mr. Symmer was endeavouring to eftablifh a theory of electricity contrary to his own, with the generofity natural to him, affifted him with his apparatus in making another experiment in purfuance of that mentioned above.

IN the middle of a paper book, of the thicknefs of a quire, Mr. Symmer put a flip of tinfoil; and in another, of the fame thicknefs, he put two flips of the fame fort of foil, including the two middle leaves of the book between them. Upon ftriking the two different books, the effects were anfwerable to what he expected. In the firft, the leaves on each fide of the foil were pierced, while the foil itfelf remained unpierced; but, at the fame time, he could perceive an impreffion had been made on each of its furfaces, at a little diftance from one another; and fuch impreffions were ftill more vifible on the paper, and might be traced, as pointing different ways. In the fecond, all the leaves of the book were pierced, excepting the two that were between the flips of foil; and in thefe two, inftead of holes, the two impreffions in contrary directions were vifible.

MR. SYMMER afterwards got an electrical apparatus of his own, formed on the model of that of Dr. Franklin, with which he frequently repeated the experiments above-mentioned, the refult of all which he comprifes in the three following obfervations.

* I. WHEN

1. When a quire of paper, without any thing between the leaves, is pierced with a stroke of electricity, the two different powers keep in the same track, and make but one hole in their passage through the paper: not but that the power from above, or that from below, sometimes darts into the paper at two or more different points, making so many holes, which, however, generally unite before they go through the paper. They seem to pass each other about the middle of the quire, for there the edges are most visibly bent different ways; whereas, in the leaves near the outside of the quire, the holes very often carry more the appearance of the passage of a power issuing out, and exploding into the air, than of one darting into the paper.

2. When any thin metallic substance, such as gilt leaf, or tinfoil, is put between the leaves of the quire, and the whole is struck; in that case, the counteracting powers deviate from the direct track, and leaving the path which they would in common have taken through the paper, only make their way in different lines to the metallic body, and strike it in two different points, distant from one another about a quarter of an inch, more or less; the distance appearing to be the least when the power is greatest: and whether they pierce, or only make impressions upon it, in either case, they leave evident marks of motion from two different parts, and in two contrary directions.   It is this deviation from a common course, and the separation of the lines of direction consequent upon it, says he, that affords a proof of the exertion of two distinct and counteracting powers.

3. When

3. When two flips of tinfoil are put into the middle of the quire, including two or more leaves between them; if the electricity be moderately ftrong, the counteracting powers only ftrike againft the flips, and leave their impreffion there. When it is ftronger, one of the flips is generally pierced, but feldom both; and from what he had obferved in fuch cafes, he fays it fhould feem, as if the power which iffued from the outfide of the phial acted more ftrongly than that which proceeded from within, for the lower flip was moft commonly pierced. But this, he adds, may be owing to the greater fpace which the power from within has to move through before it ftrikes the paper. *

In the fame paper, Mr. Symmer furnifhes a remakable inftance of the power of an hypothefis in drawing facts to itfelf, in making proofs out of facts which are very ambiguous, and in making a perfon overlook thofe circumftances in an experiment which are unfavourable to his views.

When a phial is electrified but a little, Mr. Symmer fays, if we touch the coating of it with a finger of one hand, and, at the fame time, bring a finger of the other hand to the wire, we fhall receive a pretty fmart blow upon the tip of each of the fingers, the fenfation of which reaches no farther. If the phial be electrified a degree higher, we fhall feel a ftronger blow, reaching to the wrifts, but no farther. When, again, it is electrified to a ftill higher degree, a feverer blow will be received, but will not be felt beyond the elbows. Laftly, when the phial is ftrongly charged, the ftroke may be per-

* Phil. Tranf. Vol. 51. pt. 1. p. 377, &c.

ceived

ceived in the wrifts, and elbows; but the principal fhock is felt in the breaft, as if a blow from each fide met there. This plain and fimple experiment, fays Mr. Symmer, feems obvioufly to fuggeft to obfervation the exiftence of two diftinct powers, acting in contrary directions; and, I believe, fays he, it would be held as a fufficient proof by any perfon, who fhould try the experiment, with a view to determine the queftion fimply from his own perceptions. *

IT is a fufficient anfwer to this remark of Mr. Spmmer's, that if twenty people join hands, they may all be made to feel the fhock in their wrifts, or their elbows, without having their breafts affected in the leaft. And can it be fuppofed, that the two currents of electric fire could come at all their wrifts or elbows, without paffing through their breafts. According to Mr. Symmer's hypothefis, it fhould feem, that, in a large circle, thofe perfons only who ftood near the phial, on either hand, fhould feel *a fmall fhock*; that a few perfons more, at each extremity of the circle, fhould feel one fomething ftronger; and that it could only be a very ftrong fhock, which could at all affect the perfon who ftood in the middle; and that then he fhould be affected the leaft of any perfon in the company. But all thefe confequences are contrary to fact.

THIS hypothefis of Mr. Symmer, notwithftanding he has failed in his application of it to the experiments above-mentioned, has attracted the notice of feveral electricians, both at home and abroad; and fome perfons feem inclined to adopt

* Phil. Tranf. Vol. 51. pt. 1. p. 373, &c.

it,

it, in preference to Dr. Franklin's theory. I shall therefore consider it more at large, when I come to treat of *theories* professedly; till which time, I take leave of this ingenious philosopher, and his two electric fluids.

THE experiments of Mr. Symmer excited the attention of Johannes Franciscus Cigna, and led him to a course of experiments, which throws still more light, both upon the doctrine of the two electricities, and the Leyden phial. They are also a further illustration of the discovery of Mr. Canton, improved by Mr. Wilke and Æpinus, of the mutual repellency of similar electric atmospheres.

HE took two white silk ribbons, just dried at the fire, and having extended them upon a smooth plain, either a conductor or a non-conductor, he drew over them the sharp edge of an ivory ruler, and found, that both the ribbons had acquired electricity enough to adhere to the plain; though, while they remained upon the plain, they shewed no other sign of it. If they were both taken off from the plain together, they attracted one another, the upper having acquired the resinous, and more powerful; and the lower the vitreous, and weaker electricity. If they were taken up separately, they repelled one another, having both acquired the resinous electricity.*

IN this separation of both the ribbons from the plain, as also in their separation, afterwards, from one another, electric sparks were visible between them; but if they were again put upon the plain, or joined together, no light appeared

* Memoirs of the Academy at Turin for the year 1765, p. 31.

upon

upon their fecond feparation, without another friction of the
ruler.   Alfo, when, by being taken off feparately, they had
been made to repel one another, if they were laid on the
plain again, and taken off together, they would not attract;
and if, by being taken off together, they had firft been made
to attract one another, and were laid on the plain a fecond
time, and then taken off feparately, they would not repel,
without another friction.

When, by the operation above-mentioned, they had acquired the fame electricity, if they were placed, not upon
the fmooth body on which they had been rubbed, but on a
rough one, and a conductor, as hemp or cotton, not very
dry; they would, upon being feparated, fhow contrary electricities; which, when they were joined together, would
difappear as before. *

If they had been made to repel one another, and were afterwards placed one upon the other, on the rough furface
above-mentioned, they would, in a few minutes, attract one
another; the lower of the two ribbons having changed its
refinous into a vitreous electricity.

If the two white ribbons received their friction upon the
rough furface, they always acquired contrary electricities;
the upper of the two having the refinous, and the lower the
vitreous, in whatever manner they were taken off.

The fame thing that was done by a rough furface was done
by any pointed conductor.   If two ribbons, for inftance, were
made to repel, and hang parallel to one another; and the

* Ib. p. 33.

point

point of a needle were drawn oppofite to one of them, along its whole length, they would prefently rufh together; the electricity of that ribbon to which the needle was prefented being changed into the contrary. *

IN the fame manner in which one of the ribbons changed its electricity, a ribbon not electrified would acquire electricity; viz. by putting it upon a rough furface, and laying an electrified ribbon upon it; or by holding it parallel to an electrified ribbon, and prefenting a pointed conductor to it.

HE placed a ribbon not quite dry under another that was well dried at the fire, upon a fmooth plain; and when he had given them the ufual friction with his ruler, he found that, in what manner foever they were removed from the plain, the upper of them had acquired the refinous, and the lower the vitreous electricity.†

IF both the ribbons were black, all the above-mentioned experiments fucceeded, in the fame manner as if they had been white. ‡

IF, inftead of his ivory ruler, he made ufe of any fkin, or of a piece of fmooth glafs, the event was the fame; but if he made ufe of a ftick of fulphur, the electricities were, in all cafes, the reverfe of what they were before; the ribbon which was rubbed having always acquired the vitreous electricity.

WHEN he made ufe of paper, either gilt or not gilt, the events were uncertain.

WHEN the ribbons were wrapped in paper, gilt or not gilt,

---

* Memoirs of the Academy at Turin for the year 1765, p. 34. † Ib. p. 5. ‡ Ib.

and

and the friction was made upon the paper laid on the plain above-mentioned, the ribbons acquired, both of them, the resinous electricity. *

If the ribbons were one black, and the other white, which ever of them was laid uppermost, and in whatever manner the friction was made, the black generally acquired the resinous, and the white the vitreous electricity. †

He observed, however, the following constant event; that whenever the texture of the upper piece of silk was loose, yielding, and retiform, like that of a stocking, so that it could move, and be rubbed against the lower, and the rubber was of such a nature as to impart but little electricity to glass; the electricity which the upper piece of silk acquired, did not depend upon the rubber, but upon the body it was laid upon; in which case the black was always resinous, and the white vitreous. But when the silk was of a close texture, hard and rigid, and when the rubber was such as imparted a great degree of electricity to glass; the electricity of the upper piece did not depend upon the lower, but upon the rubber. Thus a white silk stocking, rubbed with gilt paper upon glass, became resinous, and the glass vitreous; but if a piece of silk, of a firmer texture, was laid upon a plate of glass, it always acquired the vitreous electricity, and the glass the resinous, if it was rubbed with sulphur; and for the most part, if it was rubbed with gilt paper. ‡ So that the silk that was rubbed received its electricity, sometimes from the rubber, and sometimes from the substance

* Memoirs of the Academy at Turin, for the year 1765, p. 36.   † Ib. p. 38.
‡ Ib. p. 40.

O o

placed

placed under it; according as it received greater friction from the one or the other, or in proportion as one or the other was more proper to give electricity to glafs.

ANOTHER fet of experiments, which the fame Johannes Francifcus Cigna made, illuftrate the adhefion of Mr. Symmer's electrical ftockings to bodies with fmooth furfaces. He infulated a plate of lead, and bringing an electrified ribbon near it, obferved that it was attracted very feebly. Bringing his finger to the lead, a fpark iffued out of it, upon which it attracted the ribbon vigoroufly, and both together fhowed no figns of electricity. Upon the feparation of the ribbon, they again both appeared to be electrified, and a fpark was perceived between the plate and the finger. *

LAYING two plates of glafs upon a fmooth conductor communicating with the ground, and rubbing them in the fame manner as the ribbons had been rubbed, they likewife acquired electricity, and adhered firmly, both to one another, and to the conductor. If it were a plate of lead, not very thick, it would be fupported by the attraction. When they were together, they fhowed no other figns of electricity. †

WHEN the two plates of glafs were feparated from the conductor, and kept together, they fhowed, on both fides, a vitreous electricity; and the conductor, if it had been infulated, was feen to have contracted a refinous electricity.

THE two plates of glafs themfelves, when feparated, were poffeffed of the two electricities; the upper of the vitreous, and ftronger; and the lower of the refinous, and weaker.

* Memoirs of the Academy at Turin, for the year 1765, p. 43.　　† Ib. p. 52.

WITH

With a rough conductor, whether they were originally rubbed upon it, or brought to it, after they had been rubbed upon a smooth one, they scarce contracted any electricity; though, when they were separated from one another, they were affected as before.

Upon this principle, he endeavours to account for the non-excitation of a globe or tube from which the air is exhausted, or which is lined with conducting substances. In this case, he says, the vitreous electricity on the external surface of the glass is balanced by the resinous in the inward coating, or in the vacuum which serves instead of a coating; and therefore it is in the situation of the plates of glass while they lie upon the conductor above-mentioned: but when the inward coating is taken away, the electricity appears on the outside, without any fresh excitation; as when the plates were removed from the conductor. *

When he laid a number of ribbons of the same colour upon the smooth conductor, and drew his ruler over them; he found, that when he took them up singly, they all gave sparks, at the place where they were separated, as the last ribbon did with the smooth plate, and had all acquired the resinous electricity. †

If they were all taken from the plate together, they cohered in one mass, which, on both sides, appeared to be resinous. If they were laid upon the rough conductor, in the same order (whereby the opposite electricities were brought to an equilibrium) and they were all separated singly, begin-

* Ib. p. 54.   † Ib. p. 61.

O o 2   ning

ning with the lowest, sparks appeared as before; but all the ribbons had acquired the vitreous electricity; except the uppermost, which retained the resinous electricity it had received from the friction. *

If they received the friction upon the rough conductor, and were all taken up at once (in order to have a bundle in which the opposite electricities were balanced) all the intermediate ribbons acquired the electricity, either of the highest or the lowest ribbon, according as the separation was begun with the highest, or the lowest.

If two ribbons were separated from the bundle at the same time, they clung together; and, in that state, showed no sign of electricity, as one of them alone would have done. When they were separated, and the different electricities were manifest, the electricity was observed to reside in the outermost, and was opposite to that by which they had both adhered to the bundle, but much weaker. †

He placed a number of ribbons upon a plate of metal, which received electricity from the globe, while he held a pointed body to the other side of the ribbons. The consequence was, that all the ribbons became possessed of the electricity opposite to that of the plate, or of the same, according as they were taken off; except the most remote, which always kept an electricity opposite to that of the plate.

From these experiments he infers, that as electricity is propagated from the outermost ribbon to those underneath it, or else from the plate below to those next above it, when they

* Memoirs of the Academy at Turin, for the year 1765, p. 61.   † Ib.

are

are feparated; when the coating is feparated from a charged pane of glafs, it likewife depofits its electricity upon the fuperficies of the glafs, the phenomena being the fame in both, For when he put metal coatings on the fide of a plate of glafs, without any cement, they adhered firmly to the glafs when it was charged, and a light appeared upon their being feparated from it, as in the cafe of the ribbons. *

When he coated a number of ribbons in the fame manner, and charged them, the coatings adhered firmly to the ribbons; but he could never feparate one of them, but (in confequence of the loofe texture of the filk) a fpark would go to the oppofite coating, which immediately fell off, the whole being then difcharged. †

But he thought the coatings did not depofite all their electricity on the plate, when they were taken off; for though, when both were taken off, the electricities of the two fides ftill balanced one another (becaufe each retained the fame diminifhed quantity) yet, when one fuperficies of the glafs, or of the ribbons, received its electricity from friction, and the other only from the oppofite coating; he obferved, that the electricities which balanced one another while the coating was on, were no longer balanced when it was taken off; the electricity of the furface which was rubbed then prevailing, becaufe the conducting coating had, upon its feparation, taken part of its electricity along with it. ‡

To confirm this, he adds another experiment. He charged a pane of glafs, coated on one fide, while the other receiv-

*Ib. p. 63.   †Ib. p. 64.   ‡Ib. p. 65.

ed

ed electricity by a pointed conductor from the machine : He
likewise inverted the plate, and made the coated side communicate with the prime conductor, while a pointed piece of
metal was presented to the opposite side ; and, in both cases,
found, that while the coating remained, the two electricities
balanced one another ; but that when the coating was slipped
off, the electricity of the opposite side prevailed, so as to be
apparent on both sides of the plate.*

* Memoirs of the Academy at Turin, for the year 1765, p. 65.

SECTION

SECTION VII.

The history of the LEYDEN PHIAL continued.

GREAT as were the difcoveries of Dr. Franklin concerning the Leyden phial, he left fome curious particulars for this period of the hiftory of electricity; and the fubject is by no means exhaufted. Many of the properties of this wonderful bottle, as the Doctor calls it, are ftill unexplained. But as more and more light is perpetually thrown upon it, let us hope that, at length, we fhall thoroughly underftand this great experiment. The greateft difcovery concerning the properties of the Leyden phial, in this period, hath already been related in the account of Mr. Wilke's and Æpinus's method of giving the fhock by means of a plate of air; and other obfervations have, likewife, been occafionally mentioned, in places where their connection required them to be inferted. This fection, however, will contain feveral experiments of a mifcellaneous nature, which are well worth notice.

IMMEDIATELY upon the difcovery of the fhock given by glafs, all electricians attempted to charge other electric fubftances; but none of them fucceeded before Signior Beccaria.

He

He found that a very smooth plate of sealing wax, made by pouring that substance, when melted, upon an oiled marble table, would receive a considerable charge. *

After trying several other electrics, in the same manner, he found that a mixture of pitch and colophonia was charged less than sealing wax, but more than sulphur, and a great deal more than pitch alone. †

But the most curious experiment of this philosopher, relating to this subject, was made with a view to ascertain the real direction of the electric fluid in a discharge. He suspended a coated plate of glass by a silk thread, and having charged it, and kept it perfectly still; he observed that no motion was given to it, when the discharge was made by a crooked wire approaching both the sides at the same time. The experiment, in fact, proved the reaction of the glass upon the electric matter; whereby the plate was kept still, notwithstanding the fluid rushed with great violence from one side to the other. He compares the glass to an ivory ball placed between two others, which keeps its place, when, by an impulse given to one of them, the opposite ball flies off. ‡

A very curious and elegant experiment on the Leyden phial was made by professor Richman of Petersburgh, whose unfortunate death will be related in this history.

He coated both sides of a pane of glass, within two or three inches of the edge, and fastened linen threads to the upper part of the coating, on both sides; which, when the plate was not charged, hung down in contact with the coat-

* Lettere dell' elettricismo, p. 64.   † Ib. p. 66.   ‡ Ib. p. 72.

ing;

ing; but setting the plate up right, and charging it, he observed, that when neither of the sides was touched by his finger, or any other conductor communicating with the earth, both the threads were repelled from the coating, and stood at an equal distance from it; but when he brought his finger, or any other conductor, to one of the sides, the thread hanging to that side fell nearer to the coating, while the thread on the opposite side receded as much; and that when his finger was brought into contact with one of the sides, the thread on that side fell into contact with it likewise, while the thread on the opposite side receded to twice the distance at which it hung originally; so that the two threads always hung so as to make the same angle with one another. *

Æpinus shows, that it is not strictly true, that an insulated person, discharging the Leyden phial through his own body, contracts no electricity. Electrifying a large plate of air, he observed, that if the nearer plate, (by which I suppose he means that which he first touched) was electrified positively, he acquired a positive electricity by the discharge; but if it were negative, he acquired a negative electricity. He supposes that the reason why the experiment did not succeed with Dr. Franklin, was, that the surfaces with which he tried the experiment were not large enough to make the effect sensible; and that the distance of the metal plates was, likewise, too small, as it necessarily must be in charging of glass. †

JOHANNES FRANCISCUS CIGNA has invented a new me-

* Æpini Tentamen. p. 335.   † Ib. p. 27.

P p   thod

thod of charging a phial, upon the principle difcovered by
Mr. Canton and Mr. Wilke; viz. that the electricity of one
body repels that of another, efpecially if it have a flat furface,
and gives it the contrary electricity.

HE infulates a fmooth plate of lead, and while he brings
an electrified body, as a ftocking, to it, he takes a fpark
with the wire of a phial from the oppofite fide; and removing the ftocking, he takes another fpark with his finger, or
any conductor communicating with the ground. Bringing
the ftocking nearer the plate a fecond time, he takes a fecond
fpark, with the wire of the phial, as before; and, removing
it again, takes another, in the fame manner, with his finger.
This operation he continues, till the phial is charged; which,
in favourable weather, may be done with very little diminution of the electricity of the ftocking. *

IF, inftead of taking a fecond fpark with his finger, he had
taken it with the wire of another phial, that would have
been charged likewife, with no additional labour, and with
an electricity oppofite to that of the other phial. If
the fecond fpark were taken with the coating of the fame
phial, the charging would be accelerated, but the operation
would be troublefome to manage.

THE theory of this new method of charging a phial is very
eafy, upon the principle referred to above. The approach
of the electricity in the ftocking, not being able to enter the
broad fmooth furface of the metal, drives the electric fluid
out of that part of the plate which is oppofite to it, to the
other fide, which, being thereby overcharged, will part

* Memoirs of the Academy at Turin for the year 1765, F. 49.

with

with its fuperfluity to the wire of the phial.   The ftocking being taken away, the plate will have lefs than its natural fhare of the electric fluid, and will therefore readily take a fpark, either from the finger, or the wire of another phial.*

THE fame ingenious philofopher makes a confiderable difference between the electric fluid which gives the fhock, and that on which fome other phenomena of coated glafs depend. The former, which is far the greateft quantity, he fuppofes to refide, either in the coating itfelf, or on the furface of the glafs; whereas the other, he imagines, to have entered the pores, and affected the fubftance of the glafs itfelf.

HE laid two plates of glafs, well dried, one upon the other, as one piece; the lower of them being coated on the outfide; and, when they were infulated, he alternately rubbed the uppermoft plate with one hand, and took a fpark from the coating of the lower with the other, till they were charged; when the coating, and both the plates adhered firmly together.   Giving a coating to the other fide, and making a communication between that and the other coating, the ufual explofion was made.   But the plates, though thus difcharged, ftill cohered; and though, while they were in this ftate, they fhowed no other fign of electricity; yet, when they were feparated, they were each of them found to be poffeffed of electricity oppofite to that of the other.

IF the two plates were feparated before they were difcharged, and the coating of each were touched, a fpark came from

* Ib. p. 51.

P p 2   each

each; and when put together, they would cohere as before, but were incapacitated for giving a fhock. *

HE, therefore, compares the electricity which gives the fhock to the electricity of the metal plate in the former experiment; which is loft with taking one fpark, as the filk is removed from it, and is different from the electricity by which the two plates of glafs cohere. The one is difperfed at once, but the other flowly; the one exifting, as he fuppofes, in the conductors, or upon the furfaces of the electrics, and the other in the fubftance itfelf. †

AMONG experiments relating to the electric fhock, we ought to mention what has been obferved within this period of its amazing force in melting wires, and producing other furprifing effects.

THAT even artificial electricity, fays Dr. Watfon, in a paper read at the Royal Society June the 28th. 1764, when in too great a quantity, and hurried on too faft, through a fine iron wire, has a remarkable effect upon the wire, appears from a very curious experiment of Mr. Kinnerfley. This gentleman, in the prefence of Dr. Franklin, made a large cafe of bottles explode at once through a fine iron wire. The wire at firft appeared red hot, and then fell into drops, which burned themfelves into the furface of his table or floor. Thefe drops cooled in a fpherical figure like very fmall fhot, of which Dr. Franklin tranfmitted fome to Mr. Canton, who repeated the experiment. This proves the fufion to have

* Memoirs of the Academy at Turin for the year 1765, p. 55.   † Ib. p. 56.

been

been very complete; as nothing lefs than the moft perfect fluidity could give this figure to melted iron.

Mr. Canton, in a note fubfcribed to the fame paper, ob-ferves, that the diameter of a piece of Mr. Kinnerfley's wire, which he received from Dr. Franklin, was one part in 182 of an inch. He adds, that artificial lightning, from a cafe of thirty five bottles. would entirely deftroy brafs wire, of one part in 330 of an inch. At the time of the ftroke, he fays, a great number of fparks, like thofe from a flint and fteel, would fly upwards, and laterally, from the place where the wire was laid, and lofe their light, in the day time, at the diftance of about two or three inches. After the explofion, a mark appeared on the table, the whole length of the wire, and fome very round particles of brafs were difco-vered by a magnifier near the mark, but no part of the wire itfelf could be found. *

Signior Beccaria was able to melt fmall ftrips of metal, without inclofing, or covering them with pieces of glafs. But he thought that the fame colour was impreffed upon glafs by all the metals; and imagined that this circumftance was a trace of the fundamental principles being the fame in them all. †

Mr. Dalibard obferved, that, when a large pane of glafs difcharged itfelf, the polifh was taken off at the place of the difcharge, and that the track it left behind it was ufually, as he expreffes it, in tne *zig zag* form. With the

* Phil. Tranf. Vol. 54. p. 208.   * Elettricifmo artificiale, &c. p. 134, 135.

piece

piece of glaſs with which he made theſe diſcharges he pierced 160 leaves of paper.   It contained 1200 ſquare inches. *

Mr. Winkler fired the ſeeds of clubmoſs [lycopodium] by diſcharging a phial through a quantity of them.   He alſo fired the *aurum fulminans* placed upon a piece of parchment, which was torn to pieces by the exploſion. †

By the electric ſhock, Signior Beccaria could melt borax, and glaſs.   But the moſt remarkable of his experiments with the electric ſhock are thoſe by which he *revivified metals*. This he did by making the exploſion between two pieces of the calces.   In this manner he revivified ſeveral of the metals, and among others, zink.   He even produced real quickſilver from cinnabar. ‡   In this caſe of revivifications, he always obſerved ſtreaks of black beyond the coloured metalic ſtains, owing, as he imagined, to the phlogiſton driven thither from the parts that were vitrified, when the other part revivified the calx. ‖   Probably, the phlogiſton which revivified the calces was in that black duſt, which the electric ſhock will throw from metals, as will be related in its proper place.

Another curious experiment he made with the electric ſhock, by diſcharging it through ſome braſs duſt, ſprinkled between two plates of ſealing wax.   The whole was perfectly luminous, and tranſparent. §   An experiment which throws ſome light upon one of Mr. Hawkeſbee's.

With the electric ſhock he alſo made that capital experi-

* Hiſtoire abrigeé, p. 84.   † Phil. Tranſ. Vol. 38. pt. 2. p. 773.
‡ Lettere dell' elettriciſmo, p. 282.   ‖ Ib. p. 255.   § Ib. p. 257.

ment,

ment, on which he lays fo much ſtreſs in his theory of thunder ſtorms, and by which he proves, that the electric matter
forces into its paſſage all light conducting ſubſtances; by
means of which it is enabled to paſs through a quantity of
refiſting medium, which it could not otherwiſe do. He put
a narrow piece of leaf ſilver between two plates of wax, laying it acroſs the plates, but ſo as not quite to reach one
of the ſides. The diſcharge being made through this
ſtrip of metal, by bringing a wire oppoſite to the ſilver, at
the place where it was diſcontinued; the ſilver was found
melted, and part of it diſperſed all along the track that the
electric matter took, between the plates of wax, from the
ſilver to the wire. * An accident gave him occaſion to obſerve another fact of a ſimiliar nature. He once, inadvertently, received the charge of a ſmall jar, through ſome
ſmoke of ſpirit of nitre; when a hole was made in his thumb,
where the fire entered; and which he thought could only
have been made by the nitre, which was carried along with
the electric fluid. †

I ſhall cloſe the hiſtory of the Leyden phial for this period with the accounts of ſome extremely curious facts, which
Mr. Canton gives me leave to publiſh relating to this ſubject.
They certainly deſerve the utmoſt attention of philoſophers,
and may probably throw ſome light upon the electricity of
the tourmalin.

He procured ſome thin glaſs balls, of about an inch and a

* Lettere dell' elettriciſmo, p. 248.   † Ib. p. 249.

half

half diameter, with stems or tubes of eight or nine inches in length, and electrified them, some positively on the inside, and others negatively, after the manner of charging the Leyden phial, and then sealed them hermetically. Soon after, he applied the naked balls to his electrometer, and could not discover the least sign of their being electrical : but holding them at the fire, at the distance of five or six inches, they became strongly electrical, in a very short time ; and more so when they were cooling. These balls would, every time they were heated, give the electric fire to, or take it from other bodies, according to the *plus* or *minus* state of it within them. Heating them frequently, he found, would sensibly diminish their power, but keeping one of them under water a week did not appear in the least to impair it. That which he kept under water was charged the 22d. of September 1760, was several times heated before it was kept in water, and had been heated frequently afterwards ; and yet it still retained its virtue to a considerable degree, on the 31st. of October following, when he sent an account of it to Dr. Franklin. The breaking two of his balls accidentally gave him an opportunity of measuring their thickness, which he found to be between seven and eight parts in 1000 of an inch.

THE balls mentioned in the account above, which was written six years ago, still retain their virtue, but in a less degree.

SECTION

SECTION VIII.

Experiments and observations concerning
ELECTRIC LIGHT.

MY reader has been informed of the neceſſity I was under of dividing the buſineſs of this period of my hiſtory into ſeveral parts. He has already ſeen titles which he could not have expected from the diviſions of the preceding periods, but he would perhaps leaſt of all expect a diſtinct ſection upon electric light; and yet the experiments and obſervations which have been made, immediately relating to this ſubject, are ſo many, that they deſerve a place by themſelves. And I would rather err by making too many ſubdiviſions, than too few; becauſe, above all things, I would wiſh to preſerve perſpicuity, which is chiefly injured by crowding together things diſſimilar.

Many experiments had been made very early, by Mr. Hawkeſbee and others, on electricity, and particularly electric light, in vacuo; but ſo little was, at that time, known of the nature of electricity in general, that, comparatively,

Q q

little

little ufe could be made of thefe experiments. Very fortunately, Dr. Watfon happened to turn his thoughts that
way, after the great difcovery of the accumulation of electricity in the Leyden phial; and by this means he difcovered, that our atmofphere, when dry, is the agent by which,
with the affiftance of other electrics per fe, we were enabled
to accumulate electricity upon non-electrics (he might have
added electrics too) that is, to communicate to them a greater
quantity of electricity than thofe bodies naturally have. That
upon the removal of the air, the electric fluid pervaded the
vacuum to a confiderable diftance, and manifefted its effects
upon any non-electric fubftances by which it was terminated.

THIS he demonftrated by one of the moft beautiful experiments which the whole compafs of electricity yet exhibits.
He exhaufted a glafs cylinder, three feet in length, and three
inches in diameter, with a contrivance to let down a brafs
plate, as far as he pleafed, into it; in order to make it approach another plate, fixed near the bottom of the veffel.

THIS cylinder, thus prepared, he infulated, and obferved,
that when the upper plate was electrified, the electric matter would pafs from one plate, to another, at the greateft
diftance to which the brafs plates could be drawn; and that
the brafs plate at the bottom of the cylinder was ftrongly
electrified, as if a wire had connected it with the prime conductor. It was a moft delightful fpectacle, he fays, when
the room was darkened, to fee the electric matter in its paffage through this vacuum; to obferve, not as in the open air,
fmall brufhes or pencils of rays, an inch or two in length,
but corufcations of the whole length of the tube, and of a

bright

bright filver hue. Thefe did not immediately diverge, as in the open air, but frequently, from a bafe apparently flat, divided themfelves into lefs and lefs ramifications, and refembled very much the moft lively corufcations of the aurora borealis.

Sometimes he obferved, that when the tube had been exhaufted in the moft perfect manner, the electric fluid was feen to pafs between the brafs plates in one continued ftream, of the fame dimenfions throughout its whole length; which, he thought, demonftrated, that the caufe of that very powerful mutual repulfion of the particles of electric fire, which is feen in the open air, is more owing to the refiftance of the air, than to any natural tendency of the electricity itfelf. For, in the open air we obferve that thefe brufhes, when the electricity is ftrong, diverge fo much, as to form almoft a fpherical figure. *

He made this vacuum part of a circuit neceffary to make the difcharge of a phial; and, at the inftant of the explofion, there was feen a mafs of very bright embodied fire, jumping from one of the brafs plates in the tube to the other. But this did not take place when one of the plates was farther diftant from the other than ten inches. If the diftance was greater, the fire began to diverge, and lofe part of its force: and this force diminifhed in proportion to its divergency, which was nearly as the diftance of the two plates.

To find a more perfect vacuum for the paffage of the electric fluid, he had recourfe to an excellent invention of Lord

* Phil. Tranf. Vol. 47. p. 367.

Q q 2

Charles

Charles Cavendish; who, by means of a long bent tube of glass, filled with mercury, and inverted, made all the bended part of it (which was above the mercury) the most perfect vacuum that can be made by man. This vacuum Dr. Watson insulated, and one of the basons of the mercury being made to communicate with the conductor, when some non-electric substance touched the other, the electric matter pervaded the vacuum in a continued arch of lambent flame, and, as far as the eye could follow it, without the least divergency.

CONNECTING one of the basons with the machine, which was insulated, the fire was seen pervading the vacuum in a contrary direction. And this he considered as the *experimentum crucis* of two principles which he had advanced before; viz. that electricity is furnished to the conductor, not by the excited electric, but from the non-electrics in contact with the rubber; and that we are able to take from, or add to that quantity of electricity, which is naturally inherent in bodies.

HE also observed, that if, in the fore-mentioned circumstances, the hand of a person standing upon the floor was brought near the side of the glass, the coruscations would dart themselves that way in a great variety of forms, extremely curious to behold.

BUT the Doctor found, that even this vacuum, did not conduct so perfectly as metals, or water; because a person standing upon the floor, and applying his finger to the upper brass plate, received a smart stroke. This he conceived to arise from the electricity of the brass being so much more

rarified

rarefied than that of the body of the man who applied his finger. *

Mr. Wilson engaged Mr. Smeaton, the inventor of a new and more perfect kind of air pump, to make some electrical experiments in vacuo. The following is the account of them that he tranfmitted to Mr. Wilfon. They are, in feveral refpects, fimilar to thofe made by Dr. Watfon, and yet are attended with a confiderable variety of circumftances.

A glass veffel, about one foot in length, and eight inches in its greateft diameter, open at both ends, had one of its ends clofed by a brafs ferrule, which conftituted one of the centers on which it turned; the other end was clofed with a metal plate. In the center of this plate was a fquare ftem, which was applied to the arbor of a lath, by which the glafs was turned round. On one fide of this laft plate was fixed a cork, by means of which the glafs was fcrewed upon the air pump.

Upon rarefying the air within the glafs about 500 times, and afterwards turning the glafs in the lath, whilft, at the fame time, it was rubbed with his hand; a confiderable quantity of lambent flame, variegated with all the colours of the rainbow, appeared within the glafs, under the hand. This light was pretty fteady in every refpect, except that every part of it was perpetually changing colour.

When a little air was let into the glafs, the light appeared more vivid, and in a greater quantity; but was not fo fteady: for it would frequently break out into a kind of corufcations,

* Phil. Tranf. Vol. 47. p. 373.

like

like lightning, and fly all about within the glafs. When a little more air was let in, the flafhing was continual, and ftreams of bluifh light feemed to iffue from under his hand, within the glafs, in a thoufand forms, with great rapidity; and appeared like a cafcade of fire. Sometimes it feemed to fhoot out into the forms of trees, mofs, &c.

WHEN more air was let in, the quantity of light was diminifhed, and the ftreams compofing the flafhes narrower. The glafs now required a greater velocity, and harder friction. Thefe circumftances increafed as more air was let in; fo that, by fuch time as the glafs was one third full of air, thefe corufcations quite vanifhed, and a much fmaller quantity of light appeared partly within, and partly without the glafs. And when all the air was let in, the light appeared wholly without the glafs, and much lefs in quantity than when the glafs was in part exhaufted. *

MR. CANTON, in repeating Dr. Watfon's experiment with the Torricellian vacuum, obferved one circumftance attending it, which throws great light upon the Leyden phial. He obferved, that when the excited tube was brought near one of the bafons of this machine (infulated) a light was feen through more than half of the vacuum; which foon vanifhed, if the tube was not brought nearer, but which appeared again as it moved farther off; and that this appearance might be repeated feveral times, without exciting the tube afrefh.

THIS experiment he confidered as a kind of ocular demonftration of the truth of Dr. Franklin's hypothefis, that when the electric fluid was condenfed on one fide of the glafs, it was repelled from the other, if it met with no re-

* Wilfon's Effay, p. 216.

fiftance

fiftance. Thus, at the approach of the excited tube, he fup-poſed the fire to be repelled from the infide of the glaſs fur-rounding the vacuum, and to be carried off through the columns of mercury, but to return again as the tube was withdrawn. *

THIS curious experiment Mr. Canton ſhowed, and ex-plained to Mr. Wilfon; who afterwards expatiated upon it, in a book publiſhed by him and Dr. Hoadly in conjunction, entitled *Obſervations on a ſeries of electrical experiments*; in a note of which, p. 28, he ſays, Mr. Canton has taken no-tice of this vaniſhing and returning of the light.

MR. CANTON has ſince diverſified this beautiful experi-ment, by bringing the excited tube to another glaſs tube, exhauſted, and hermetically ſealed; by which means, he ex-hibits the perfect appearance of an aurora borealis. The flame from one of its extremities, which is in a manner coat-ed by the hand which holds it, will dart to the other extre-mity, at uncertain intervals of time, for near a quarter of an hour together, without repeating the application of the ex-cited tube.

WHEN it was generally agreed among electricians, that what had been called vitreous and refinous electricity were in reality a redundancy of the electric fluid in one caſe, and a deficiency in the other; and when, in conſequence of this ſuppoſition, the one was called poſitive, and the other nega-tive electricity; there ſtill remained ſome doubt which of the two was poſitive, and which negative. Mr. Wilfon, in a

---

* Phil. Tranſ. Vol. 48. pt. 1. p. 356.

paper

paper read at the Royal Society, December the 6th. 1759, recites an experiment which, he thought, put the matter beyond all difpute, and abfolutely determined, that what had been called vitreous was really pofitive, and what had been called refinous was negative; as, indeed, had generally been fuppofed, though, as Mr. Wilfon thought, without fufficient reafon, notwithftanding what had been advanced by Dr. Franklin, and Mr. Canton upon that fubject.

REPEATING the beautiful experiment mentioned before, as firft contrived by Lord Charles Cavendifh, he fays he attended to a circumftance which feemed to have been overlooked by Dr. Watfon, who publifhed the account of it. This was a fingular appearance of light upon one of the furfaces of the quickfilver. To obferve this remarkable appearance to more advantage, Mr. Wilfon let a fmall quantity of air into the tube, by which means four columns of quickfilver were obtained, and confequently fix vifible furfaces, in one of the legs of the inverted tube. He then electrified the mercury in the other leg, while the mercury on the oppofite fide had a communication with the earth; when, the room being dark, the ftream of electric light was vifible through the whole length of the vacuum, and its general appearance was of a feeming uniform denfity; except at the upper furfaces of each column, where about one tenth of an inch above the furface, the light was always confiderably brighter; whereas the under furfaces exhibited no fuch an appearance, the light being rather lefs bright in thofe places than in the general appearance of the whole illuminated vacuum.

THIS

This luminous appearance Mr. Wilson afcribed to the re-fiftance the fluid met with at the upper furface of the quick-filver, in endeavouring to get into it. He therefore inferred that excited glafs electrified bodies pofitively, or gave them a greater quantity of the electric fluid than they had.

Electrifying, in the fame place, with a cylinder of rofin, inftead of glafs, the luminous appearances were all on the under furfaces of the columns of quickfilver; from which he inferred, that rofin electrified bodies negatively, depriving them of part of the electric fluid which they naturally had; or, as he expreffes it, occafioning a current of electric fluid to fet the contrary way.

These luminous appearances, Mr. Wilson alfo confidered, as a ftrong confirmation of the exiftence of a *medium*, at or near the furfaces of bodies, which hindered the entrance or exit of the electric fluid. A doctrine which Mr. Wilson had advanced, and laid great ftrefs upon on feveral other oc-cafions. *

The arguments which to Mr. Wilson appeared conclufive, in proof of what is commonly fuppofed, that glafs electrifies *plus*, and fulphur, &c. *minus*, did not appear fo to Æpinus; though he acknowledges that the knobs of light in the vacuum did, in common with many other appearances, prove a real difference between the two electricities; and thought that it was very eafy to conceive, that when an elaftic fluid iffues from a body, it fhould be denfer near the furface from whence it iffues, than where it finds more liberty to expand

* Phil. Tranf. Vol. 51. pt. 1. p. 308.

R r   itfelf.

itself. He might have added, that this might have been expected, from the mutual attraction which is supposed to subsist between the electric fluid, and other bodies. But Æpinus did not expressly mention this circumstance. Mr. Wilson, therefore, makes light of the objection; and adds, that when he related the experiment with the bent tube, in his letter to Dr. Heberden, he omitted some phenomena attending the fact; which greatly favoured the doctrine he advanced. If, says he, when glass is electrified, and applied to the first column, we suffer the electric fluid to pass along the tube in small quantities only, and at short intervals, little luminous streams will be seen to move from the first to the second column of quicksilver, and consequently from the glass. The like appearances happen, but in a contrary direction, when rosin or amber is made use of, and applied to the same column. Glass, therefore, he concludes, electrifies *plus*, or fills bodies with more of this fluid than belongs to them naturally, and rosin, &c. *minus.* *

IT will excite no disagreeable sensation in the less grave and saturnine of my readers to be informed, that I have been amusing them with a controversy which took its rise from a mere deception: for Mr. Canton informs me, and gives me leave to inform the public, that the light which Mr. Wilson takes notice of, as appearing on one surface of the mercury in the double barometer of Lord Charles Cavendish, and which Mr. Wilson takes to be a proof of the existence of a medium on the surfaces of bodies, which hinders the entrance or exit

* Phil. Transf. Vol. 53. pt. 438. p. 441.

of

of the electric fluid to some degree, he found to be caused by nothing but common air. For if the Torricellian vacuum be properly made, no difference of light can be seen on the surfaces of the columns of mercury; but if as much air be let into the vacuum, as will make each column of mercury a quarter of an inch shorter than that of a good barometer, the light will appear much as Mr. Wilson has described it. When Mr. Wilson supposed that Dr. Watson, when he made the experiment of the Torricellian vacuum, did not attend to the singular appearance of light on one of the surfaces of the mercury, he little suspected that if the vacuum Dr. Watson made was free from air, there was no such singular appearance of light to be attended to. Air, Mr. Canton adds, must be condensed near the surface of all bodies that attract it; and will, therefore, be some hindrance to the exit, or entrance of the electric fluid, except the bodies be very sharply pointed.

Some curious observations relating to electric light were made by Mr. Wilke. Rubbing two pieces of glass together in the dark, he observed a vivid phosphoreal light: which, however, threw out no rays, but adhered to the place where it was excited. It was attended with a strong phosphoreal smell, but with no attraction, or repulsion. From this experiment he inferred, that friction alone would not excite electricity, so as to be accumulated upon any body; and that to produce this effect, the bodies rubbed together must be of different natures, with respect to their attracting the electric fluid. He, moreover, imagined, that all examples of phosphoreal light, without attraction, were owing to the same

R r 2   excitation

excitation of electricity, without the accumulation of it. Such he imagined to be the case of light emitted by the Bolognian stone, cadmea fornacum, rotten wood, pounded sugar, and glass of all kinds. *

A TUBE excited with a woollen cloth, on which white wax or oil had been put, he says, threw out flames; each of which when examined, appeared to rise out of a little protuberance of fire. The flame was one, and very narrow at the bottom, but farther from the tube it divided into several ramifications; which always leaned to those parts of the tube which were the least excited, or to conductors in the neighbourhood †

HE says that, upon presenting a finger or other non-electric to an excited negative electric, a cone of light is formed; the base of which is at the finger, or other non-electric, and the apex at the electric, on the surface of which it spreads to a considerable distance all round. ‡

SOMETIMES, he had seen fiery particles thrown laterally from an irregular electric spark, which shone like stars, and were very like those which are produced by the collision of flint and steel. ||

SUSPENDING various balls from his conductor, and presenting others to them, which were sometimes of glass and sometimes of metal, and varying them in every manner possible, he always found (except when two metal balls were used) that the light between them formed a cone, the base of which was always on the body which was positive, and the apex on

* Wilke, p. 123, 124.   † Ib. p. 125.   ‡ Ib. p. 127. || Ib. p. 130.

that

that which was negative. He fays that this criterion is fufficient to diftinguifh the two electricities from one a-nother.

He obferves that, at the apex of a cone iffuing from point-ed bodies, electrified pofitively, there is a cylindrical fpark, out of which lucid rays, like a river, are darted. Thefe rays, he fays, form a lucid cone, the apex of which is turned towards the point from which the fire proceeds. Sometimes from the apex, or at fome diftance from it, there is a lucid point, which, he fays, Haufenius calls *the fire of the fecond kind*, out of which flew ftreams of fire. The ftreams never iffue from the electrified body itfelf, but always from this lucid point. He fays, moreover, that this lucid point at the extremity of an electrified body, and which throws out lucid rays, forms the diftinctive character of the pofitive cone. *

A negative cone, he fays, is fmall, confifting of very flender filaments, which immediately adhere to the point at which the light enters, or to its fides; and, if accurately ex-amined, feems to form little cones, the bafes of which reft upon the body.

When he afterwards comes to confider the caufe of nega-tive cones of light, he owns himfelf to be at a great lofs how to do it.

Mr. Wilke put Englifh phofphorus upon a pointed body, which, in the dark, rendered the whole vifible; and when he fufpended this pointed body perpendicularly, the phofpho-real vapours were feen to afcend; but upon electrifying it, as

* Wilke, p. 132.

it

it hung in the fame direction, the vapours were carried down-wards, and formed a very long cone, extending out of the middle of the cone of electric light, which was feen per-fectly diftinct from it.   When the electrification was difcon-tinued, the phofphoreal vapour afcended as at firft.   From this depreffion of the phofphoreal effluvia Mr. Wilke infers the efflux of the electric fluid from the point, and upon the furface, and not only through the fubftance of the point-ed body.   It is pity that he did not try this curious experi-ment with pointed bodies electrified negatively.   He would certainly have found the fame depreffion of the phofphoreal effluvia, and would, probably, have retracted his conclufion concerning this proof of the efflux. *

Mr. Wilke alfo thought it to be a proof, that the elec-tric matter did not only flow out of the fubftance of elec-trified bodies, but upon the furface of them, that a metallic ring, projecting ever fo little beyond the point of a wire on which it had been put, prevents the appearance of the lucid point.

The laft obfervation which I fhall recite of Mr. Wilke concerning electric light is, that if a point not electrified, be oppofed to a point electrified pofitively, the cones of light, which, in other circumftances, would appear upon both of them, difappear; but that if a pofitive cone be oppofed to a negative cone they both preferve their own characteriftic properties. †

* Wilke, p. 134.   † Ib. p. 140.

SIGNIOR

SIGNIOR BECCARIA was of opinion, that the direction of the electric fluid may be determined from the phenomena of pointed bodies. The *pencil* (by which he means the electric fire at a point electrified positively) he says, contracts as it approaches a flat piece of metal not electrified; whereas the *star* (by which he means the electric fire at a point electrified negatively) expands in the same circumstances, and has a small cavity near the point towards the large superficies. The pencil is attended with a snapping noise, the star makes little or no noise. He hardly gives any reason for the first of these phenomena; he only says, that such is the necessary consequence of a fluid issuing out of, or entering into a point. But the greater noise made by the pencil, he thought was made by the impulse given by the electric matter to the air, causing it to vibrate: and this must be greater when the fluid is thrown from the point into the air, than when it comes through different portions of the air, and meets in one point. *

WHEN two points are opposed to one another, he says, the phenomena are much the same in both. †

SIGNIOR BECCARIA observed that hollow glass vessels, of a certain thinness, exhausted of air, gave a light when they were broken in the dark. By a beautiful train of experiments, he found, at length, that the luminous appearance was not occasioned by the breaking of the glass, but by the dashing of the external air against the inside, when it was broke. He covered one of these exhausted vessels with a receiver, and let-

* Elettricifmo artificiale, p. 63.   † Ib.

ting

ting the air suddenly on the outside of it, observed the very same light. This he calls his *new invented phosphorus.* *

THAT electric light is more subtle and penetrating, if one may say so, than light produced in any other way is manifest from several experiments, particularly the remarkable one of Mr. Hawkesbee; but none prove it so clearly as some made by the ingenious Mr. Lane, who has likewise made several other original experiments, with which, it is hoped, he will soon oblige the public. He gives me leave to mention these.

WHEN he had, for some different purpose, made the electric shock pass over the surface of a piece of marble, in the dark; he observed, that the part over which the fire had passed was luminous, and retained that appearance for some time. No such effect of the electric shock having ever been observed before, he repeated the experiment with a great variety of circumstances, and found it always answered with all calcarious substances, whether animal or mineral, and especially if they had been burnt into a lime. And, as far as he had tried, many more substances would retain this light than would not do it; among others several vegetable substances would do it, particularly white paper. Tiles, and brick were luminous, but not tobacco-pipe clay, though well burnt.

THAT gypseous substances, when calcined, were luminous, appeared from bits of images made of plaister of Paris; and of this class, he says, is the famous Bolognian stone. But many bodies, he found, were luminous after the electric stroke,

* Lettere dell' elettricismo, p. 365, &c.

which

which were not apparently fo, when expofed to the rays of the fun.

He made thefe curious experiments by placing the chains, or wires that led from the conductor to the outer coating of his jar, within one, two, or three inches (according to the ftrength of the charge) from one another, on the furface of the body to be tried, and difcharging a fhock through them. If the ftone was thin, he found, that if one chain was placed at the top, and the other at the bottom, it would appear luminous on both fides after the explofion.

Mr. Canton, to whom thefe experiments were communicated, clearly proved, that it was the *light* only that the fubftances retained, and nothing peculiar to electricity; and, moreover, after frequent trials, difcovered a compofition, which retains both common light, and that of electricity, much more ftrongly than either the Bolognian ftone, or any other known fubftance whatever. With this new phofphorus he makes a great number of moft beautiful experiments. The flafh made by the difcharge of a common jar, within an inch of a circular piece of it, of about two inches and a half in diameter, will illuminate it fo much, that the figures on a watch plate may be eafily diftinguifhed by it in a darkened room, and it will retain the light half an hour.

S f   SECTION

S E C T I O N   IX.

THE ELECTRICITY OF THE TOURMALIN.

THIS period of my hiſtory furniſhes an intirely new ſubject of electrical inquiries; which, if properly purſued, may throw great light upon the moſt general properties
of electricity. This is the *Tourmalin:* though, it muſt be
acknowledged, the experiments which have hitherto been
made upon this foſſil ſtand like exceptions to all that was before known of the ſubject.

THE tourmalin, as Dr. Watſon has in a manner de-
monſtrated, was known to the antients under the name of
the *lyncurium.* All that Theophraſtus ſays concerning the
lyncurium agrees with the tourmalin, and with no other
foſſil that we are acquainted with. He ſays, that it was uſed
for ſeals, that it was very hard, that it was endued
with an attracting power like amber, and that it was
ſaid, particularly by Diocles, to attract not only ſtraws, and
ſmall pieces of wood, but alſo copper and iron, if beaten

very

very thin; that it was pellucid, of a deep red colour, and required no small labour to polish it. The account which passed current among the ancients concerning the origin of this stone was fabulous, which made Pliny think that all that was said of it was fabulous too.

This stone, though not much attended to by European philosophers, till very lately, is common in several parts of the East Indies, and more particularly in the island of Ceylon, where it is called by the natives *tournamal*. In this island the Dutch became acquainted with it, and by them it is called *aschentrikker*, from its property of attracting ashes, when it is thrown into the fire.

The first account we have had, of late years, concerning this extraordinary stone is in the History of the Royal Academy of sciences at Paris for the year 1717; where we are told, that Mr. Lemery exhibited a stone, which was not common, and came from Ceylon. This stone, he said, attracted and repelled small light bodies, such as ashes, filings of iron, bits of paper, &c.

Linnæus, in his Flora Zeylonica, mentions this stone under the name of *lapis electricus*, and takes notice of Mr. Lemery's experiments.

Notwithstanding this, no further mention was made of this stone and its effects till some years after; when the Duc De Noya, in his letter to Mr. Buffon, presented to the Royal Society, informed us, that when he was at Naples, in the year 1743, the Count Pichetti, secretary to the king, assured him, that, during his stay at Constantinople, he had seen a

S f 2   small

small stone called *tourmalin*, which attracted and repelled ashes. This account the Duc De Noya had quite forgotten, but being in Holland in the year 1758, he saw, and purchased two of those stones. With these, in company with Messrs. Daubenton and Adamson, he made a great number of experiments, of which he favoured the public with a particular account. *

BUT prior to the Duc De Noya's experiments, Mr. Leehman had acquainted Æpinus with the attractive power of the tourmalin, and furnished him with two of them, on which he made many experiments; the result of which he published in the History of the Academy of sciences and belles lettres at Berlin for the year 1756. The substance of the memoir is as follows.

THE tourmalin has always, at the same time, a positive and a negative electricity; one of its sides being in one state, and the other in the opposite; and this does not depend on the external form of the stone. These electricities he could excite in the strongest degree, by plunging the stone in boiling water.

IF one side of the tourmalin be heated more than the other (as if it be laid upon a hot cake of metal) each of the sides acquire an electricity opposite to that which is natural to it; but if left to itself, it will return to its natural state.

IF one of the sides of the tourmalin be rubbed, while the other is in contact with some conductor communicating with

* Phil. Transf. Vol. 51. pt. 1. p. 396.

the

the ground; the rubbed side is always positive, and the other negative. If neither side be in contact with a conductor, both become positive. If, in the former of these cases, the tourmalin be rubbed, so as to acquire a sensible heat, and the side which is naturally positive be made negative, it will, upon standing to cool, return to its natural state; but if it have acquired no sensible heat, it will not return to its natural state while any kind of electricity remains. If it be heated, even when it is rubbed and insulated, (in which case both sides become positive) it will still return to its natural state upon cooling.

The Duc De Noya mentions these experiments of Æpinus, but does not admit of a *plus* and *minus* electricity belonging to the tourmalin when heated. On the contrary, he says, that both the sides are electrified plus, but one of them more than the other, and that it was the difference between those degrees which led Æpinus into his mistake. *

The tourmalin was introduced to the notice of the English philosophers by Dr. Heberden, who fortunately recollecting to have seen one of them, many years before, in the possession of Dr. Sharpe at Cambridge (and it was the only one known in England at that time) procured it for Mr. Wilson, who, though it was but a small one, repeated with it most of the experiments of Æpinus, so far as to satisfy himself that his opinion of its positive and negative power was well founded.

Afterwards, Dr. Heberden, ever desirous of extending the bounds of science, procured some of these stones from

* Phil. Transf. Vol. 57. pt. 1. p. 315.

Holland,

Holland, and put them into the hands of thofe perfons who were likely to make the beft ufe of them, particularly Mr. Wilfon and Mr. Canton; in whofe hands they were not lodged in vain, as will appear by the brief account I fhall fubjoin of their experiments upon them.

Mr. Wilson's obfervations are too many, and too particular to be all inferted in this work. The refult of them was, in the main, the fame with that of Æpinus, eftablifhing the opinion of the two different powers of this ftone; but, contrary to Æpinus, he afferts, that when the fides of the tourmalin are unequally heated, it exhibits that fpecies of electricity which is natural to the hotter fide, that is, the tourmalin is *plus* on both fides, when the *plus* fide is the hotter; and *minus* on both fides, when the *minus* fide is the hotter.

Upon this Æpinus repeated all his former experiments, and ftill found the refult of them agreeable to his former conclufion, and contrary to that of Mr. Wilfon. Mr. Wilfon alfo repeated his, without any variation in the event, and imagined the difference between him and Æpinus might arife from the different fizes of the tourmalins they made ufe of, or from their different manner of making the experiments. And it is evident, from the defcription of both their apparatufes, that that of Mr. Wilfon was much better calculated for the purpofe of accurate experiments than that of Æpinus. Mr. Wilfon alfo ufed a greater variety of methods of communicating heat to his tourmalins. He both plunged them in boiling water, held them to the flame of a candle, and expofed them to heated infulated electrics. *

* Phil. Tranf. Vol. 53. p. 436, &c.

Though

THOUGH the detail of all Mr. Wilfon's experiments would be, as I obferved before, much too long for my purpofe, I cannot help relating one of them, which was made with the laft mentioned method of treating. He heated one end of a glafs tube red hot, and when he had expofed what both he and Æpinus call the negative fide of the tourmalin to it; he obferved that about three inches of the heated part of the glafs were electrified *minus*, though the glafs beyond that was electrified *plus*, and continued fo even after the glafs was cold, the electric fluid having paffed from the tourmalin to the glafs ; fince thefe were the fame appearances that were produced by prefenting an excited tube to the heated glafs.

HE then applied the *plus* fide of the tourmalin to the fame heated glafs, and found that the tube was electrified *minus*, above a foot in length, without the leaft appearance of a *plus* electricity beyond the *minus* one, as in the other experiment; and this *minus* electricity appeared when the tube was nearly cold.   In this cafe he judged that the electric fluid had paffed from the glafs to the tourmalin.

MR. WILSON imagined that the tourmalin, as well as glafs, was permeable to the electric fluid, and that the refiftance to its entering the fubftance of it was lefs on what he calls the negative than on the pofitive fide. Thefe conclufions he drew from the two following experiments. Rubbing the pofitive fide of the ftone flightly, he found both fides electrified *plus*, but rubbing the negative fide in the fame manner, both fides were electrified *plus* more ftrongly than before. *

* Phil. Tranf. Vol. 51. pt. 1. p. 327.

SEVERAL

SEVERAL experiments led Mr. Wilſon to conclude that the tourmalin reſiſted the exit and entrance of the electric fluid conſiderably leſs than glaſs, or even than amber; and upon the whole he infers, that the tourmalin differs in nothing from other electric bodies but in acquiring electricity by heat. *

EXAMINING a great number of tourmalins, he found that a line drawn from the *plus* part, through the center of the ſtone, would always paſs through the *minus* part.

ALMOST all theſe tourmalins he greaſed over, and whilſt they were warm enough to preſerve the greaſe liquid, he tried each tourmalin ſeparately, but found no alteration in the virtue of the ſtone, except that it was a little weakened; though it is well known, that moiſture of any ſort readily conducts the electric fluid. If, therefore, the tourmalin had not a fixed kind of electricity, the *plus* and *minus* electricity, obſervable on the two ſides of the ſtone, muſt, by this treatment, have united, and deſtroyed each other. From this circumſtance Mr. Wilſon concluded, that the tourmalin ſuffered the electric fluid to paſs through it only in one direction, and that in this it bore ſome analogy to the loadſtone; having, as it were, two electric poles, which are not eaſily deſtroyed or altered. †

THIS induced him to try whether, like the loadſtone, the tourmalin would loſe its virtue after being made red hot; but, though he kept two of them in a ſtrong fire for half an hour, he could not perceive the leaſt alteration in them : but plunging one in water, while it was red hot entirely deſtroyed its

* Phil. Tranſ. Vol. 51. pt. 1. p. 329.   † Ib. p. 337, 338.

virtue,

virtue, and gave it the appearance of having been shivered in many parts without breaking. *

Notwithstanding the attention given to this subject by Æpinus and Mr. Wilson, the most important discovery relating to the electricity of the tourmalin was reserved for Mr. Canton; who, in a paper read at the Royal Society in the same month with that above-mentioned of Mr. Wilson, viz. December 1759, observes, that the tourmalin emits and absorbs the electric fluid only by the increase or diminution of its heat. For if the tourmalin, he says, be placed on a plain piece of heated glass or metal, so that each side of it, by being perpendicular to the surface of the heating body, may be equally heated; it will, while heating, have the electricity of one of its sides positive, and that of the other negative. This will, likewise, be the case when it is taken out of boiling water; but the side which was positive while it was heating will be negative while it is cooling, and the side which was negative will be positive.

In this paper Mr. Canton refers to the Gentleman's Magazine for the month of September before, in which he had published the result of some experiments he had made on a tourmalin which he had procured from Holland. The propositions he there lays down are so few, and comprise the principal part of what is known upon this subject in so concise and elegant a manner, that I shall recite them all in this place.

1. When the tourmalin is not electrical or attractive,

* Phil. Tranf. Vol. 51. pt. 1. p. 338.

T t

heating

heating it, without friction, will make it so; and the electricity of one side of it, (distinguished by A) will be positive, and that of the other side (B) will be negative.

2. The tourmalin not being electrical will become so by cooling; but with this difference, that the side A will be negative, and the side B positive.

3. If the tourmalin, in a non-electrical state, be heated, and suffered to cool again, without either of its sides being touched; A will be positive, and B negative, the whole time of the increase and decrease of its heat.

4. Either side of the tourmalin will be positive by friction, and both may be made so at the same time.

These, says he, are the principal laws of the electricity of this wonderful stone; and he adds, if air be supposed to be endued with similar properties, i. e. of becoming electrical by the increase or diminution of its heat (as is probable, if its state before and after a thunder storm be attended to) thunder clouds both positive and negative, as well as thunder gusts, may easily be accounted for.

These capital discoveries were made before Mr. Canton had received of Dr. Heberden the tourmalins above-mentioned. When those came to his hands he was enabled to make several new and curious experiments, which I have leave to publish.

He put one of the tourmalins, which was of the common colour into the fire, and burnt it white; when he found that its electrical property was intirely destroyed. The electricity of another was only in part destroyed by the fire. Two other tourmalins he joined together, when they were made soft by

fire,

fire, without deſtroying their electrical property.   The vir
tue of another was improved by its being melted at one end ;
and he found (contrary to what Mr. Wilſon had obſerved of
another tourmalin, which he treated in the ſame manner)
that one tourmalin retained its electrical property after it had
been frequently made red hot, and, in that ſtate, put into
cold water.

But the moſt curious of his experiments were made upon
a large irregular tourmalin, about half an inch in length,
which he cut in three pieces ; taking one part from the poſitive, and another from the negative end.   Trying theſe pieces
ſeparately, he found the outer ſide of the piece which he cut
from the end that was negative when cooling was likewiſe
negative when cooling, and that the outer ſide of that piece
which was cut from the end that was poſitive when cooling
was likewiſe poſitive when cooling ; the oppoſite ſides of both
pieces being, agreeable to the general law of the electricity of
the tourmalin, in a contrary ſtate.

The middle part of the ſame tourmalin was affected juſt as
it had been when it was intire ; the poſitive end remained poſitive, and the negative end continued negative.   The ſame
he had alſo obſerved of two other tourmalins, each about the
ſize of this, which were alſo cut out of a large one.

On January the 8th. 1762, Mr. Canton took the Doctor's
large tourmalin (the ſame of which Mr. Wilſon has given a
drawing and deſcription, in the 51ſt. vol. of the Philoſophical
Transactions, p. 316) and having placed a ſmall tin cup of
boiling water on one end of his electrometer, which was ſupported by warm glaſs, while the pith balls were ſuſpended at the

other

other end; he dropped it into the water, and obferved, that during the whole time of its being heated, and alfo while it was cooling in the water, the balls were not at all electrified.

THE refult of this experiment was contrary to one of a fimilar nature made by Mr. Wilfon on the fame tourmalin; who fays, as has been related above, that when the tourmalin was covered over with liquid greafe, it ftill retained its electric virtue.

TILL the year 1760 it had been fuppofed, that, of all electric fubftances, the tourmalin alone, poffeffed the property of being excited by heating and cooling, but in the beginning of that year, Mr. Canton having had an opportunity of examining a variety of gems, by the favour of Mr. Nicolas Crifp a jeweller in Bow church yard, firft found the *Brazil topaz* to have the electrical properties of the tourmalin. The largeft he met with he put into the hands of Dr. Heberden, who returned it November the 27th. 1760, and fent with it the tourmalins above-mentioned.

IN September 1761, Mr. Wilfon (who had been informed of Mr. Canton's difcovery) met with feveral other gems, of different fizes and colours, that refembled the tourmalin with refpect to electrical experiments. The moft beautiful of them were fomething like the ruby, others were more pale, and one inclining to an orange colour. In point of hardnefs and luftre, they were nearly the fame with the topaz.

FROM all his experiments upon thefe gems he thought it was abundantly evident, that the direction of the fluid did not depend upon the external figure of the gem, but upon fome

particular

particular internal make and conftitution thereof. And that there is fome fuch natural difpofition in all gems affording thefe appearances may be collected, he fays, from another curious fpecimen of the tourmalin kind, which is green, and formed of long flender chryftals with feveral fides; many of which are found fticking together, and are brought from South America.

These gems, numbers of which were furnifhed him by Mr. Emanuel Mendes Da Cofta, he found not only, like the tourmalin, in regard to electric appearances; but that the direction of the electric fluid moving therein was always along the grain or fhootings of the chryftals, one end thereof being electrified *plus*, and the other end *minus*. And that the fluid is more difpofed to pafs in that direction than in another, he thought, might be farther collected from what has been obferved on the grain of the load ftone by Dr. Knight; who found that though the magnetic poles of the natural loadftone might be varied in any direction, yet that the fame loadftone admitted of being made much more magnetical along the grain than acrofs it.

From thefe experiments and obfervations Mr. Wilfon inferred by anology, that the electric fluid, flowing through all thefe ftones and gems, moves in that direction in which the grain happens to lie; and that the reafon of this is, that the refiftance the fluid meets with in paffing through the gem is lefs in that direction than in any other. *

In a fubfequent paper of Mr. Wilfon's, read at the Royal

* Phil. Tranf. Vol. 52. pt. 2. p. 443.

Society

Society December the 23d. 1763, and March 1764, he recites several curious experiments on the effects of removing
the tourmalin from one room to another, in which there was
some difference of heat; the result of which exactly confirms
Mr. Canton's discovery, that the side which is positive when
heating is negative when cooling, and *vice versa*. Upon a
very nice examination, and during some favourable circumstances, Mr. Wilson says he has observed the tourmalin to
be feebly electrified, when the thermometer varied up and
down only one degree. *

* Phil. Transf. Vol. 52. pt. 2. p. 457.

SECTION

SECTION X.

DISCOVERIES THAT HAVE BEEN MADE SINCE THOSE OF
DR. FRANKLIN, WITH RESPECT TO THE SAMENESS
OF LIGHTNING AND ELECTRICITY.

THE year 1752 makes an aera in electricity no less famous than the year 1746, in which the Leyden phial
was discovered. In the year 1752, was verified the hypothesis of Dr. Franklin, of the identity of the matter of
lightning and of the electric fluid; and his great project
of trying the experiment, by real lightning actually brought
down from the clouds, was carried into execution.

THE French philosophers were the first to distinguish themselves upon this memorable occasion, and the most active
persons in the scene were Mr. Dalibard and Delor, both zealous partisans (as Mr. Nollet calls them) of Dr. Franklin.
The former prepared his apparatus at Marly La Ville, situated
five or six leagues from Paris, the other at his own house, on
some of the highest ground in that capital. Mr. Dalibard's
machine consisted of an iron rod forty feet long, the lower

extremity

extremity of which was brought into a centry box, where the rain could not come; while, on the outfide, it was faftened to three wooden pofts by long filk ftrings defended from the rain. This machine happened to be the firft that was favoured with a vifit from this ethereal fire. The philofopher himfelf happened not to be at home at that time; but, in his abfence, he entrufted the care of his apparatus to one Coiffier a joiner, a man who had ferved fourteen years among the dragoons, and on whofe underftanding and courage he could depend. This artifan had all proper inftructions given him both how to make obfervations, and alfo to guard himfelf from any harm there might be from them; befides being exprefly ordered to get fome of his neighbours to be prefent, and particularly to fend for the curate of Marly whenever there fhould be any appearance of the approach of a thunder ftorm. At length the long expected event arrived.

On wednefday the 10th. of May 1752, between two and three o'clock in the afternoon, Coiffier heard a pretty loud clap of thunder. Immediately he flies to the machine, takes a phial furnifhed with a brafs wire, and prefenting one end of the wire to the rod, he fees a fmall bright fpark iffue from it, and hears the fnapping that it made. Taking a fecond fpark ftronger than the former, and attended with a louder report, he calls his neighbours, and fends for the curate. The curate runs with all his might, and the parifhoners feeing the precipitation of their fpiritual guide, imagine that poor Coiffier had been killed with lightning. The alarm fpreads through the village, and the hail which came on did not prevent the flock from following their fhepherd. The hone

ecclefiaft

ecclefiaftic arriving at the machine, and feeing there was no danger, took the wire into his own hand and immediately drew feveral ftrong fparks, which were moft evidently of an electrical nature, and compleated the difcovery for which the machine was erected. The thunder cloud was not more than a quarter of an hour in paffing over the zenith of the machine, and there was no thunder heard befides that fingle clap. As foon as the ftorm was over, and no more fparks could be drawn from the bar, the curate wrote a letter to Mr. Dalibard, containing an account of this remarkable experiment, and fent it immediately by the hands of Coiffier himfelf.

He fays, that he drew fparks from the bar of a blue colour, an inch and a half in length, and which fmelt ftrong of fulphur. He repeated the experiment at leaft fix times in the fpace of about four minutes, in the prefence of many perfons, each experiment taking up the time, as he, in the ftyle of a prieft, expreffeth himfelf, of a *pater* and an *ave*. In the courfe of thefe experiments he received a ftroke on his arm, a little above the elbow, but he could not tell whether it came from the brafs wire inferted into the phial, or from the iron bar. He did not attend to the ftroke at the time he received it, but, the pain continuing, he uncovered his arm when he went home, in the prefence of Coiffier; and a mark was perceived round his arm, fuch as might have been made by a blow with the wire on his naked fkin; and afterwards feveral perfons, who knew nothing of what had happened, faid that they perceived a fmell of fulphur when he came near them.

U u   Coiffier

COIFFIER told Mr. Dalibard, that for about a quarter of an hour before the curate arrived, he had, in the prefence of five or fix perfons, taken much ftronger fparks than thofe which the curate mentioned. *

EIGHT days after Mr. Delor faw the fame thing at his own houfe, although only a cloud paffed over, without either thunder or lightning. †

THE fame experiments were afterwards repeated by Mr. Delor, at the requeft of the king of France, who, it is faid, faw them with the greateft fatisfaction, and expreffed a juft fenfe of the merit of Dr. Franklin. Thefe applaufes of the king excited in Mr. De Buffon, Dalibard, and Delor, a defire of verifying Dr. Franklin's hypothefis more completely, and of purfuing his fpeculations upon the fubject.

MR. DELOR's apparatus in Paris confifted of a bar of iron ninety nine feet high, and anfwered rather better than that of Mr. Dalibard, which, as was obferved before, was only forty feet high. But as the quantity of electricity which they could procure from the clouds, in thefe firft experiments, was but fmall, they added to this apparatus what they called a *magazine* of electricity, confifting of many bars of iron infulated, and communicating with the pointed iron rod. This magazine contained more of the electric matter, and gave a more fenfible fpark, upon the approach of the finger, than the pointed bar.

A MAGAZINE of this kind Mr. Mazeas had in an upper room of his houfe, into which he brought the lightning, by

* Dalibard's Franklin, Vol. 2. p. 109, &c.   † Nollet's Letters, Vol. 1. p. 9.

means

means of a wooden pole projecting out of his window, at the extremity of which a glaſs tube, filled with roſin, received a pointed iron rod, twelve feet long. But all this while the electrics, which they made uſe of to ſupport theſe iron rods, were too much expoſed to the open air, and conſequently were liable to be wet, which would infallibly defeat their experiments.

THE moſt accurate experiments, made with theſe imperfect inſtruments, were thoſe of Mr. Monnier. He was convinced that the high ſituation in which the bar of iron had commonly been placed was not abſolutely neceſſary for this purpoſe: for he obſerved a common ſpeaking trumpet, ſuſpended upon ſilk five or ſix feet from the ground, to exhibit very evident ſigns of electricity. He alſo found that a man placed upon cakes of roſin, and holding in his hand a wooden pole, about eighteen feet long, about which an iron wire was twiſted, was ſo well electrified when it thundered, that very lively ſparks were drawn from him; and that another man, ſtanding upon non-electrics, in the middle of a garden, and only holding up one of his hands in the air, attracted, with the other hand, ſhavings of wood which were held to him.

HE ſays, that he obſerved a continual diminution of the electricity when the rain came on, though the thunder was ſtill very loud, and though the cake of roſin which ſupported his conductor was not wet. But he afterwards found, that this was not univerſally true.

HE once obſerved, that when the conducting wire was ſurrounded with drops of rain, only ſome of them were electri-

U u 2   fied;

fied; as was evident from the conic figure they had, while the reft remained round as before. It was alfo perceived, that the electrified and non-electrified drops generally fucceeded each other alternately; which made Mr. Monnier call to mind a fingular phenomenon, which happened fome years before to five peafants who were paffing through a corn field, near Franckfort upon the Oder, during a thunder ftorm; when the lightning killed the firft, the third, and the fifth of them, without injuring the fecond, or the fourth.*

It was not owing to any want of attention to this fubject that the Englifh philofophers were not the firft to verify the theory of Dr. Franklin. They happened to have few opportunities of trying experiments. In the few they had, they were difappointed by the rain wetting their apparatus, which was not better conftructed than that of the French.

At length fuccefs crowned the affiduity and happier contrivance of Mr. Canton, who, to the lower end of his conducting wire, had had the precaution to faften a tin cover, to keep the rain from the glafs tube which fupported it. By this means, on the 20th. of July 1752, he got fparks at the diftance of half an inch, but the whole appearance ceafed in the fpace of two minutes. †

Mr. Wilson, who took great pains in thefe purfuits, as he did in every thing elfe relating to electricity, perceived feveral electrical fnaps, on the 12th. of Auguft following, from no other apparatus than an iron curtain rod; one end of which he put into the neck of a glafs phial, which he held in his hand, while the other was made to point into the air.

* Phil. Tranf. Vol. 47. p. 551.   † Ib. p. 568.

Dr.

Dr. Bevis also, at this time, viz. on the 12th. of August, observed nearly the same appearances that Mr. Canton had done before. *

Mr. Canton afterwards, resuming his observations on lightning, with the assiduity and accuracy with which he observes every thing, found, by a great number of experiments, that some clouds were in a positive, and some in a negative state of electricity. And that, by this means, the electricity of his conductor would sometimes change from one state to the other five or six times in less than half an hour. †

This observation of Mr. Canton, on the different electricity of the clouds was made, and the account of it published in England, before it was known that Dr. Franklin had made the same discovery in America.

When the air was dry, he observed, that the apparatus would continue electrified for ten minutes, or a quarter of an hour after the clouds had passed the zenith, and sometimes till they were more than half way towards the horizon; that rain, especially when the drops were large, generally brought down the electric fire; and hail, in summer, he believed, never failed. The last observation he had made before the time of his writing this paper, his apparatus had been electrified by a fall of thawing snow. This was on the 12th. of November 1753, which, he says, was the 26th. day, and the sixty first time, it had been electrified since it was first set up viz. about the middle of the preceding May.

Only two thunder storms had happened at London during that whole summer, and his apparatus was so strongly electrified by one of them, that the ringing of the bells (which

* Phil. Transf. Vol. 47. p. 569.   † Ib. Vol. 48. pt. 1. p. 356.

he

he fufpended to his apparatus, to fignify when the electrification was begun, and which were frequently rung fo loud as
to be heard in every room of the houfe) was ftifled by the
almoft conftant ftream of denfe electric fire between each
bell and brafs ball, which would not fuffer it to ftrike.

UPON a farther occafion, he obferves, that, in the fucceeding months of January, February, and March, his apparatus was electrified no lefs than twenty five times, both
pofitively and negatively, by fnow as well as by hail and rain;
and almoft to as great a degree when Fahrenheits thermometer was between twenty eight and thirty four, as he had ever
known it in fummer, except in a thunder ftorm. *

MR. CANTON concludes his paper with propofing the two
following queries. 1. May not air fuddenly rarefied give
electric fire to, and may not air fuddenly condenfed receive
electric fire from, clouds and vapours paffing through it?
2. Is not the aurora borealis the flafhing of electric fire from
pofitive towards negative clouds at a great diftance, through
the upper part of the atmofphere, where the refiftance is
leaft? †

MR. CANTON not only obferved the different ftates of pofitive and negative electricity in the clouds, but alfo noted
the proportion that the one bore to the other for a confiderable time. In the firft period he had obferved the clouds had
been pofitively electrical 83 times, and negative 101. In this
period he had punctually fet down how often the powers had
fhifted, and the whole time that the apparatus continued to

* Phil. Tranf. Vol. 48. pt. 2. p. 785.   † Ib. pt. 1. p. 358.

be

be electrified, but he had intirely neglected to note the time that each power lasted. But this last circumstance he afterwards carefully attended to, for about two months, viz. from the 28th. of June to the 23d. of August 1754; and found the apparatus to be electrified positively thirty one times, which taken together lasted three hours thirty five minutes; and negatively forty five times, the whole duration of which was ten hours thirty nine minutes. He also observed that the positive power was generally the strongest. This account he wrote the 31st. of August 1754.

THESE observations which Mr. Canton gives me leave to make public are extremely curious, and must have required great attention; but they are hardly sufficient to authorize any general conclusion.

ONE of the effects of lightning and electricity is the melting of metals. This was first thought to be a cold fusion; but that opinion is refuted, in a very sensible manner, by Dr. Knight, in a paper read at the Royal Society November 22d. 1759. He observes, that the instances most generally given of cold fusion are two; viz. that of a sword being melted in its scabbard, and that of money being melted in a bag, both the scabbard and the bag remaining unhurt.

A GREAT number of authors, he says, have mentioned both the facts, but without giving either their own testimony, or that of any one else for the truth of them, or describing any of the other concomitant circumstances. And it seemed to him very possible, that lightning might produce effects similar to those above-mentioned, without our being obliged to have recourse to a cold fusion to account for them.

IF,

If, says he, the edge or external surface of a sword had been melted, whilst the main part of the blade remained entire, it would have afforded sufficient ground to assert, in general terms, that the sword was melted, and yet the scabbard might have remained unhurt; because either the edge or surface of a sword might be instantly melted by lightning and cooled so suddenly, as to make no impression of burning upon the scabbard. Metals, as well as other bodies, he observes, will both heat and cool sooner, in proportion as they are thin and slender; that very small wire will instantly become red hot, and even melt, and run into a round globule in the flame of a common candle; though it is no sooner removed out of the flame, but it is instantly cold. He therefore concludes, that the edge of a sword, or even its surface, might be instantly melted by lightning; and being in contact with, or rather still united to the rest of the blade, which might be cold, it would part with its heat too suddenly to produce any appearance of burning.

He was confirmed in this reasoning, by examining some fragments and particles of wire melted by lightning, which were sent him by Mr. Mountaine. Amongst them appeared globules of various sizes, which had undergone very different degrees of fusion. The largest of them had not been fluid enough to put on a spherical figure, but they approached nearer to it, in proportion as they were smaller; so that, in the smallest granulæ, the fusion was most perfect, the globules being very round and smooth. Their sizes continued diminishing till they became invisible to the naked eye, and

some

some of them, when viewed with a microscope, required a third or fourth magnifier to see them distinctly.

Some of the bits of wire were rough and scaly, like burnt iron, and were swelled in those places where they were beginning to melt. Others continued straight, and of an equable thickness; but their outward surface seemed to have undergone a perfect fusion, so that there were two or more pieces adhering together, as if joined by a thin solder.

In the Philosophical Transactions, Dr. Knight says, there are two or three relations which seem at first to favour a cold fusion, but when duly considered prove nothing conclusive. *

But that there is really no such thing as cold fusion, either by electricity or lightning, was most clearly demonstrated by Mr. Kinnersley, in a letter to Dr. Franklin, dated Philadelphia March the 12th. 1761.

He suspended a piece of small brass wire, about twenty four inches long, with a pound weight at the lower end; and, by sending through it the charge of a case of bottles, containing above thirty feet of coated glass, he discovered what he calls a new method of wire drawing. The wire was red hot, the whole length well annealed, and above an inch longer than before. A second charge melted it so that it parted near the middle, and measured, when the ends were put together, four inches longer than at first.

This experiment, he says, was proposed to him by Dr. Franklin, in order to find whether the electricity, in passing

* Phil. Transf. Vol. 51. pt. 1. p. 294, &c.

through

through the wire, would so relax the cohesion of its constituent particles, as that the weight might produce a separation; but neither of them had the least suspicion that any heat would be produced.

THAT he might have no doubt of the wire being *hot*, as well as *red*, he repeated the experiment on another piece of the same wire, encompassed with a goose quill, filled with loose grains of gun-powder; which took fire, as readily as if it had been touched with a red hot poker. Also tinder, tied to another piece of the wire, kindled by it; but when he tried a wire about twice as big, he could produce no such effects.

HENCE, says he, it appears, that the electric fire, though it has no sensible heat when in a state of rest, will, by its violent motion, and the resistance it meets with, produce heat in other bodies, when passing through them, provided they be small enough. A great quantity will pass through a large wire, without producing any sensible heat; when the same quantity, passing through a very small one, being there confined to a narrower passage, the particles crowding closer together, and meeting with greater resistance, will make it red hot, and even melt it.

HENCE, he concludes, that lightning does not melt metal by a cold fusion, as Dr. Franklin and himself had formerly supposed; but that, when it passed through the blade of a sword, if the quantity was not very great, it might heat the point so as to melt it, while the broadest and the thickest part might not be sensibly warmer than before.

WHEN trees or houses are set on fire by the dreadful quantity,

tity, which a cloud, or fometimes the earth difcharges, muft not the heat, fays he, by which the wood is firft kindled, be generated by the lightning's violent motion through the refifting combuftible matter?

If lightning, by its rapid motion, produced heat in itfelf as well as in other bodies (which Mr. Kinnerfley imagined was evident from fome experiments made with his electrical thermometer, mentioned before) he thought that its fometimes fingeing the hair of animals killed by it might eafily be accounted for; and that the reafon of its not always doing fo might be, that the quantity, though fufficient to kill a large animal, might not be great enough, or not have met with refiftance enough, to become by its motion burning hot.

We find, fays he, that dwelling houfes, ftruck with lightning, are feldom fet on fire by it; but that when it paffes through barns, with hay or ftraw in them, or ftore houfes containing large quantities of hemp, or fuch like matter, they feldom, if ever, efcape a conflagration. This, he thought, might be owing to fuch combuftibles being apt to kindle with a lefs degree of heat than was neceffary to kindle wood. *

All that was done by the French and Englifh electricians, with refpect to lightning and electricity, fell far fhort of what was done by Signior Beccaria at Turin. His attention to the various ftates of the atmofphere, his affiduity in making experiments, his apparatus for making them, the ex-

* Phil. Tranf. Vol. 53. p. 92, &c.

X x 2

tent

tent of his views in making them, the minute exactnefs with
which he has recorded them, and his judgment in applying
them to a general theory, far exceeded every thing that had
been done by philofophers before him, or that has been done
by any perfon fince. And though I fhall give a confiderable
fcope to my account of his experiments and obfervations, I
fhall be able to give my reader but a faint idea of the extent,
variety, and value of his labours in this great field.

He made ufe both of kites, and pointed rods, and of a
great variety of both, at the fame time, and in different
places. Some of the ftrings of his kites had wires in them,
and others had none. Some of them flew to a prodigious
height, and others but low; and he had a great number of
affiftants to note the nature, time, and degree of appearances,
according as his views required.

To keep his kites conftantly infulated; and at the fame to
give them more or lefs ftring, and for many other purpofes,
he had the ftring rolled upon a reel, which was fupported by
pillars of glafs; and his conductor had a communication with
the axis of the reel. *

To diftinguifh the pofitive and negative ftate of the clouds,
when the electricity was vigorous, with more certainty, and
with more fafety than it could be done by prefenting an excited ftick of glafs, or fealing wax to threads diverging from
his conductor; he inclofed a pointed wire and a flat piece of
lead oppofite to it within a cylindrical glafs veffel, wrapped
in pafteboard, fo that the infide could have no communication

* Lettere dell' elettricifmo, p. 112.

with

with the external light.   Into this cover, and oppofite to the
point of the wire, he inferted a very long tube of pafteboard ;
through which he could look from a confiderable diftance,
and fee the form of the electric light at the end of the wire ;
which is the fureft indication of its quality. *

From Signior Beccaria's extremely exact and circumftan-
tial account of the external appearances of thunder clouds,
which he prefixes to his obfervations on their probable caufes,
I fhall draw a general outline of the moft remarkable particulars, in the ufual progrefs of a thunder ftorm.

The firft appearance of a thunder ftorm (which generally
happens when there is little or no wind) is one denfe cloud,
or more, increafing very faft in fize, and rifing into the higher
regions of the air.   The lower furface is black, and nearly
level ; but the upper finely arched, and well defined. Many
of thefe clouds often feem piled one upon another, all arched
in the fame manner ; but they keep continually uniting,
fwelling, and extending their arches.

At the time of the rifing of this cloud, the atmofphere is
generally full of a great number of feparate clouds, motionlefs, and of odd and whimfical fhapes.   All thefe, upon the
appearance of the thunder cloud, draw towards it, and become more uniform in their fhapes as they approach ; till,
coming very near the thunder cloud, their limbs mutually
ftretch towards one another ; they immediately coalefce, and
together make one uniform mafs.   Thefe he calls *adfcititious*
clouds, from their coming in, to enlarge the fize of the

* Ib. p. 107.

thunder

thunder cloud. But, fometimes, the thunder cloud will fwell, and increafe very faft without the conjunction of any adfcititious clouds, the vapours in the atmofphere forming themfelves into clouds wherever it paffes. Some of the ad-fcititious clouds appear like white fringes, at the fkirts of the thunder cloud, or under the body of it, but they keep con-tinually growing darker and darker, as they approach to unite with it.

WHEN the thunder cloud is grown to a great fize, its low-er furface is often ragged, particular parts being detached towards the earth, but ftill connected with the reft. Some-times the lower furface fwells into various large protuber-ances, bending uniformly towards the earth. And fome-times one whole fide of the cloud will have an inclination to the earth, and the extremity of it will nearly touch the earth * When the eye is under the thunder cloud, after it is grown large, and well formed, it is feen to fink lower, and to darken prodigioufly; at the fame time that a number of fmall adfcititious clouds (the origin of which can never be perceived) are feen in a rapid motion, driving about in very uncertain directions under it. While thefe clouds are agitated with the moft rapid motions, the rain generally falls in the greateft plenty, and if the agitation be exceeding great, it commonly hails. †

WHILE the thunder cloud is fwelling, and extending its branches over a large tract of country, the lightning is feen to dart from one part of it to another, and often to illuminate

* Lettere dell' elettricifmo, p. 151.   † Ib. p. 155.

its

its whole mafs. When the cloud has acquired a fufficient extent, the lightning ftrikes between the cloud and the earth, in two oppofite places, the path of the lightning lying through the whole body of the cloud and its branches. The longer this lightning continues, the rarer does the cloud grow, and the lefs dark is its appearance; till, at length, it breaks in different places, and fhows a clear fky. When the thunder cloud is thus difperfed, thofe parts which occupy the upper regions of the atmofphere are equally fpread, and very thin; and thofe that are underneath are black, but thin too; and they vanifh gradually, without being driven away by any wind. *

Having feen what this philofopher obferved abroad, and in the air, let us fee what he took notice of at his apparatus within doors. This never failed to be electrified upon every approach of a thunder cloud, or any of its branches; and the ftream of fire from it was generally perpetual, while it was directly over the apparatus. †

That thunder clouds were fometimes in a pofitive as well as negative ftate of electricity, Signior Beccaria had difcovered, before he heard of its having been obferved by Dr. Franklin, or any other perfon. ‡ The fame cloud, in paffing over his obfervatory, electrified his apparatus, fometimes pofitively, and fometimes negatively. * The electricity continued longer of the fame kind, in proportion as the thunder cloud was fimple, and uniform in its direction; but when the lightning changed its place, there commonly happened a

* Lettere dell' ellettricifmo, p. 146, 176.   † Ib. p. 167.   ‡ Ib. p. 138. || Ib. p. 172.

change

change in the electricity of his apparatus. It would change suddenly after a very violent flash of lightning, but the change would be gradual when the lightning was moderate, and the progress of the thunder cloud flow.

IT was an immediate inference from his observations of the lightning abroad, and his apparatus within, that the quantity of electric matter, in an usual storm of thunder, is almost inconceivably great; considering how many pointed bodies, as trees, spires, &c. are perpetually drawing it off, and what a prodigious quantity is repeatedly discharged to, or from the earth. *

AFTER this summary view of appearances, I shall, in the same succinct manner, represent the manner in which this excellent philosopher accounts for them, and some other principal and well known phenomena of thunder storms.

CONSIDERING the vast quantity of electric fire that appears in the most simple thunder storms, he thinks it impossible that any cloud, or number of clouds should ever contain it all, so as either to discharge or receive it. Besides, during the progress and increase of the storm, though the lightning frequently struck to the earth, the same clouds were the next moment ready to make a still greater discharge, and his apparatus continued to be as much affected as ever. The clouds must, consequently, have received at one place, the moment that a discharge was made from them in another. † In many cases, the electricity of his apparatus, and consequently of the clouds, would instantly change from one kind to another several times; an effect which cannot be accounted for by any simple dis-

* Lettere dell' ellettricismo, p. 180.   † Ib. p. 183, 188.

charge,

charge or recruit.   Both muſt have taken place in a very quick ſucceſſion. *

THE extent of the clouds doth not leſſen this difficulty : for, be it ever ſo great, ſtill the quantity ought to be leſſened by every diſcharge : and, beſides, the points, by which the ſilent diſcharges are made, are in proportion to the extent of the clouds. † Nor is the difficulty leſſened by ſuppoſing that freſh clouds bring recruits ; for beſides that the clouds are not ripe for the principal ſtorm, till all the clouds, to a great diſtance, have actually coaleſced, and formed one uniform maſs, thoſe recruits bear no ſort of proportion to the diſcharge, and whatever it was, it would ſoon be exhauſted.

THE fact, therefore, muſt be, that the electric matter is continually darting from the clouds in one place, at the ſame time that it is diſcharged from the earth in another. And it is a neceſſary conſequence from the whole, that the clouds ſerve as conductors to convey the electric fluid from thoſe places of the earth which are overloaded with it, to thoſe which are exhauſted of it. ‡

To aſcertain this fact in the moſt complete manner, he propoſes that two obſervatories be fixed, about two leagues aſunder, in the uſual path of the thunder clouds ; and that obſervations be made, whether the apparatus be not often poſitive at one place, when it is negative at the other. ||

THAT great quantities of electric matter do ſometimes ruſh out of particular parts of the earth, and riſe through the air, into the higher regions of the atmoſphere, he thinks is evi-

* Ib. p. 220.   † Ib. p. 185.   ‡ Ib. p. 193. || Ib. p. 194.

Y y

dent

dent from the great quantities of fand, afhes, and other light fubftances, which have often been carried up into the air, and fcattered uniformly over a large tract of country. * No other known efficient caufe of this phenomenon can be affigned, except the wind; and it has been obferved when there was no wind ftirring; and the light bodies have even been carried againft the wind. † He fuppofes, therefore, that thefe light bodies are raifed by a large quantity of electric matter, iffuing out of the earth, where it was overcharged with it, and (by that property of it which he had demonftrated) attracting, and carrying with it every fubftance that could ferve as a conductor in its paffage. All thefe bodies, being poffeffed of an equal quantity of the electric fluid, will be difperfed equally in the air, and confequently over that part of the earth where the fluid was wanting, and whither they ferve to convey it. ‡ Had thefe bodies been raifed by the wind, they would have been difperfed at random, and in heaps.

THIS comparatively rare phenomenon (but of which he had been more than once a fpectator) he thinks exhibits both a perfect image, and demonftration, of the manner in which the vapours of the atmofphere are raifed to form thunder clouds. The fame electric matter, wherever it iffues, attracts to it, and carries up into the higher regions of the air, the watery particles that are difperfed in the atmofphere. The electric matter afcends to the higher regions of the atmofphere, being folicited by the lefs refiftance it finds there than in the common mafs of the earth; which, at thofe times, is generally

*. Lettere dell' elettricifmo, p. 199.   † Ib. p. 225.   ‡ Ib. p. 202.

very

very dry, and confequently highly electric. The uniformity
with which thunder clouds fpread themfelves, and fwell into
arches, muft be owing to their being affected by fome caufe
which, like the electric matter, diffufes itfelf uniformly
wherever it acts, and to the refiftance they meet with in
afcending through the air. * As a proof of this, fteam, rifing
from an electrified eolipile, diffufes itfelf with the fame uniformity, and in fimilar arches, extending itfelf towards any
conducting fubftance. †

The fame caufe which firft raifed a cloud, from vapours
difperfed in the atmofphere, draws to it thofe that are already
formed, and continues to form new ones ; till the whole collected mafs extends fo far, as to reach a part of the earth
where there is a deficiency of the electric fluid. ‡   Thither
too, will thofe clouds, replete with electricity, be ftrongly
attracted, and there will the electric matter difcharge itfelf
upon the earth. A channel of communication being, in this
manner, found, a frefh fupply of electric matter will be raifed
from the overloaded part, and will continue to be conveyed
by the medium of the clouds, till the equilibrium of the fluid
between the two places of the earth be reftored. When the
clouds are attracted in their paffage by thofe parts of the earth
where there is a deficiency of the fluid, thofe detached fragments are formed, and alfo thofe uniform depending protuberances, which will be fhown to be, in fome cafes, the caufe
of water-fpouts, and hurricanes. ‖

That the electric matter, which forms and animates the

* Ib. p. 205.   † Ib. p. 206.   ‡ Ib. p. 212.   ‖ Ib. p. 214.

thunder clouds, iffues from places far below the furface of the earth, and that it buries itfelf there, is probable from the deep holes that have, in many places, been made by lightning. * Flafhes of lightning have, alfo, been feen to arife from fubterraneous cavities, and from wells. † Violent inundations have accompanied thunder ftorms, not occafioned by rain, but by water burfting from the bowels of the earth, from which it muft have been diflodged by fome internal concuffion. Deep wells have been known to fill fafter in thunder ftorms,‡ and others have conftantly grown turbid at the approach of thunder. ‖

THIS very rife, as well as the whole progrefs of thunder clouds, has fometimes been, in a manner, vifible. Exhalations have been frequently feen to rife from particular caverns, attended with a rumbling noife, and to afcend into the higher regions of the air, with all the phenomena of thunder ftorms defcribed above, according to the defcription of perfons who lived long before the connection between electricity and lightning was fufpected. §

THE greateft difficulty attending this theory of the origin of thunder ftorms relates to the collection, and infulation of electric matter within the body of the earth. With refpect to the former, he has nothing particular to fay. Some operations in nature are certainly attended with a lofs of the equilibrium in the electric fluid, but no perfon has yet affigned a more probable caufe of the redundancy of the electric matter which, in fact, often abounds in the clouds, than what we

---

* Lettere dell' ellettricifmo, p. 227.   † Ib. p. 228.   ‡ Ib. p. 233.   ‖ Ib. p. 360.
§ Ib. p. 231.

may

may fuppofe poffible to take place in the bowels of the earth. And fuppofing the lofs of the equilibrium poffible, the fame caufe that produced the effect would prevent the reftoring of it ; fo that not being able to force a way, at leaft one fufficiently ready, through the body of the earth, it would iffue at the moft convenient vent into the higher regions of the air, as the better paffage. His electrical apparatus, though communicating with the earth, has frequently, in violent thunder ftorms, given evident fparks to his finger. *

In the enumeration of the effects of thunder ftorms, he obferves that a wind always blows from the place from which the thunder cloud proceeds ; that this is agreeable to the obfervations of all mariners, and that the wind is more or lefs violent in proportion to the fuddennefs of the appearance of the thunder cloud, the rapidity of its expanfion, and the velocity with which the adfcititious clouds join it. The fudden condenfation of fuch a prodigious quantity of vapours muft difplace the air, and repel it on all fides. †

He, in fome meafure, imitated even this effect of thunder, at leaft produced a circulation of all the air in his room, by the continued electrification of his chain. ‡

Among other effects of lightning, he mentions the cafe of a man rendered exceeding ftiff, prefently after he was ftruck dead in a ftorm of thunder. But the moft remarkable circumftance, in this cafe, was the lightning (chufing the beft conductor) having ftruck one particular vein, near his neck, and followed it through its minuteft ramifications ; fo that

* Ib. p. 236.   † Ib. p. 339, 340.   ‡ Ib. p. 343.

the

the figure of it appeared through the skin, finer than any pencil could have drawn it. *

HE cautions persons not to depend upon the neighbourhood of a higher, or, in all cases, a better conductor than their own body; since, according to his repeated observations, the lightning by no means descends in one undivided track; but bodies of various kinds conduct their share of it, at the same time, in proportion to their quantity and conducting power. †

A GREAT number of observations, relating to the descent of lightning, confirm his theory of the manner of its ascent: for, in many cases, it throws before it the parts of conducting bodies, and distributes them along the resisting medium through which it must force its passage. ‡

UPON this principle it is, that the longest flashes of lightning seem to be made, by its forcing into its way part of the vapours in the air. ‖ One of the principal reasons why those flashes make so long a rumbling, is their being occasioned by the vast length of a vacuum, made by the passage of the electric matter. For though the air collapses the moment after it has passed, and the vibration (on which the sound depends) commences at the same moment, through the whole length of the track; yet, if the flash was directed towards the person who hears the report, the vibrations excited at the nearer end of the track will reach his ear much sooner than those excited at the more remote end; and the sound will, without

* Lettere dell' ellettricismo, p. 242.   † Ib. p. 246.   ‡ Ib. p. 247.   ‖ Ib. p. 851.

any

any repercuffion or echo, continue till all the vibrations have fuccefflively reached him. *

ONE of the moft remarkable effects of lightning is that it gives polarity to the magnetic needle, and to all bodies that have any thing of iron in them, as bricks, &c; and by obferving which way the poles of thefe bodies lye, it may be known, with the utmoft certainty, in what direction the ftroke paffed. †   In one cafe he actually afcertained the direction of the lightning in this manner. ‡

SINCE a fudden ftroke of lightning gives polarity to magnets, he conjectures that a regular and conftant circulation of the whole mafs of the fluid, from North to South, may be the original caufe of magnetifm in general. ‖ This is a truly great thought; and, if juft, will introduce greater fimplicity into our conceptions of the laws of nature.

THAT this ethereal current is infenfible to us, is no proof of its non-exiftence, fince we ourfelves are involved in it. He had feen birds fly fo near a thunder cloud, as he was fure they would not have done, if they had been affected by its atmofphere. §

THIS current he would not fuppofe to arife from one fource, but from feveral, in the northern hemifphere of the earth. The aberration of the common center of all thefe currents from the North point may be the caufe of the variation of the needle, the period of this declination of the center of the currents may be the period of the variation, and the obliquity with which the currents ftrike into the earth

* Ib. p. 252.   † Ib. p. 262.   ‡ Ib. p. 263.   ‖ Ib. p. 268.   § Ib.

may

may be the cauſe of the dipping of the needle, and alſo why bars of iron more eaſily receive the magnetic virtue in one particular direction. *

He thinks that the *Aurora Borealis* may be this electric matter performing its circulation, in ſuch a ſtate of the atmoſphere as renders it viſible, or approaching nearer to the earth than uſual. Accordingly very vivid appearances of this kind have been obſerved to occaſion a fluctuation in the magnetic needle. †

Stones and bricks ſtruck by lightning are often vitrified. He ſuppoſes that ſome ſtones in the earth, having been ſtruck in this manner, firſt gave occaſion to the vulgar opinion of the thunder bolt. ‡

Signior Beccaria was very ſenſible that heat contributes much to the phenomena of thunder, lightning, and rain ; but he could not find, by any experiment, that it tended to promote electricity. He, therefore, rather thought that heat operated, in this caſe, by exhaling the moiſture of the air, and thereby cutting off the communication of the electric fluid between one place and another, particularly between the earth and the higher regions of the air, whereby its effects were more viſible. ||

Having entertained my reader with the obſervations of this great Italian genius, I muſt once more conduct him to France, where he will ſee ſeveral experiments well worth his notice. In this country we have ſeen that Dr. Franklin's theory of the identity of electricity and the matter of light-

* Lettere dell' elettriciſmo, p. 269.   † Ib. p. 272.   ‡ Ib. p. 263. || Ib. p. 359.

ning

ning was firſt verified, and we ſhall now ſee it verified in the grandeſt and moſt conſpicuous manner.

THE greateſt quantity of electricity that was ever brought from the clouds, by any apparatus prepared for that purpoſe, was by Mr. De Romas, aſſeſſor to the preſideal of Nerac. This gentleman was the firſt who made uſe of a wire interwoven in the hempen cord of an electrical kite, which he made ſeven feet and a half high, and three feet wide, ſo as to have eighteen ſquare feet of ſurface. This cord was found to conduct the electricity of the clouds more powerfully than an hempen cord would do, even though it was wetted; and, being terminated by a cord of dry ſilk, it enabled the obſerver (by a proper management of his apparatus) to make whatever experiments he thought proper, without danger to himſelf.

BY the help of this kite, on the 7th. of June 1753, about one in the afternoon, when it was raiſed 550 feet from the ground, and had taken 780 feet of ſtring, making an angle of near forty five degrees with the horizon; he drew ſparks from his conductor three inches long and a quarter of an inch thick, the ſnapping of which was heard about 200 paces. Whilſt he was taking theſe ſparks, he felt, as it were, a cobweb on his face, though he was above three feet from the ſtring of the kite; after which he did not think it ſafe to ſtand ſo near, and called aloud to all the company to retire, as he did himſelf about two feet.

THINKING himſelf now ſecure enough, and not being incommoded by any body very near him, he took notice of what paſſed among the clouds which were immediately over the kite; but could perceive no lightning, either there, or any

Z z   where

where elfe, nor fcarce the leaft noife of thunder, and there was no rain at all. The wind was Weft, and pretty ftrong, which raifed the kite 100 feet higher, at leaft, than in the other experiments.

AFTERWARDS, cafting his eyes on the tin tube, which was faftened to the ftring of the kite, and about three feet from the ground, he faw three ftraws, one of which was a-bout one foot long, a fecond four or five inches, and the third three or four inches, all ftanding erect, and performing a circular dance, like puppets, under the tin tube, without touching one another.

THIS little fpectacle, which much delighted feveral of the company, lafted about a quarter of an hour; after which, fome drops of rain falling, he again perceived the fenfation of the cobweb on his face, and at the fame time heard a con-tinual ruftling noife, like that of a fmall forge bellows. This was a further warning of the increafe of electricity; and from the firft inftant that Mr. De Romas perceived the dancing ftraws, he thought it not advifable to take any more fparks even with all his precautions; and he again intreated the company to fpread themfelves to a ftill greater diftance.

IMMEDIATELY after this came on the laft act of the enter-tainment, which Mr. De Romas acknowledged made him tremble. The longeft ftraw was attracted by the tin tube, upon which followed three explofions, the noife of which greatly refembled that of thunder. Some of the company compared it to the explofion of rockets, and others to the violent crafhing of large earthen jars againft a pavement. It

is

is certain that it was heard into the heart of the city, not-withstanding the various noises there.

The fire that was seen at the instant of the explosion had the shape of a spindle eight inches long and five lines in dia-meter.   But the most astonishing and diverting circumstance was produced by the straw, which had occasioned the explo-sion, following the string of the kite.   Some of the company saw it at forty five or fifty fathoms distance, attracted and re-pelled alternately, with this remarkable circumstance, that every time it was attracted by the string, flashes of fire were seen, and cracks were heard, though not so loud as at the time of the former explosion.

It is remarkable, that, from the time of the explosion to the end of the experiments, no lightning at all was seen, nor scarce any thunder heard.   A smell of sulphur was per-ceived, much like that of the luminous electric effluvia issu-ing out of the end of an electrified bar of metal.   Round the string appeared a luminous cylinder of light, three or four inches in diameter; and this being in the day time Mr. De Romas did not question but that, if it had been in the night, that electric atmosphere would have appeared to be four or five feet in diameter.   Lastly, after the experiments were over, a hole was discovered in the ground, perpendicularly under the tin tube, an inch deep, and half an inch wide, which was probably made by the large flashes that accom-panied the explosions.

An end was put to these remarkable experiments by the falling of the kite, the wind being shifted into the East, and rain mixed with hail coming on in great plenty. Whilst the

kite

kite was falling, the ſtring came foul of a penthouſe; and it was no ſooner diſengaged, than the perſon who held it felt ſuch a ſtroke in his hands, and ſuch a commotion through his whole body, as obliged him inſtantly to let it go; and the ſtring, falling on the feet of ſome other perſons, gave them a ſhock alſo, though much more tolerable. *

THE quantity of electric matter brought by this kite from the clouds at another time is really aſtoniſhing. On the 26th. of Auguſt 1756, the ſtreams of fire iſſuing from it were ob-ſerved to be an inch thick, and ten feet long. This amazing flaſh of lightning, the effect of which on buildings or animal bodies, would perhaps have been equally deſtructive with any that are mentioned in hiſtory, was ſafely conducted by the cord of the kite to a non-electric body placed near it, and the report was equal to that of a piſtol.

MR. ROMAS had the curioſity to place a pigeon in a cage of glaſs, in a little edifice, which he had purpoſely placed, ſo as that it ſhould be demoliſhed by the lightning brought down by his kite. The edifice was, accordingly, ſhattered to pieces, but the cage and the pigeon were not ſtruck. †

THE Abbé Nollet, who gives this account, adds, that if a ſtroke of this kind had gone through the body of Mr. De Romas, the unfortunate profeſſor Richman had not probably been the only martyr to electricity, and adviſes, that great caution be uſed in conducting ſuch dangerous experiments. ‡

WHEN we conſider how many ſevere ſhocks the moſt cau-tious and judicious electricians often receive through inadver-

* Gent. Magaz. for Auguſt 1756, p. 378.   † Nollet's Letters, Vol. 2. p. 259.
‡ Phil. Tranſ. Vol. 52. pt. i. p. 342.

tence,

tence, we fhall not be furprifed, that when philofophers firft began to collect and make experiments upon real lightning, it fhould fometimes have proved a little untractable in their hands, and that they were obliged to give one another frequent cautions how to proceed with it.

The Abbé Nollet, as early as the year 1752, advifes that thefe experiments be made with circumfpection ; as he had been informed, by letters from Florence and Bologna, that thofe who had made them there had had their curiofity more than fatisfied, by the violent fhocks they had fuftained in drawing fparks from an iron bar electrifed by thunder. One of his correfpondents informed him, that once, as he was endeavouring to fasten a fmall chain, with a copper ball at one of its extremities, to a great chain, which communicated with the bar at the top of the building (in order to draw off the electric fparks by means of the ofcillations of this ball) there came a flafh of lightning, which he did not fee, but which affected the chain with a noife like wild fire. At that inftant, the electricity communicated itfelf to the chain of the copper ball, and gave the obferver fo violent a commotion, that the ball fell out of his hands, and he was ftruck backwards four or five paces. He had never been fo much fhocked by the experiment of Leyden. *

Mr. Romas received a fevere ftroke when he firft raifed his kite : and Mr. Dalibard fays, that Mr. Monnier, a phyfician of St. Germain en Laye, member of the academy of fciences at Paris ; and Mr. Bertier of the oratory at Montmorency,

* Phil. Tranf. Vol. 48. pt. 1. p. 205.

a correspondent of the academy, were both struck down by strokes of lightning, as they were taking sparks from their apparatus. *

But the greatest sufferer by experiments with lightning, since mankind have introduced so dangerous a subject of their inquiries, was professor Richman of Petersburgh before mentioned. He was struck dead, on the 6th. of August 1753, by a flash of lightning drawn by his apparatus into his own room, as he was attending to an experiment he was making with it. There were two accounts of this fatal accident communicated to the Royal Society, one by Dr. Watson who had it from the best authority; † and the other translated from the High Dutch. ‡ From both these the following is extracted.

The professor had provided himself with an instrument which he called an electrical gnomon, the use of which was to measure the strength of electricity. It consisted of a rod of metal terminating in a small glass vessel, into which (for what reason I do not know) he put some brass filings. At the top of this rod, a thread was fastened, which hung down by the side of the rod when it was not electrified; but when it was, it avoided the rod, and stood at a distance from it, making an angle at the place where it was fastened. To measure this angle, he had the arch of a quadrant fastened to the bottom of the iron rod.

He was observing the effect of the electricity of the clouds, at the approach of a thunder storm, upon this gnomon; and,

* Dalibard's Franklin, Vol. 2. p. 129.   † Phil. Transf. Vol. 48. pt. 2. p. 765.
‡ Ib. Vol. 49. pt. 1. p. 61.

of

of courfe, ftanding with his head inclined towards it, accompanied by Mr. Solokow (an engraver, whom he frequently took with him, to be a joint obferver of his electrical experiments, in order to reprefent them the better in cuts) when this gentleman, who was ftanding clofe to his elbow, obferved a globe of blue fire, as he called it, as big as his fift, jump from the rod of the gnomon towards the head of the profeffor, which was, at that inftant, at about a foot diftance from the rod. This flafh killed Mr. Richman: but Mr. Solokow could give no account of the particular manner in which he was immediately affected by it : for, at the fame time that the profeffor was ftruck, there arofe a fort of fteam, or vapour, which intirely benumbed him, and made him fink down upon the ground; fo that he could not remember even to have heard the clap of thunder, which was very loud.

The globe of fire was attended with a report as loud as that of a piftol: a wire, which brought the electricity to his metal rod, was broken to pieces, and its fragments thrown upon Mr. Solokow's cloaths. Half of the glafs veffel in which the rod of the gnomon ftood was broken off, and the filings of metal that were in it were thrown about the room.

Upon examining the effects of the lightning in the profeffor's chamber, they found the door cafe half fplit through, the door torn off, and thrown into the room. * They opened a vein of the breathlefs body twice, but no blood followed, and endeavoured to recover fenfation by violent chafing, but in vain. Upon turning the corps with the face downwards, during the

* Ib. Vol. 48. pt. 2. p. 763.

rubbing,

rubbing, an inconfiderable quantity of blood ran out of the mouth. There appeared a red fpot on the forehead, from which fpirted fome drops of blood through the pores, without wounding the fkin. The fhoe belonging to the left foot was burft open, and, uncovering the foot at that place, they found a blue mark; from which it was concluded, that the electrical force of the thunder, having entered the head, made its way out again at that foot.

Upon the body, particularly on the left fide, were feveral red and blue fpots, refembling leather fhrunk by being burnt. Many more blue fpots were afterwards vifible over the whole body, and in particular on the back. That upon the forehead changed to a brownifh red, but the hair of the head was not finged, notwithftanding the fpot touched fome of it. In the place where the fhoe was unripped, the ftocking was intire; as was the coat every where, the waiftcoat only being finged on the foreflap, where it joined the hinder; but there appeared on the back of Mr. Solokow's coat long narrow ftreaks, as if red hot wires had burned off the nap, and which could not be well accounted for.

When the body was opened the next day, twenty four hours after he was ftruck, the cranium was very intire, having no fiffure, nor crofs opening; the brain as found as it poffibly could be, but the tranfparent pellicles of the windpipe were exceffively tender, gave way, and eafily rent. There was fome extravafated blood in it, as likewife in the cavities below the lungs; thofe of the breaft being quite found, but thofe towards the back of a brownifh black colour, and filled with more of the above-mentioned blood: otherwife, none of the

entrails

entrails were touched; but the throat, the glands, and the thin inteſtines were all inflamed. The ſinged leather-colour-ed ſpots penetrated the ſkin only. Twice twenty four hours being elapſed, the body was ſo far corrupted that it was with difficulty they got it into a coffin. *

* Phil. Tranf. Vol. 49. pt. 1. p. 67.

A a a   SECTION

SECTION XI.

OBSERVATIONS ON THE GENERAL STATE OF ELECTRICITY
IN THE ATMOSPHERE, AND ITS MORE USUAL EFFECTS.

ELECTRICIANS, after obferving the great quantity of
electric matter with which the clouds are charged
during a thunder ftorm, began to attend to the leffer quantities of it which might be contained in the common ftate of
the atmofphere, and the more ufual effects of this great and
general agent in nature. Mr. Monnier, whofe obfervations
on the electricity of the air during a thunder ftorm have been
already mentioned, was the firft who found that there was
very often, and perhaps always, a quantity of electric matter
in the atmofphere, when there was no appearance of thunder.
This he confirmed by decifive experiments, made at St. Germain en Laye, and publifhed in a memoir read at the Royal
Academy of fciences at Paris November the 15th. 1752.*
BUT more accurate experiments upon the electricity of the

* Phil. Tranf. Vol. 48. pt. 1. p. 203.

air

air were made by the Abbé Mazeas, at Chateau de Maintenon, during the months of June, July, and October 1753, and communicated to the Royal Society, in a letter to Dr. Stephen Hales.

The Abbé's apparatus confifted of an iron rod 370 feet long, raifed ninety feet above the horizon. It came down from a very high room in the caftle, where it was faftened to a filken cord fix feet long; and was carried from thence to the fteeple of the town, where it was likewife faftened to another filken cord of eight feet long, and fheltered from rain. And a large key was fufpended, by the end of this wire, in order to receive the electric fluid.

On the 17th. of June, when he began his experiments, the electricity of the air was fenfibly felt every day, from fun rife till feven or eight in the evening, except in moift weather, when he could perceive no figns of electricity. In dry weather, the wire attracted minute bodies at no greater diftance than three or four lines. He repeated the experiment carefully every day, and conftantly obferved, that, in weather void of ftorms, the electricity of a piece of fealing wax of two inches long was above twice as ftrong as that of the air. This obfervation inclined him to conclude, that in weather of equal drinefs the electricity of the air was always equal.

It did not appear to him that hurricanes and tempefts increafed the electricity of the air, when they were not accompanied with thunder; for that, during three days of a very violent continual wind, in the month of July, he was obliged to put fome duft within four or five lines of the conductor, before any fenfible attraction could be perceived.

　　　　　　　　　　　　　　The

THE direction of the winds, whether Eaſt, Weſt, North, or South, made no ſenſible alteration in the electricity of the air, except when they were moiſt.

IN the drieſt nights of that ſummer, he could diſcover no ſigns of electricity in the air; but it returned in the morning, when the ſun began to appear above the horizon, and vaniſh-ed again in the evening, about half an hour after ſun ſet.

THE ſtrongeſt common electricity of the atmoſphere, du-ring that ſummer, was perceived in the month of July, on a very dry day, the heavens being very clear, and the ſun ex-tremely hot. The diſtance of ten or twelve lines was then ſufficient for the approach of the duſt to the conductor, in order to ſee the particles riſe in a vertical direction, like the filings of iron on the approach of a magnet.

ON the 27th. of June, at two in the afternoon, he per-ceived ſome ſtormy clouds riſing above the horizon, and im-mediately went up to his apparatus; and, having applied the duſt to the key, it was attracted with a force which increaſed in proportion as the clouds reached the zenith. When they had come nearly over the wire, the duſt was ſo impetuouſly repelled, as to be entirely ſcattered from the paper. He drew conſiderable ſparks from it, though there was neither thunder nor lightning. When the ſtormy clouds were in the zenith of his wire, he obſerved that the electricity was increaſed to ſuch a degree, that even the ſilken thread at-tracted light bodies at the diſtance of ſeven or eight inches.

THESE ſtormy clouds remained about two hours above the horizon, without either thunder or lightning; nor did a very

heavy

heavy rain diminish the electricity, except about the end, when the clouds began to be dissipated. *

Mr. Kinnersley observed, that when the air was in its driest state, there was always a considerable share of electricity in it, and which might be easily drawn from it. Let a person, he says, in a negative state, standing out of doors, in the dark, when the air is dry, hold, with his arm extended, a long sharp needle, pointing upwards, and he will soon be convinced that electricity may be drawn out of the air; not indeed very plentifully, for, being a bad conductor, it seems loth to part with it, yet some will evidently be collected. The air near the person's body, having less than the natural quantity, will have none to spare; but his arm being extended, as above, some will be collected from the remoter air, and will appear luminous as it converges to the point of the needle.

Let a person electrised negatively, he says, present the point of a needle horizontally, to a cork ball suspended by silk, and the ball will be attracted towards the point, till it has parted with so much of its natural quantity of electricity, as to be in a negative state, in the same degree with the person who holds the needle; then it will recede from the point, being, as he supposes, attracted the contrary way by the electricity of greater density in the air behind it. But as this opinion, he pleasantly says, seems to deviate from *electrical orthodoxy*, he would be glad to see these phenomena better accounted for by the superior and more penetrating genius of his friend Dr. Franklin, to whom he is writing.

* Phil. Transf. Vol. 48. pt. 1. p. 377, &c.

WHETHER

WHETHER the electricity in the air, in clear dry weather, be of the same denfity at the heighth of 200 or 300 yards, as near the furface of the earth, he thought might be fatisfactorily determined by Dr. Franklin's old experiment of the kite.

THE twine, he fays, fhould have throughout a very fmall wire in it, and the ends of the wire, where the feveral lengths are united, ought to be tied down with a waxed thread, to prevent their acting in the manner of points.

WHEN he wrote this letter, he had tried the experiment twice, when the air was as dry as it ever is in that country, and fo clear that not a cloud had been feen, and found the twine each time in a fmall degree electrifed pofitively. *

THE preceding obfervations of Mr. Monnier, Mr. Mazeas, and Mr. Kinnerfley, fall far fhort of the extent and accuracy of thofe of Signior Beccaria; whofe obfervations on the general ftate of electricity in the atmofphere I have referved for the laft place of the fection, becaufe they are the moft confiderable, though they were all made independent of, and, many of them, prior to thofe mentioned before.

HE obferved that, during very high winds, his apparatus gave no figns of being electrified. †   Indeed he found that in three different ftates of the atmofphere, he could find no electricity in the air.   1. In windy and clear weather.   2. When the fky was covered with diftinct and black clouds, that had a flow motion.   3. In moft moift weather, not ac-

* Phil. Tranf. Vol. 53. pt. 1. p. 87.   † Lettere dell' elettricifmo, p. 106.

tually

tually raining. * In a clear sky, when the weather was calm, he always perceived signs of a moderate electricity, but interrupted. In rainy weather, without lightning, his apparatus was always electrified a little time before the rain fell, and during the time of the rain, but it ceased to be affected a little before the rain was over.

THE higher his rods reached, or his kites flew, the stronger signs they gave of their being electrified. † Also longer strings or cords, extended and insulated in the open air, acquired electricity sooner than those which were shorter. A cord 1500 Paris Feet long, stretched over the river Po, was as strongly electrified during a shower, without thunder, as a metalic rod, to bring lightning into his house, had been in any thunder storm. ‡

HAVING two rods for bringing the lightning into his house, 140 feet asunder, he observed, that if he took a spark from the higher of them, the spark from the other, which was thirty feet lower, was at that instant lessened; but, what is remarkable, is that its power revived again, though he kept his hand upon the former. ||

HE imagined that the electricity communicated to the air might sometimes furnish small sparks to his apparatus; since the air parts with the electricity it has received very slowly, and therefore the equilibrium of the electric matter in the air will not be restored so soon as in the earth and clouds. §

AMONG the effects of a moderate electricity in the atmosphere, Signior Beccaria reckons *rain, hail* and *snow*.

* Lettere dell' elletricismo, p. 166.   † Ib. p. 114.   ‡ Ib. p. 165.
|| Ib. p. 176   §. Ib. p. 347.

CLOUDS

CLOUDS that bring rain, he thought, were produced in the same manner as thunder clouds, only by a more moderate electricity. He deſcribes them at large, and the reſemblance which all their phenomena bear to thoſe of thunder clouds is indeed very ſtriking. *

HE notes ſeveral circumſtances attending rain without lightning, which make it very probable, that it is produced by the ſame cauſe as when it is accompanied with lightning. Light has been ſeen among the clouds by night in rainy weather; and even by day rainy clouds are ſometimes ſeen to have a brightneſs evidently independent of the ſun. † The uniformity with which the clouds are ſpread, and with which the rain falls, he thought were evidences of an uniform cauſe, like that of electricity. ‡ The intenſity of electricity in his apparatus generally correſponded very nearly, to the quantity of rain that fell in the ſame time. ‖ Nor is any thing to be inferred to the contrary of this ſuppoſition from the apparatus not being always electrified during rain. It has ſometimes failed during thunder. Indeed it follows from his general theory, that the electricity of his apparatus could not always correſpond to the electricity of the clouds; ſince it muſt in ſome meaſure depend upon the ſituation of the obſervatory, with reſpect to thoſe parts of the earth or clouds which are giving or taking electric fire. This was confirmed by an obſervation which he made upon one thunder cloud, which paſſed over his obſervatory. At its approach his apparatus was electrified poſitively, when it was directly over him all ſigns of electri-

* Lettere dell' ellettriciſmo, p. 284.   † Ib. p. 288.   ‡ Ib. p. 299.
‖ Ib. p. 307.

city

city ceafed, and when it was paffed, his apparatus was electrified negatively. *   This obfervation very much favours his
general theory of thunder clouds.

SOMETIMES all the phenomena of thunder, lightning, hail,
rain, fnow, and wind, have been obferved at one time ; which
fhows the connection they all have with fome common caufe. †

SIGNIOR BECCARIA, therefore, fuppofes that, previous to
rain, a quantity of electric matter efcapes out of the earth, in
fome place where there was a redundancy of it ; and, in its
afcent to the higher regions of the air, collects and conducts
into its path a great quantity of vapours. The fame caufe
that collects, will condenfe them more and more : till, in the
places of the neareft intervals, they come almoft into contact,
fo as to form fmall drops ; which uniting with others as they
fall, come down, in rain. The rain will be heavier in proportion as the electricity is more vigorous, and the cloud approaches more nearly to a thunder cloud. ‡

HE imitated the appearance of clouds that bring rain by
infulating himfelf between the rubber and conductor of his
electrical machine, and with one hand dropping *colophonia* into a fpoon faftened to the conductor, and holding a burning
coal, while his other hand communicated with the rubber.
In thefe circumftances, the fmoke fpread along his arm, and,
by degrees, all over his body ; till it came to the other hand
that communicated with the rubber. The lower furface of
this fmoke was every where parallel to his cloaths, and the
upper furface was fwelled and arched like clouds replete with

* Ib. p. 310.   † Ib. p. 290, 345.   ‡ Ib. p. 305.

B b b   thunder

thunder and rain. *   In this manner, he supposes, the clouds that bring rain diffuse themselves from over those parts of the earth which abound with electric fire, to those parts that are exhausted of it; and, by letting fall their rain, restore the equilibrium between them.

SIGNIOR BECCARIA thought that the electricity communicated to the air, which both receives and parts with it slowly, would account for the retention of vapours in a clear sky; for small disjoined clouds, not dispersed into rain; for the smaller and lighter clouds in the higher regions of the air, which are but little affected by electricity; and also for the darker, heavy, and sluggish clouds in the lower regions, which retain more of it. †   The degree of electricity which he could communicate to the air of his room, notwithstanding its being in contact with the floor, the walls, &c. made this appear to him both possible, and probable. ‡

HE even imagined, that some alteration in the weight of the air might be made by this electricity of it. ‖   He observed his barometer to fall a little immediately upon a flash of lightning; but he acknowledges that this circumstance is no sufficient foundation to suppose that electricity will account for *much* variation of the height of the barometer. §   But he thought that the phenomena of rain favoured the supposition, that the electric matter in the air did, in some measure, lessen its pressure.   For when the electric matter is actually in the air, collecting and condensing the vapours, the barometer is lowest.   When the communication is made between the earth

* Lettere dell' elettricismo, p. 294. † Ib. p. 348, 349. ‡ Ib. p. 350. ‖ Ib. § Ib. p. 353.

and

and the clouds by the rain, the quickfilver begins to rife; the electric matter, which fupported part of the preffure, being difcharged. And this, he fhows, will be the cafe whether the electricity in the air be pofitive or negative. *

Hail, this ingenious philofopher fuppofes to be formed in the higher regions of the air, where the cold is intenfe, and where the electric matter is very copious. In thefe circumftances, a great number of particles of water are brought near together, where they are frozen, and in their defcent collect other particles: fo that the denfity of the fubftance of the hail ftone grows lefs and lefs from the center; this being formed firft, in the higher regions, and the furface being collected in the lower. Agreeable to this, it is obferved, that, in mountains, hail-ftones, as well as drops of rain, are very fmall; there being but fmall fpace through which to fall, and thereby increafe their bulk. Drops of rain and hail agree alfo in this circumftance, that the more intenfe is the electricity that forms them, the larger they are. † Motion is known to promote freezing, and fo the rapid motion of the electrified clouds may promote that effect in the air. ‡

Clouds of fnow differ in nothing from clouds of rain, but in the circumftance of cold, which freezes them. Both the regular diffufion of fnow, and the regularity in the ftructure of the parts of which it confifts (particularly fome figures of fnow or hail, which he calls *rofette*, and which fall about Turin) fhow the clouds of fnow to be actuated by fome uniform caufe like electricity. ‖ He even endeavours, very par-

* Ib. p. 354.   † Ib. p. 314.   ‡ Ib. p. 318.   ‖ Ib. p. 320, 322, 325.

ticularly,

ticularly, to fhow in what manner certain configurations of snow are made, by the uniform action of electricity. * All thefe conjectures about the caufe of hail and fnow were confirmed by obferving, that his apparatus never failed to be electrified by fnow, as well as by rain.

A MORE intenfe electricity unites the particles of hail more clofely than the more moderate electricity does thofe of fnow. In like manner, we fee thunder clouds more denfe than thofe which merely bring rain, and the drops of rain are larger in proportion, though they often fall not from fo great a height. †

* Lettere dell' ellettricifmo, p. 325, 331, 333.   † Ib. p. 328.

SECTION

SECTION XII.

THE ATTEMPTS THAT HAVE BEEN MADE TO EXPLAIN
SOME OF THE MORE UNUSUAL APPEARANCES IN
THE EARTH AND HEAVENS BY ELECTRICITY.

IN the two preceding sections of this period, relating to the
electricity of the atmosphere, the experiments and observations of Signior Beccaria have made a principal figure;
and the materials I have collected from him make a no less
considerable part of this. They who may have thought he
indulged too much to imagination before, will think him
absolutely extravagant here; but his extravagancies, if they
be such, are those of a great genius; and had he a thousand
more such extravagancies, I should, with pleasure, have followed him though them all.

THE meteor usually called a *falling star* has hitherto
puzzled all philosophers. Signior Beccaria makes it pretty
evident, that it is an electrical appearance; and the fact
which he relates as a proof of it, is exceeding curious and
remarkable.

As

As he was one time fitting with a friend in the open air, an hour after fun fet, they faw what is called a falling ftar directing its courfe towards them; and apparently growing larger and larger, till it difappeared not far from them; when it left their faces, hands, and cloaths, with the earth, and all the neighbouring objects, fuddenly illuminated, with a diffufed and lambent light, attended with no noife at all. While they were ftarting up, ftanding, and looking at one another, furprifed at the appearance, a fervant came running to them out of a neighbouring garden, and afked them if they had feen nothing; for that he had feen a light fhine fuddenly in the garden, and efpecially upon the ftreams which he was throwing to water it. *

ALL thefe appearances were evidently electrical; and Signior Beccaria was confirmed in his conjecture, that electricity was the caufe of them, by the quantity of electric matter which, as was mentioned before, he had feen gradually advancing towards his kite; for that, he fays, had very much the appearance of a falling ftar. Sometimes alfo he faw a kind of *glory* round the kite, which followed it when it changed its place, but left fome light, for a fmall fpace of time, in the place which it had quitted. †

THAT appearances, which bear evident marks of electricity, have a very fenfible progreffive motion, is demonftrated from a variety of meteorological obfervations. I fhall relate one made by Mr. Chalmers, when he was on board the Montague under the command of Admiral Chambers. The account

* Lettere dell' elettricifmo, p. 111.   † Ib. p. 130.

of

of it was read at the Royal Society March the 22d. 1749.

On the 4th. of November 1749, in lat. 42°. 48, long. 9°. 3, he was taking an obfervation on the quarter deck, about ten minutes before twelve, when one of the quartermafters defired he would look to the windward; upon which he obferved a large ball of blue fire rolling on the furface of the water, at about three miles diftance from them. They immediately lowered their top fails. &c. but it came down upon them fo faft, that before they could raife the main tack, they obferved the ball to rife almoft perpendicular, and not above forty or fifty yards from the main chains; when it went off with an explofion as if hundreds of cannon had been fired at one time, and left fo great a fmell of brimftone, that the fhip feemed to be nothing but fulphur. After the noife was over, which, he believed, did not laft longer than half a fecond, they found their main top-maft fhattered into above a hundred pieces, and the main maft rent quite down to the heel. There were fome of the fpikes which nail the fifh of the main maft drawn with fuch force out of the maft, and they ftuck fo faft in the main deck, that the carpenter was obliged to take an iron crow to get them out. There were five men knocked down, and one of them greatly burnt by the explofion. They believed, that when the ball, which appeared to them to be of the bignefs of a large mill-ftone, rofe, it took the middle of the main top maft, as the head of the maft above the hounds was not fplintered. They had a hard gale of wind from the N. by W. to the N. N. E. for two days before the accident, with a great deal of rain and hail, and a large fea. From the northward they had no
thunder

thunder or lightning, neither before nor after the explosion. The ball came down from the North East, and went to the South West.

THAT the *Aurora Borealis* is an electrical phenomenon was, I believe, never disputed, from the time that lightning was proved to be one. To the circumstances of resemblance which had before been taken notice of between this phenomenon and electricity; Signior Beccaria adds, that when the aurora borealis has extended lower than usual into the atmosphere, various sounds, as of rumbling, and hissing, have been heard. *

MR. BERGMAN says, he has often observed the magnetic needle to be disturbed by a high aurora borealis, but that he could never procure any electricity from them, either with pointed metallic rods, or by means of a kite. †

MR. CANTON (besides his conjecture, mentioned before, p. 334, that the aurora borealis may be the flashing of electric fire from positive towards negative clouds at a great distance, through the upper part of the atmosphere, where the resistance is least) supposes that the aurora borealis, which happens at the time that the needle is disturbed by the heat of the earth, is the electricity of the heated air above it; and this, he says, will appear chiefly in the northern regions, as the alteration in the heat of the air in those parts will be the greatest. This hypothesis, he adds, will not seem improbable, if it be considered, that electricity is now known to be the cause of thunder and lightning, that it has been extracted from the air

* Elettricismo artificiale and naturale, p. 221.   † Phil. Transf. Vol. 52. pt. 2. p. 485.

at

at the time of an aurora borealis; that the inhabitants of the northern countries obferve the aurora to be remarkably ftrong, when a fudden thaw happens after fevere cold weather; and that the curious in thefe matters are now acquainted with a fubftance that will, without friction, both emit and abforb the electric fluid, only by the increafe or diminution of its heat; meaning the tourmalin, in which he had difcovered that property. *

IN a paper dated November the 11th. 1754, he fays he has fometimes known the air to be electrical in clear weather, but never at night, except when there has appeared an aurora borealis, and then but to a fmall degree, which he had feveral opportunities of obferving that year. How far pofitive and negative electricity in the air, with a proper quantity of moifture between, to ferve as a conductor, will account for this, and other meteors, fometimes feen in a ferene fky, he leaves to be inquired into. †

SIGNIOR BECCARIA takes fome pains to fhow that *water fpouts* have an electrical origin. To make this more evident, he firft defcribes the circumftances attending their appearance, which are the following.

THEY generally appear in calm weather. The fea feems to boil, and fend up a fmoke under them, rifing in a hill towards the fpout. At the fame time, perfons who have been near them have heard a rumbling noife. The form of a water fpout is that of a fpeaking trumpet, the wider end being in the clouds, and the narrower end towards the fea. The

* Phil. Tranf. Vol. 51. pt. 1. p. 403.   † Ib. Vol. 48. pt. 2. p. 784.

fize

size is various, even in the same spout. The colour is sometimes inclining to white, and sometimes to black. Their
position is sometimes perpendicular to the.sea, sometimes oblique; and sometimes the spout itself is in the form of a
curve. Their continuance is very various, some disappearing
as soon as formed, and some continuing a considerable.time.
One that he had heard of continued a whole hour. But they
often vanish, and presently appear again in the same place. *

THE very same things that water spouts are at sea are some
kinds of *whirlwinds* and *hurricanes* by land. They have been
known to tear up trees, to throw down buildings, make caverns in the earth; and, in all these cases, to scatter earth,
bricks, stones, timber, &c. to a great distance in every direction. † Great quantities of water have been left, or raised
by them, so as to make a kind of deluge; and they have always been attended with a prodigious rumbling noise.

THAT these phenomena depend upon electricity cannot but
appear very probable from the nature of several of them; but
the conjecture is made more probable from the following additional circumstances. They generally appear in months
peculiarly subject to thunder storms, and are commonly preceded, accompanied, or followed by lightning, rain, or hail;
the previous state of the air being similar. Whitish or yellowish flashes of light have sometimes been seen moving with
prodigious swiftness about them. And, lastly, the manner
in which they terminate exactly resembles what might be expected from the prolongation of one of the uniform protube-

---

* Elettricismo artificiale e naturale, p. 206, &c.   † Ib. p. 210.

rances

rances of electrified clouds, mentioned before, towards the sea; the water and the cloud mutually attracting one another: for they suddenly contract themselves, and disperse almost at once; the cloud rising, and the water of the sea under it falling to its level. But the most remarkable circumstance, and the most favourable to the supposition of their depending upon electricity is, that they have been dispersed by presenting to them sharp pointed knives or swords. This, at least, is the constant practice of mariners, in many parts of the world where these water spouts abound; and he was assured by several of them, that the method has often been undoubtedly effectual. *

THE analogy between the phenomena of water spouts and electricity, he says, may be made visible, by hanging a drop of water to a wire communicating with the prime conductor, and placing a vessel of water under it. In these circumstances, the drop assumes all the various appearances of a water spout, both in its rise, form, and manner of disappearing. Nothing is wanting but the smoke, which may require a great force of electricity to become visible.

MR. WILKE also considers the water spout as a kind of great electrical cone, raised between the cloud strongly electrified, and the sea or the earth. *

To Signior Beccaria's theory of water spouts and hurricanes, I shall add a description of a hurricane in the West Indies, from the *Account of the European settlements in America*. It was evidently written without the most distant view to any

---

* Elettricismo artificiale e naturale, p. 213.   † Wilke, p. 142.

C c c 2

philo-

philosophical theory, and least of all that of electricity; and yet those who are disposed to favour this hypothesis may perceive several circumstances, which tend to strengthen it. I need not point them out.

" IT is in the rainy season, principally, in the month of " August, more rarely in July and September, that they are " assaulted by *hurricanes*, the most terrible calamity to " which they are subject from the climate. This destroys, " at one stroke, the labours of many years, and frustrates " the most exalted hopes of the planter; and often just at " the moment when he thinks himself out of the reach of " fortune. It is a sudden and violent storm of wind, rain, " thunder, and lightning; attended with a furious swelling " of the sea, and sometimes with an earthquake; in short, " with every circumstance which the elements can assemble " that is terrible and destructive.

" FIRST they see, as a prelude to the ensuing havock, " whole fields of sugar canes whirled into the air, and scat- " tered over the face of the country. The strongest trees of " the forest are torn up by the roots, and driven about like " stubble. Their wind-mills are swept away in a moment. " Their works, the fixtures, the ponderous copper boilers, " and stills, of several hundred weight, are wrenched from " the ground, and battered to pieces. Their houses are no " protection. The roofs are torn off at one blast, whilst the " rain, which in an hour rises five feet, rushes in upon them " with an irresistible violence.

" THERE are signs, which the Indians of these islands " taught our planters, by which they can prognosticate the
approach

" approach of a hurricane. It comes on either in the
" quarters, or at the full or change of the moon. If it
" comes at the change, obferve thefe figns. That day
" you will fee the fky very turbulent. You will ob-
" ferve the fun more red than at other times. You
" will perceive a dead calm, and the hills clear of all
" thofe clouds and mifts which ufually hover about them.
" In the clefts of the earth, and in the wells, you will hear
" a hollow rumbling found, like the rufhing of a great
" wind. At night the ftars feem much larger than ufual,
" and furrounded with a fort of burs. The North Weft fky
" has a black and menacing look, and the fea emits a ftrong
" fmell, and rifes into vaft waves, often without any wind.
" The wind itfelf now forfakes its ufual fteady Eafterly
" ftream, and fhifts about to the Weft; from whence it
" fometimes blows, with intermiffions, violently and irre-
" gularly, for about two hours at a time. You have the
" fame figns at the full of the moon. The moon itfelf is
" furrounded with a great bur, and fometimes the fun has
" the fame appearance." *

THE firft perfon who advanced that earthquakes were
probably caufed by electricity, was DR. STUKELEY, upon
occafion of the earthquakes at London, on February the 8th.
and on March the 8th. 1749; and another which affected
various other parts of England, the center being about Daventry in Northamptonfhire, on the 30th. of September 1750.

---

* Account of the European Settlements in America, Vol. 2. p. 96, &c.

The

The papers which the Doctor delivered to the Royal Society
on these occasions, and which were read March the 22d. 1749,
and December the 6th. 1750, are very valuable, and well
deserve the attention of all philosophers and electricians. I
shall here give the substance of both; only abridging, and
differently arranging the materials of them.

THAT earthquakes are not owing to subterraneous winds,
fires, vapours, or any thing that occasions an explosion, and
heaves up the ground, he thought might easily be concluded
from a variety of circumstances. In the first place, he
thought there was no evidence of any remarkable cavernous
structure of the earth; but that, on the contrary, there is
rather reason to presume, that it is, in a great measure, solid;
so as to leave little room for internal changes, and fermentations within its substance; nor do coal-pits, he says, when on
fire, ever produce any thing resembling an earthquake.

IN the second earthquake at London, there was no such
thing as fire, vapour, smoke, smell, or an eruption of any
kind observed, though the shock affected a circuit of thirty
miles in diameter. This consideration alone, of the extent
of surface shaken by an earthquake, he thought was sufficient
to overthrow the supposition of its being owing to to the expansion of any subterraneous vapours. For it could not possibly be imagined, that so immense a force, as could act upon
that compass of ground instantaneously, should never break
the surface of it, so as to be discoverable to the sight or smell;
when small fire balls, bursting in the air, have instantly propagated a sulphureous smell all around them, to the distance
of several miles.

BESIDES

BESIDES, the operation of this great fermentation, and production of elaſtic vapours, &c. ought to be many days in continuance, and not inſtantaneous; and the evaporation of ſuch a quantity of inflammable matter would require a long ſpace of time.

HE thought that if vapours and ſubterraneous fermentations, explosions, and eruptions were the cauſe of earthquakes, they would abſolutely ruin the whole ſyſtem of ſprings and fountains, wherever they had once been: which is quite contrary to fact, even where they have been frequently repeated. Mentioning the great earthquake which happened A. D. 17, when no leſs than thirteen great cities of Aſia Minor were deſtroyed in one night, and which may be reckoned to have ſhaken a maſs of earth 300 miles in diameter, he aſks; How can we poſſibly conceive the action of any ſubterraneous vapours to produce ſuch an effect ſo inſtantaneouſly? How came it to paſs, that the whole country of Aſia Minor was not at the ſame time deſtroyed, its mountains reverſed, its fountains and ſprings broken up, and ruined for ever, and the courſe of its rivers quite changed? Whereas, nothing ſuffered but the cities. There was no kind of alteration in the ſurface of the country, which, indeed, remains the ſame to this day.

To make the hypotheſis of ſubterraneous vapours, &c. being the cauſe of earthquakes the more improbable, he obſerves, that any ſubterraneous power, ſufficient to move a ſurface of earth thirty miles in diameter, as in the earthquakes which happened at London, muſt be lodged at leaſt fifteen or twenty miles below the ſurface of the earth, and therefore muſt move an inverted cone of ſolid earth, whoſe baſis is

thirty

thirty miles in diameter, and axis fifteen or twenty miles; an effect which, he fays, no natural power could produce.

Upon the fame principle, the fubterraneous caufe of the earthquake in Afia Minor muft have moved a cone of earth of 300 miles in bafe and 200 in the axis; which, he fays, all the gunpowder which has ever been made fince the invention of it would not have been able to ftir, much lefs any vapours, which could be fuppofed to be generated fo far below the furface.

It is not upon the principles of any fubterraneous explo-fion that we can, in the leaft, account for the manner in which fhips, far from any land, are affected during an earth-quake; which feem as if they ftruck upon a rock, or as if fomething thumped againft their bottoms. Even the fifhes are affected by an earthquake. This ftroke, therefore, muft be occafioned by fomething that could communicate motion with unfpeakably greater velocity than any heaving of the earth under the fea, by the elafticity of generated vapours. This could only produce a gradual fwell, and could never give fuch an impulfe to the water, as would make it feel like a ftone.

Comparing all thefe circumftances, he fays, he had al-ways thought, that an earthquake was an electrical fhock, of the fame nature with thofe which are now become familiar in electrical experiments. And this hypothefis he thought was confirmed by the phenomena preceding and attending earthquakes, particularly thofe which occafioned this publi-cation.

THE

THE weather, for five or fix months before the firft of thefe earthquakes, had been dry and warm to an extraordinary degree, the wind generally South and South Weft, and that without rain ; fo that the earth muft have been in a ftate of electricity ready for that particular vibration in which electrification confifts. On this account, he obferves, that the Northern regions of the world are but little fubject to earthquakes in comparifon with the Southern, where the warmth and drynefs of the air, fo neceffary to electricity, are common. All the flat country of Lincolnfhire before the earthquake in September, though, underneath, it is a watery bog, yet, through the whole preceding fummer and autumn, (as they can have no natural fprings in fuch a level) the drought had been fo great on the furface of the earth, that the inhabitants were obliged to drive their cattle feveral miles to water. This, he fays, fhows how fit the dry furface was for an electrical vibration ; and alfo, which is of great importance, that earthquakes reach but very little below the furface of the earth.

BEFORE the earthquake at London, all vegetables had been uncommonly forward. At the end of February, in that year, all forts of garden ftuff, fruits, flowers, and trees were obferved to be as forward as, in other years, about the middle of April ; and electricity is well known to quicken vegetation.

THE Aurora Borealis had been very frequent about the fame time, and had been twice repeated juft before the earthquake, of fuch colours as had never been feen before. It had alfo removed to the South, contrary to what is common in

D d d　　　　　　　England ;

England; so that some Italians, and people from other places where earthquakes are frequent, observing these lights, and the peculiar temperature of the air, did actually foretell the earthquake. For a fortnight before the earthquake in September, the weather was serene, mild, and calm; and, one evening, there was a deep red Aurora Borealis, covering the cope of heaven, very terrible to behold.

THE whole year had been exceedingly remarkable for fire balls, thunder, lightning, and coruscations, almost throughout all England. Fire balls were more than once seen in Ireland, and Lincolnshire, and particularly observed. And all these kinds of meteors, the Doctor says, are rightly judged to proceed from the electrical state of the atmosphere.

IN these previous circumstances of the state of the earth and air, nothing, he says, is wanting to produce the wonderful effect of an earthquake, but the touch of some non-electric body; which must necessarily be had *ab extra*, from the region of the air, or atmosphere. Hence, he infers that, if a non-electric cloud discharge its contents upon any part of the earth, in that highly electrical state, an earthquake must necessarily ensue. As the discharge from an excited tube produces a commotion in the human body, so the discharge of electric matter from the compass of many miles of solid earth must needs be an earthquake, and the snap from the contact be the horrid uncouth noise attending it.

THE Doctor had been informed, by those who were up and abroad the night preceding the earthquake, and early in the morning, that coruscations in the air were extremely

frequent;

frequent; and that, a little before the earthquake, a large and black cloud fuddenly covered the atmofphere, which probably occafioned the fhock, by the difcharge of a fhower Dr. Childrey, he fays, obferves, that earthquakes are always preceded by rain, and fudden tempefts of rain in times of great draught.

A sound was obferved to roll from the river Thames to-wards Temple Bar, before the houfes ceafed to nod, juft as the electrical fnap precedes the fhock. This noife, an ob-ferver faid, was much greater than any he had ever heard. Others, who write upon earthquakes, commonly obferve, that the noife precedes the fhock: whereas it muft have been quite the contrary, if the concuffion had depended upon a fubterraneous eruption. This noife attending earthquakes, the Doctor thought, could not be accounted for, but upon the principles of electricity. The earthquake in September was attended with a rufhing noife, as if houfes were falling, and people, in fome places, were fo univerfally frighted, as to run out of their houfes, imagining that their own, and thofe of their neighbours were tumbling on their heads. In fome villages, the people, being at divine fervice, were much a-larmed with the noife; which they faid, beyond all compari-fon, exceeded all the thunder they had ever heard.

The flames and fulphureous fmells, which are fometimes obferved during earthquakes, the Doctor thought were more eafily accounted for, from the fuppofition of their being elec-trical phenomena, than from their being occafioned by the eruption of any thing from the bowels of the earth.

D d d 2

The

THE impreſſion made by an earthquake upon land and water, to the greateſt diſtances is, as was obſerved before, inſtantaneous; which could only be effected by electricity. In
the earthquake in September, the concuſſion was felt through
a ſpace of 100 miles in length, and forty in breadth; and,
as far as could be judged, at the ſame inſtant of time. That
this tract of ground, which amounted to 4000 ſquare miles
in ſurface, ſhould be thrown into ſuch agitation in a moment,
is ſuch a prodigy, the Doctor ſays, as we could never believe,
or conceive, did we not know it to be fact from our ſenſes.
But if we ſeek the ſolution of it, we cannot think any natural power equal to it, but that of electricity; which acknowledges no ſenſible tranſition of time, no bounds.

THE little damage generally done by earthquakes, the
Doctor thought to be an argument of their being occaſioned
by a ſimple vibration, or tremulous motion of the ſurface of
the earth, by an electrical ſnap. This vibration, he ſays,
impreſſed on the water, meeting with the ſolid bottoms of
ſhips, and lighters, occaſions that thump which is ſaid to be
felt by them: yet, of the millions of ordinary houſes, over
which it paſſed, not one fell. A conſideration which ſufficiently points out what ſort of a motion this was not; alſo
what ſort of a motion it was, and whence derived; not a
convulſion in the bowels of the earth, but an uniform vibration along its ſurface, like that of a muſical ſtring, or what
we put a drinking glaſs into, by rubbing one's finger on the
edge; which yet, being brought to a certain pitch, breaks
the glaſs; undoubtedly, he adds, an electrical repulſion of its
parts.

THAT

THAT earthquakes are electrical phenomena, is further evident from their chiefly affecting the sea coast, places along rivers, and, I may add, eminences. The earthquake in September spread mostly to the North and South; which the Doctor says is the direction of the Spalding river, whereby it was conveyed to the sea shore, where it was particularly sensible; thence up Boston channel, and so up Boston river to Lincoln. The greatest part of this earthquake displayed its effects along, and between the two rivers Welland, and Avon, and that from their sources down to their mouths. It likewise reached the river Witham, which directed the electrical stream that way also to Lincoln; for which reason, meeting the same coming from Boston, it was most sensibly felt there. It reached, likewise, to the Trent at Nottingham, which conveyed it to Newark.

THE first electrical stroke in this earthquake seemed to the Doctor to have been made on the high ground about Daventry in Northamptonshire. From thence it descended chiefly Eastward, and along the river Welland, from Harborough, to Stamford, Spalding, and the sea; and along the river Avon and Nen to Northampton, Peterborough, Wisbich, and the sea. It spread itself all over the vast level of the isle of Ely, promoted by a great number of canals; natural and artificial, made for draining that country. It was still conducted Eastward, by Mildenhall river in Suffolk, to Bury, and the parts adjacent. All these circumstances duly considered were to him a confirmation of the doctrine he advanced on this subject.

LASTLY,

LASTLY, the Doctor adds, as a further argument in favour of his hypothesis, that pains in the back, rheumatic, hysteric, and nervous cases; head aches, cholics, &c. were felt by many people of weak constitutions, for a day or two after the earthquake; just as they would after electrification; and, to some, these disorders proved fatal.

IN what manner the earth and atmosphere are put into that electrical and vibratory state, which prepares them to give or receive that snap and shock, which we call an earthquake, and whence it is that this electric matter comes, the Doctor does not pretend to say; but thinks it as difficult to account for as magnetism, gravitation, muscular motion, and many other secrets in nature. *

To these valuable observations of Dr. Stukeley, I shall add some circumstances which were observed by Dr. Hales, in the earthquake at London, on March the 8th. 1749; as tending to strengthen the hypothesis of its being caused by electricity; though the Doctor, who relates them, thought that the electric appearances were only occasioned by the great agitation which the electric fluid was put into, by the shock of so great a mass of the earth.

AT the time of the earthquake, about twenty minutes before six in the morning, the Doctor, being awake in bed, on a ground floor, at a house near the church of St. Martin's in the fields, very sensibly felt his bed heave, and heard an obscure rushing noise in the house, which ended in a loud explosion up in the air, like that of a small cannon. The whole

* Phil. Transf. abridged, Vol. 10. p. 526, 535, and p. 541, 551.

duration,

duration, from the beginning to the end, feeming to be about four feconds.

THIS great noife, the Doctor conjectured, was owing to the rufhing, or fudden expanfion of the electric fluid at the top of St. Martin's fpire, where all the electric effluvia, which afcended along the large body of the tower, being ftrongly condenfed, and accelerated at the point of the wea‐thercock, as they rufhed off, made fo much the louder ex‐panfive explofion.

THE Doctor further fays, that the foldiers, who were upon duty in St. James's park, and other perfons who were then up, faw a blackifh cloud, and a confiderable lightning, juft before the earthquake began. *

MY reader, who has feen to how great an extent Signior Beccaria has already carried the principles of electricity, will have no doubt but that he fuppofes *earthquakes* to be derived from that caufe. And indeed, without any knowledge of what Dr. Stukeley had done, he did fuppofe them to be elec‐trical phenomena; but, contrary to the Doctor, imagined the electric matter which occafioned them to be lodged deep in the bowels of the earth, agreeable to his hypothefis con‐cerning the origin of lightning.

IT is certain that if Signior Beccaria's account of the origin of thunder clouds be admitted, there will be little difficulty in admitting further, that *earthquakes* are to be reckoned a‐mong the effects of electricity. For if the equilibrium of the electric matter can, by any means, be loft in the bowels

* Phil. Tranf. abridged, Vol. 10. p. 540, 541.

of

of the earth; fo that the beft method of reftoring it fhall be by the fluid burfting its way into the air, and traverfing feveral miles of the atmofphere, to come to the place where it is wanted; it may eafily be imagined, that violent concuffions may be given to the earth, by the fudden paffage of this powerful agent. And feveral circumftances attending earthquakes he thought rendered this hypothefis by no means improbable.

VOLCANOS are known to have a near connection with earthquakes; and flafhes of light, exactly refembling lightning, have frequently been feen to rufh from the top of mount Vefuvius, at the time that afhes and other light matter have been carried out of it into the air, and been difperfed uniformly over a large tract of country. Of thefe he produces a great number of inftances, from the beft authority. *

A RUMBLING noife, like thunder, is generally heard during an earthquake. At fuch times, alfo, flafhes of light have been feen rifing out of the ground, and darting up into the air. Real lightning hath fometimes occafioned fmall fhakings of the earth, at leaft has been attended by them. But the ftrongeft circumftance of refemblance which he obferved is the fame that Dr. Stukeley lays fo much ftrefs on, viz. the amazing fwiftnefs with which the earth is fhaken in earthquakes. An earthquake, fays he, is by no means a gradual heaving, as we might have expected from other caufes, but an inftantaneous concuffion, fo that the fluidity of the water is no fecurity againft the blow. The very fhips, many leagues off the coaft, feel as if they ftruck againft a rock.

* Lettere dell' ellettricifmo, p. 226, 362, &c.

THIS

THIS admirable philosopher, having imitated all the great phenomena of natural electricity in his own room, doth not let the earthquake escape him. He says, that if two pieces of glass, inclosing a thin piece of metal, be held in the hand, while a large shock is sent through them, a strong vibration, or concussion will be felt; which sometimes, as in Dr. Franklin's experiments, breaks them to pieces.

SIGNIOR BECCARIA thinks, that there are traces of electrical operations in the earthquake, that happened at Julian's attempt to rebuild the temple of Jerusalem. *

THAT the electric fluid is sometimes collected in the bowels of the earth, he thought was probable from the appearance of *ignes fatui* in mines, which sometimes happens; and is very probably an electrical phenomenon. †

WHICH of these two philosophers have advanced the more probable opinion concerning the seat of the electric matter which occasions earthquakes, I shall not pretend to decide. I shall only observe that, perhaps, a more probable general hypothesis than either of them may be formed out of them both. Suppose the electric matter to be, some way or other, accumulated on one part of the surface of the earth, and, on account of the dryness of the season, not easily to diffuse itself; it may, as Signior Beccaria supposes, force itself a way into the higher regions of the air, forming clouds in its passage out of the vapours which float in the atmosphere, and occasion a sudden shower, which may further promote the

* Lettere dell' elettricismo, p. 363.   † Dell' Elettricismo artificiale e naturale, p. 223.

passage

paſſage of the fluid.   The whole ſurface, thus unloaded, will receive a concuſſion, like any other conducting ſubſtance on parting with, . or receiving a quantity of the electric fluid. The ruſhing noiſe will, likewiſe, ſweep over the whole extent of the country.   And, upon this ſuppoſition, alſo, the fluid, in its diſcharge from the country, will naturally follow the courſe of the rivers, and alſo take the advantage of any eminencies, to facilitate its aſcent into the higher regions of the air.

I SHALL cloſe this account of the theory of lightning, and other phenomena of the atmoſphere, with an enumeration of the principal appearances of natural electricity obſerved by the antients, and which were never underſtood before the diſcovery of Dr. Franklin.   It will be very eaſy for me to do this, as I find them already collected to my hands by Dr. Watſon. *

A LUMINOUS appearance, which muſt have been of an electrical nature, is mentioned by Plutarch in his life of Lyſander.   He conſidered it as a meteor.

PLINY, in his ſecond book of natural hiſtory, calls thoſe appearances *ſtars*, and tells us, that they ſettled not only upon the maſts, and other parts of ſhips ; but alſo upon men's heads.   *Exſiſtunt*, ſays that hiſtorian, *ſtellæ et in mari terriſque.   Vidi nocturnis militum vigiliis inhærere pilis pro vallo fulgorem effigie ea : et antennis navigantium, aliiſque navium partibus, ceu vocali quodam ſono inſiſtunt, ut volucres, ſedem ex ſede mutantes.———Geminæ autem ſalutares et proſperi curſus præ-*

* Phil. Tranſ. Vol. 48. pt. 1. p. 210.

nuncia ;

nunciæ ; quarum adventu, fugari diram illam ac minacem appel-latamque Helenam ferunt. Et ob id Polluci et Castori id numen assignant, eosque in mari deos invocant. Hominum quoque capiti vespertinis horis, magno præsagio circumfulgent. But, adds he, these things are *incerta ratione et in naturæ majestate abdita.*

" STARS make their appearance both at land and sea. I
" have seen a light in that form on the spears of soldiers,
" keeping watch by night upon the ramparts. They are seen
" also on the sailyards, and other parts of ships, making an au-
" dible sound, and frequently changing their places. Two of
" these lights forebode good weather, and a prosperous voyage ;
" and drive away *one* that appears single, and wears a threat-
" ning aspect. This the sailors call Helen, but the *two* they
" call Castor and Pollux, and invoke them as Gods. These
" lights do sometimes, about the evening, rest on men's
" heads, and are a great and good omen. But these are a-
" mong the awful mysteries of nature. "

SENECA in his Natural Questions, chap. I. takes notice of the same phenomenon. *Gylippo, Syracusas petenti visa est stella supra ipsam lancem constitisse. In Romanorum castris visa sunt ardere pila, ignibus scilicet in illis delapsis.*

" A STAR settled on the lance of Gylippus, as he was
" sailing to Syracuse : and spears have seemed to be on fire in
" the Roman camp. "

IN Cæsar de Bello Africano, cap. 6. edit. Amstel. 1686. We find them attending a violent storm. *Per id tempus fere Cæsaris exercitui res accidit incredibilis auditu ; nempe Virgili-arum signo confecto, circiter vigilia secunda noctis, nimbus cum*

saxea

ſaxea grandine ſubito eſt coortus ingens.   Eadem noĉte legionis V.
pilorum cacumina ſua ſponte arſerunt.

" ABOUT that time, there was a very extraordinary appear-
" ance in the army of Cæſar.   In the month of February,
" about the ſecond watch of the night, there ſuddenly aroſe
" a thick cloud, followed by a ſhower of ſtones ; and the
" ſame night, the points of the ſpears belonging to the fifth
" legion ſeemed to take fire. "

LIVY, cap. 32. mentions two ſimilar facts.   *In Sicilia mi-*
litibus aliquot ſpicula, in Sardinia muro circumeunti vigilias equi-
ti, ſcipionem, quem in manu tenuerat, arſiſſe ; et litora crebris
ignibus fulſiſſe.

" THE ſpears of ſome ſoldiers in Sicily, and a walking
" ſtick, which a horſeman in Sardinia was holding in his
" hand, ſeemed to be on fire.   The ſhores were alſo luminous
" with frequent fires. "

THESE appearances are called, both by the French and Spani-
ards, inhabiting the coaſts of the Mediterranean, St. Helme's,
or St. Telme's fires ; by the Italians the fires of St. Peter, and
St. Nicholas ; and are frequently taken notice of by the
writers of voyages.

IF ſome late accounts from France, adds the Doctor, are
to be depended upon, this phenomenon, has been obſerved at
Plauzet for time immemorial, and Mr. Binon, the curé of
the place, ſays, that for twenty ſeven years, which he has
reſided there, in great ſtorms accompanied with black clouds,
and frequent lightnings, the three pointed extremities of the
croſs of the ſteeple of that place appeared ſurrounded with a
body of flame ; and that when this phenomenon has been ſeen,

the

the ſtorm was no longer to be dreaded, and calm weather returned ſoon after.

MODERN hiſtory furniſhes a great number of examples of
flames appearing at the extremities of pointed metallic bodies,
projecting into the air.   Little notice was taken of theſe,
while the cauſe of them was unknown; but ſince their near
affinity with lightning has been diſcovered, they have been
more attended to, and collected.

SECTION

SECTION XIII.

OBSERVATIONS ON THE USE OF METALLIC CONDUC-
TORS TO SECURE BUILDINGS, &c. FROM THE EFFECTS
OF LIGHTNING.

THE former fections of this period relate chiefly to the
theory of electricity.   In the two next I fhall confider
what has been done towards reducing this fcience into practice.   And, in the firft place, I fhall recite the obfervations
that have been made refpecting the ufe of metallic conductors, to fecure buildings from lightning, as having the neareft connecton with the fubject of the fections immediately
preceding.

DR. FRANKLIN's propofal to preferve buildings from the
dreadful effects of lightning was by no means a matter of
mere theory.   Several ftriking facts, which occurred within
the period of which I am treating, demonftrate its utility.

INNUMERABLE obfervations fhow how readily metallic rods
actually conduct lightning, and how fmall a fubftance of metal is fufficient to difcharge great quantities of it.   Mr. Ca-
lendrini, who afterwards applied to Dr. Watfon, to be informed

formed of the beft methods of fecuring powder magazines, fays that he himfelf was an eye witnefs of the effect of a flafh of lightning, where he obferved it had ftruck the wire of a bell, and had been completely conducted by it, from one room of a houfe to another, through a very fmall hole in the partition. This obfervation was prior to the difcoveries of Dr. Franklin, but was recollected and recorded afterwards.*

Dr. Franklin himfelf, in a letter to Mr. Dalibard, dated Philadelphia, June the 29th. 1755, relating what had been fhown him of the effects of lightning on the church of New-bury in New England, obferves; that a wire, not bigger than a common knitting needle, did in fact conduct a flafh of lightning, without injuring any part of the building as far as it went; though the force of it was fo great that, from the termination of the wire down to the ground, the fteeple was exceedingly rent and damaged, fome of the ftones, even in the foundation, being torn out, and thrown to the diftance of twenty or thirty feet. No part of the wire, however, could be found, except about two inches at each extremity, the reft being exploded, and its particles diffipated in fmoke and air, as the Doctor fays, like gunpowder by a common fire. It had only left a black fmutty track upon the plaifter of the wall along which it ran, three or four inches broad, darkeft in the middle, and fainter towards the edges. From the circumftances of this fact it was very evident, that, had the wire been continued to the foot of the building, the whole fhock would have been conducted without the leaft injury to it, though the wire would have been deftroyed. *

* Phil. Tranf. Vol. 54. pt. 1. p. 203.

But

BUT the moſt complete demonſtration of the real uſe of Dr. Franklin's method of ſecuring buildings from the effects of lightning, is Mr. Kinnerſley's account of what happened to the houſe of Mr. Weſt, a merchant of Philadelphia in Penſilvania, which was guarded by an apparatus conſtructed according to the directions of Dr. Franklin. It conſiſted of an iron rod, which extended about nine feet and an half above a ſtack of chimnies, to which it was fixed. It was more than half an inch in diameter in the thickeſt part, and went tapering to the upper end, in which there was an hole that received a braſs wire about three lines thick and ten inches long, terminating in a very acute point. The lower part of the apparatus joined to an iron ſtake, driven four or five feet into the ground.

MR. WEST judging, by the dreadful flaſh of lightning and inſtant crack of thunder, that the conductor had been ſtruck, got it examined; when it appeared, that the top of the pointed rod was melted, and the ſmall braſs wire reduced to ſeven inches and a half in length, with its top very blunt. The ſlendereſt part of the wire, he ſuſpected, had been diſſipated in ſmoke; but ſome of it, where the wire was a little thicker, being only melted by the lightning, ſunk down (while in a fluid ſtate) and formed a rough irregular cap, lower on one ſide than on the other, round the upper end of what remained, and became intimately united with it. It is remarkable, that, notwithſtanding the iron ſtake, in which the apparatus terminated, was driven three or four feet into the ground, yet

* Phil. Tranſ. Vol. 49. pt. 1. p. 309.

the

the earth did not conduct the lightning fo faft, but that, in thunder ftorms, the lightning would be feen diffufed near the ftake two or three yards over the pavement, though at that time very wet with rain. *

In order to fecure fhips from fuftaining damage by lightning, Dr. Watfon, in a letter to Lord Anfon, read at the Royal Society, December the 16th. 1762, advifes, that a rod of copper, about the thicknefs of a goofe quill, be connected with the fpindles and iron work of the maft; and, being continued down to the deck, be from thence, in any convenient direction, fo difpofed, as always to touch the fea water. †

With refpect to powder magazines, Dr. Watfon advifed Mr. Calandrini above-mentioned, that the apparatus to conduct the lightning from them be detached from the buildings themfelves, and conveyed to the next water.

What lately happened to St. Bride's church in London is a fufficient proof of the utility of metallic conductors for lightning. Dr. Watfon, who publifhed an account of this fact in the Philofophical Tranfactions, obferves, that the lightning firft took a weathercock, which was fixed at the top of the fteeple; and was conducted without injuring the metal, or any thing elfe, as low as where the large iron bar or fpindle which fupported it (and which came down feveral feet into the top of the fteeple) terminated.   There, the metallic communication ceafing, part of the lightning exploded,

* Phil. Tranf. Vol. 53. pt. 1. p. 96.   † Ib. Vol. 52. pt. 2. p. 633.

cracked,

cracked, and fhattered the obelifk, which terminated the
fpire of the fteeple, in its whole diaméter, and threw off at
that place feveral large pieces of Portland ftone, of which the
fteeple was built.   Here it likewife removed a ftone from its
place, but not far enough to be thrown down.   From thence
the lightning feemed to have rufhed upon two horizontal iron
bars, which were placed within the building crofs each o-
ther, to give additional ftrength to the obelifk, almoft at the
bafe of it, and not much above the upper ftory.   At the end
of one of thefe iron bars, on the Eaft and North Eaft fide, it
exploded again, and threw off a confiderable quantity of ftone.
Almoft all the damage done to the fteeple, except near the top,
was confined to the Eaft and North Eaft fide, and generally,
where the ends of the iron bars had been inferted into the
ftone, or placed under it; and, in fome places, by its violence in the ftone, its paffage might be traced from one iron
bar to another.

IT is very remarkable, that, to leffen the quantity of ftone
in this beautiful fteeple, cramps of iron had been employed
in feveral parts of it; and upon thefe, ftones of no great
thicknefs had been placed, both by way of ornament, and to
cover the cramped joint.   In feveral places thefe ftones had,
on account of their covering the iron, been quite blown off,
and thrown away.   A great number of ftones, fome of them
large ones, were thrown from the fteeple, three of which fell
upon the roof of the church, and did great damage to it; and
one of them broke through the large timbers which formed
the roof, and lodged in the gallery.

UPON

Upon the whole, the steeple was found, on a survey, to be so much damaged in several of its parts, that eighty five feet were taken down, in order to restore it substantially; and the manner in which this steeple was damaged completely indicated, as Dr. Watson observes, the great danger of insulated masses of metal from lightning; and, on the contrary, evinced the utility and importance of masses of metal continued, and properly conducted, in defending them from its direful effects. The iron and lead employed in this steeple, in order to strengthen and preserve it, did almost occasion its destruction; though, after it was struck by the lightning, had it not been for these materials keeping the remaining parts together, a great part of the steeple must have fallen.

This building suffered the more, on account of the thunder storm having been preceded by several very warm days. The nights had scarce furnished any dew, the air was quite dry, and in a state perfectly unfit to part with its highly accumulated electricity, without violent efforts. This great dryness had made the stones of St. Bride's steeple, and all other buildings under the like circumstances, far less fit, than if they had been in a moist state, to conduct the lightning, and prevent the mischief. For, though this thunder storm ended in a heavy shower of rain, none, except a few very large drops, fell till after the church was struck. And Dr. Watson had no doubt, but that the succeeding rain prevented many accidents of a similar kind, by bringing down, with every drop of it, part of the electric matter, and thereby restoring the equilibrium between the earth and clouds.

It is frequently observed, he says, that, in attending to

the

the apparatus for collecting the electricity of the clouds, though the sky is much darkened, and there have been several claps of thunder, at no great distance, yet the apparatus will scarcely be affected by it; but that, as soon as the rain begins and falls upon so much of the apparatus as is placed in the open air, the bells belonging to it will ring, and the electrical snaps succeed each other in a very extraordinary manner. This, as he observes, demonstrates, that every drop of rain brings down part of the electric matter of a thunder cloud and dissipates it in the earth and water, thereby preventing the mischiefs of its violent and sudden explosion. Hence when the heavens have a menacing appearance, a shower of rain is much to be wished for.

FROM all these considerations, Dr. Watson had no doubt, but that the mischief done to St. Bride's steeple was owing to the efforts of the lightning, after it had possessed the apparatus of the weathercock, endeavouring to force itself a passage from thence to the iron work employed in the steeple. As this must be done *per saltum*, as he expresses it, there being no regular metallic communication, it was no wonder, when its force was vehement, that it rent every thing which was not metallic that obstructed its easy passage; and that, in this particular instance, the ravages increased, as the lightning, to a certain distance, came down the steeple.

THE Doctor advises that, in order to have ocular demonstration when these metallic conductors do really discharge the lightning, that they be discontinued for an inch or two, in some place convenient for observation; in which case the fire will be seen to jump from one extremity of the wire to the
other.

other. If any danger be apprehended from this difcontinuance of the metallic conductor, he fays that a loofe chain may
be ready to hang on, and complete the communication. *

MR. DELAVAL, who alfo gives an account of the fame
accident, obferved, that, in every part of the building that
was damaged, the lightning had acted as an elaftic fluid, endeavouring to expand itfelf where it was accumulated in the
metal; and that the effects were exactly fimilar to thofe which
would have been produced by gunpowder pent up in the fame
places, and exploded.

IN the fame paper Mr. Delaval gives it as his opinion, that
a wire, or very fmall rod of metal, did not feem to have been
a canal fufficiently large to conduct fo great a quantity of
lightning as ftruck this fteeple; efpecially if any part of it, or
of the metal communicating with it; was inclofed in the
ftone work; in which cafe, he thought, the application of
it would tend to increafe its bad effects, by conducting it to
parts of the building which it might otherwife not have
reached.

UPON the whole, he thought that a conductor of metal,
lefs than fix or eight inches in breadth, and a quarter of an
inch in thicknefs (or an equal quantity of metal in any other
form that might be found more convenient) cannot with
fafety be depended on, where buildings are expofed to the
reception of a great quantity of lightning. †

MR. WILSON, in a paper written upon the fame occafion,
advifes, that pointed bars or rods of metal be avoided in all
conductors of lightning.

* Phil. Tranf. Vol. 54. p. 201, &c.　　† Ib. Vol. 54. p. 234.

As

As the lightning, he says, muft vifit us fome way or other, from neceffity, there can be no reafon to invite it at all; but, on the contrary, when it happens to attack our buildings, we ought only fo to contrive our apparatus, as to be able to carry the lightning away again, by fuch fuitable conductors, as will very little, if at all, promote any increafe of its quantity.

To attain this defirable end, in fome degree at leaft, he propofes, that the feveral buildings remain as they are at the top; that is, without having any metal above them, either pointed or not, by way of conductor; but that, on the infide of the higheft part of the building, within a foot or two of the top, a rounded bar of metal be fixed, and continued down along the fide of the wall, to any kind of moifture in the ground. *

Signior Beccaria whofe obfervations and experience with refpect to lightning give a weight to his opinion fupe-rior to that of any other man whatever, feems to think very differently from Mr. Wilfon on this subject. He fays that no metallic apparatus can attract more lightning than it can conduct. And fo far is he from thinking one conductor, rounded at the top, and a foot or two under the roof fuffici-ent; that if the building be of any extent, he advifes to have feveral of the ufual form; that is, pointed, and higher than the building. One conductor he thought fufficient for one tower, fteeple, or fhip; but he thought, two neceffary for the wing of a building 200 feet long, one at each extremity; three for two fuch wings, the third being fixed in the mid-

* Phil. Tranf. Vol. 54. p. 249.

dle;

dle; and four for a square palace of the same front, one at each corner. *

My readers at a distance from London will hardly believe me, when I inform them, that the elegant spire which has been the subject of a great part of this section, and which has been twice damaged by lightning (for it is now very probable, that a damage it received in the year 1750, was owing to the same cause) is now repaired, without any metallic conductor, to guard it in case of a third stroke.

* Lettere dell' ellettricismo, p. 278.

SECTION

S E C T I O N   XIV.

OF MEDICAL ELECTRICITY.

THE fubject of medical electricity falls almoft wholly within the period of which I am now treating. For, though fome effects of electricity upon animal bodies had been noted by the Abbé Nollet, and a few difeafed perfons had faid they had received benefit from being electrified; yet very little had been done this way, and phyficians had fcarcely attended to it, till within this period; whereas electricity is now become a confiderable article in the *materia medica*.

THERE is, however, one celebrated inftance of the cure of a palfy before this period; which is that performed by Mr. Jallabert, profeffor of Philofophy and Mathematics at Geneva, on a lockfmith, whofe right arm had been paralytic fifteen years, occafioned by a blow of a hammer. He was brought to Mr. Jallabert on the 26th of December 1747, and was completely cured by the 28th of February 1748. In

this

this interval he was frequently electrified, sparks being taken from the arm, and sometimes the electric shock sent through it. *

THE report of this cure performed at Geneva engaged Mr. Sauvages of the Academy in Montpelier to attempt the cure of paralytics, in which he had considerable success. In one case it occasioned a salivation, and in another a profuse sweat. Many paralytics, however, were electrified without any success. Indeed the prodigious concourse of patients of all kinds, which the report of these cures brought together, was so great, that few of them could be electrified, except very imperfectly. For two or three months together, twenty different patients were electrified every day. It is not surprising to find, that the neighbouring populace considered these cures as an affair of witchcraft, and that the operators were obliged to have recourse to their priests to undeceive them. † In the course of these experiments it was found, by very accurate observations made with a pendulum, that electrification increases the circulation of the blood about one sixth.

ONE of the first who attended to electricity in a medical way was Dr. Bohadtch, a Bohemian; who, in a treatise upon medical electricity, communicated to the Royal Society, gave it as his opinion, after the result of much experience, that of all distempers, the *hemiplegia* seemed to be the most proper object of electricity. He also thought it might be of use in intermitting fevers. ‡

THE palsy having happened to be the first disorder in which

* Histoire, pt. 3. p. 36.   † Ib. p. 97.   ‡ Phil. Transf. Vol. 47. p. 351.

electricity

electricity gave relief, there was a confiderable number of cafes publifhed pretty early, in which paralytics were faid to have found benefit from this new method of treatment. In the year 1757 Mr. Patric Brydone performed a complete cure of a hemiplegia, and, indeed, an almoft univerfal paralytic affection, in about three days. The patient was a woman, aged thirty three, and the palfy was of about two years continuance. * And John Godfrey Tefke, very nearly cured a young man, of the age of twenty, of a paralytic arm, of which he had not had the leaft ufe from the age of five years.†

THE Abbé Nollet's experiments upon paralytics had no permanent good effect. ‡ He obferves however that, during fifteen or fixteen years that he had electrified all forts of perfons, he had known no one bad effect to have arifen from it to any of them. ∥

DR. HART, in a letter to Dr. Watfon, dated Salop March the 20th. 1756, mentions a cure performed by electricity upon a woman of twenty three years of age, whofe hand and wrift had for fome time been rendered ufelefs by a violent contraction of the mufcles. She was not fenfible of the firft fhock that was given her; but, as the fhocks were repeated, the fenfation increafed, till fhe was perfectly well. She was alfo cured a fecond time, after a relapfe occafioned by a cold. §

BUT perhaps the moft remarkable cafe that has yet occurred of the ufe of electricity in curing a diforder of this kind; and indeed of any that is incident to the human body was of

* Phil. Tranf. Vol. 50. pt 1. p. 392. † Ib. Vol. 51. pt. 1. p. 179. ‡ Recherches, p. 412. ∥ Ib. p. 416. § Phil. Tranf. Vol. 49. pt. 2. p. 558.

that

that dreadful diſorder, an univerſal *tetanus*. It is related by Dr. Watſon in the Philoſophical Tranſactions; and the account was read at the Royal Society the 10th. of February 1763. The patient was a girl belonging to the Foundling hoſpital, about ſeven years of age, who was firſt ſeized with a diſorder occaſioned by the worms, and at length by an univerſal rigidity of her muſcles; ſo that her whole body felt more like that of a dead animal than a living one. She had continued in this diſmal condition above a month, and about the middle of November 1762, after all the uſual medicines had failed, Dr. Watſon began to electrify her; and continued to do it by intervals, till the end of January following; when, every muſcle of her body was perfectly flexible, and ſubſervient to her will, ſo that ſhe could not only ſtand upright, but could walk, and even run like other children of her age.*

THAT electricity may be hurtful, and even in ſome caſes in which analogy would lead us to promiſe ourſelves it might be of uſe, is evident from many caſes, and particularly from one related by Dr. Hart of Shrewſbury, in a letter to Dr. Watſon, which was read at the Royal Society November the 14th. 1754.

A YOUNG girl about ſixteen, whoſe right arm was paralytic, and greatly waſted in compariſon of the other, on being electrified twice, became univerſally paralytic, and remained ſo above a fortnight; when the new palſy was removed by proper medicines, though the firſt diſeaſed arm remained as before. However Dr. Hart, notwithſtanding this bad acci-

* Phil. Tranſ. Vol. 53. p. 10.

dent,

dent, had a mind to try electricity again. The girl submitted to it; but after having been electrified about three or four days, she became a second time universally paralytic, and even lost her voice, and the use of her tongue; so that it was with great difficulty she could swallow. She was relieved of this additional palsy a second time by a proper course of medicines, continued about four months; but was discharged out of the hospital as incurable of her first palsy. It is said that the Doctor would have tried electricity a third time; but the girl, being more nearly concerned in the experiment than her physician, thought proper to decline it. *

DR. Franklin's account of the effects of electricity, in the manner in which he applied it, is by no means favourable to its use in such cases. He says, in a letter to Dr. Pringle, read at the Royal Society January the 12th. 1758, that some years before, when the news papers made mention of great cures performed in Italy and Germany by electricity, a number of paralytics were brought to him, from different parts of Pensilvania and the neighbouring provinces, to be electrised, and that he performed the operation at their request. His method was, first to place the patient in a chair, or upon an electric stool, and draw a number of large strong sparks from all parts of the affected limb or side. He then fully charged two six gallon glass jars, and sent the united shock of them through the affected limb or limbs, repeating the stroke commonly three times each day.

THE first thing he observed was an immediate greater sen-

* Phil. Transf. Vol. 48. pt. 2. p. 786.

sible

fible warmth in the affected limbs, which had received the stroke, than in the others ; and the next morning the patients ufually faid, that, in the night, they had felt a pricking fenfation in the flefh of the paralytic limbs ; and would fometimes fhow a number of fmall red fpots, which they fuppofed were occafioned by thofe prickings. The limbs too were found more capable of voluntary motion, and feemed to receive ftrength. A man, for inftance, who could not, the firft day, lift his lame hand from off his knee, would the next day raife it four or five inches, the third day higher ; and on the fifth was able, but with a feeble languid motion, to take off his hat. Thefe appearances, the Doctor fays, gave great fpirits to the patients, and made them hope for a perfect cure ; but he did not remember that he ever faw any amendment after the fifth day ; which the patients perceiving, and finding the fhocks pretty fevere, became difcouraged, went home, and in a fhort time relapfed ; fo that he never knew any permanent advantage from electricity in palfies.

PERHAPS, fays he, fome permanent advantage might be obtained, if the electric fhocks had been accompanied with proper medicine and regimen, under the direction of a fkilful phyfician. He thought too, that many fmall fhocks might have been more proper than the few great ones which he gave ; fince, in an account from Scotland, a cafe was mentioned in which 200 fhocks from a phial were given daily, and a perfect cure had been made. *

THAT there is an intimate connection between the ftate of

* Phil. Tranf. Vol. 50. pt. 2. p. 481.

electricity

electricity in the air and the human body, is evident from
several facts, particularly a very remarkable one related by the
Abbé Mazeas, in a letter to Dr. Hales. He was electrifying
a person who was subject to epileptic fits, by his apparatus to
make observations upon the electricity of the common atmosphere. At first this person bore the sparks very well,
but in two or three minutes the Abbé, perceiving his countenance to change, begged he would retire, lest any accident
should happen; and he was no sooner returned home, than
his senses failed him, and he was seized with a most violent
fit. His convulsions were taken off with spirits of hartshorn,
but his reason did not return in an hour and an half. He
went up and down the stairs like one who walks in his sleep,
without speaking to, or knowing any person, settling his papers, taking snuff, and offering chairs to all who came in.
When he was spoken to, he pronounced inarticulate words,
which had no connection.

When this poor man recovered his reason, he fell into
another fit; and his friends told the Abbé, he was more affected with that distemper when it thundered than at any o-
ther time; and if it ever happened, which it rarely did,
that he then escaped; his eyes, his countenance, and the
confusion of his expressions, sufficiently demonstrated the
weakness of his reason.

The next day, the Abbé learned from the person himself,
that the fear of thunder was not the cause of his disease; but
that, however, he found a fatal connection between that phenomenon and his distemper. He added, that when the fit
seized him, he perceived a vapour rising in his breast, with

so

fo much rapidity, that he loft all his fenfes before he could call for help. *

Mr. Wilson cured a woman of a deafnefs of feventeen years ftanding. He alfo obferves, that fhe had a very great cold when fhe began to be electrified; but that the inflammation ceafed the firft time, and the cold was quite gone when the operation had been performed again the fecond day. But he acknowledges, that he had tried the fame experiment upon fix other deaf perfons without any fuccefs. †

The fame perfon obferves that one gentleman, near feventy years old, could never be made to receive a fhock except in his wrifts. He fays that he himfelf could once have borne very great fhocks without inconvenience, but that he could not bear them at the time that he wrote.

Medical electricity is very much obliged to the labours and obfervations of Mr. Lovet, Lay-clerc of the cathedral church at Worcefter, who has for many years been indefatigable in the application of electricity to a great variety of difeafes. His fuccefs has been very confiderable, and all the cafes he has publifhed feem to be well authenticated.

According to Mr. Lovet, electricity is almoft a fpecific in all cafes of violent pains, of however long continuance, in every part of the body; as in obftinate head-aches, the fciatica, the cramp, and diforders refembling the gout. He had no trials of the proper gout, but only on thofe who were flightly attacked, and who received immediate relief.

The tooth ach, he fays, is generally cured inftantly, and

* Phil. Tranf. Vol. 48. pt. 1. p. 383.   † Wilfon's, Effay, p. 207.

he

he scarce ever remembered any one who complained of its raging a minute after the operation. *

It has seldom failed, he says, to cure rigidities, or a wasting of the muscles, and hysterical disorders, particularly if they be attended with coldness in the feet. According to him, it cures inflammations, it has stopped a mortification, cured a fistula lachrymalis, and dispersed extravasated blood.† He also says it has been of excellent use in bringing to a suppuration, or in dispersing without suppuration, obstinate swellings of various kinds, even those that were scrophulous. In his hands it cured the falling sickness, and fits of various kinds; though the patients had been subject to them for many years, and one cure he mentions of a hemiplegia. ‡ Lastly, he relates a well attested case, from Mr. Floyer, surgeon at Dorchester, of a complete cure of what seemed to be a *gutta serena*. The same Mr. Floyer, he also says, cured with it two young women of obstructions, one of whom had taken medicines a year to no purpose. ||

In the rheumatism, Mr. Lovet candidly confesses, it has failed; but says it was seldom in the case of young persons, if they were taken in time.

The manner in which electricity operated in these cures, Mr. Lovet imagined to be, by removing secret obstructions; which are probably the cause of those disorders. In all his practice he never knew an instance of harm being done by it, and thinks that, in all the cases in which it has done harm, the manner of administering it has been injudicious. In ge-

* Lovet's Essay, p. 112.　† Ib. p. 76.　‡ Ib. p. 101.　|| Ib. p. 119.

neral,

neral, he thinks the shocks have been made too great.   Thi,
he imagined to have been the case of the patient before mentioned of Dr. Hart, who was made more paralytic by electric shocks.   Mr. Lovet advises to begin, in general, with
simple electrification, especially in hysterical cases ;   then to
proceed to taking sparks, and lastly to giving moderate shocks,
but hardly ever any that are violent, or painful.

THE account of the application of electricity by Dr. Zetzel of Upsal, which may be seen in Mr. Lovet's treatise, a-
grees in the main with the result of his own practice ;   and
where there is any difference between them,   Mr. Lovet
thinks there are evident marks of unfairness in the Swedish
account.   And a subsequent account from Sweden mentions
several cures being made in those very cases, in which Dr.
Zetzel says that no relief was to be had from electricity.

THE Rev. Mr. J. Wesley has followed Mr. Lovet in the same
useful course of medical electricity ;   and recommends the use
of it to his numerous followers, and to all people.   Happy it is
when an ascendency over the minds of men is employed to
purposes favourable to the increase of knowledge, and to the
best interests of mankind.   Mr. Wesley's account of cures
performed by electricity agrees very well with that of Mr.
Lovet, whom he often quotes.   He adds, that he has scarce
ever known an instance in which shocks all over the body
have failed to cure a quotidian or tertian ague. *   He mentions cases of blindness cured or relieved by it ; and says that
he has known hearing given by it to a man that was born

* Wesley's Desideratum, p. 3.

H h h   deaf.

deaf. * He mentions cures in cases of bruises, running sores, the dropsy, gravel in the kidnies (causing the patient to part with it) a palsy in the tongue, and lastly in the genuine consumption. But Mr. Boisser says it is of disservice in phthisical complaints.†

MR. WESLEY candidly says, he has not known any instance of the cure of an hemiplegia; and though many paralytics have been helped by electricity, he scarcely thinks that any palsy of a year's standing has been thoroughly cured by it. He asserts, however, that he has never yet known any person, man, woman, or child, sick or well, who has found (what Mr. Wilson says, that some persons complained of) an unusual pain some days after the shock. Mr. Wesley had only known that rheumatic pains, which were afterwards perfectly cured, had increased on the first or second application. ‡

MR. WESLEY directs the same method of administration with Mr. Lovet. In deep hysterical cases, he advises that the patients be simply electrified, sitting on cakes of rosin, at least half an hour morning and evening; when, after some time, small sparks may be taken from them, and afterwards shocks given to them, more or less strong, as their disorder requires; which, he says, has seldom failed of the desired effect. ||

THIS account of the medical use of electricity by Mr. Lovet and Mr. Wesley is certainly liable to an objection, which

* Wesley's Desideratum, p. 48.   † Carmichael Tentamen inaugurale medicum de Paralysi, p. 34. ex Act. Ups.   ‡ Wesley's Desideratum, p. 50. || Ib. p. 56.

will always lie againſt the accounts of thoſe perſons who, not being of the faculty, cannot be ſuppoſed capable of diſtinguiſhing with accuracy either the nature of the diſorders, or the conſequences of a ſeeming cure.   But, on the other hand, this very circumſtance of their ignorance of the nature of diſorders, and conſequently of the beſt method of applying electricity to them, ſupplies the ſtrongeſt argument in favour of its innocence, at leaſt.   If in ſuch unſkilful hands it has produced ſo much good, and ſo little harm ; how much more good, and how much leſs harm would it probably have produced in more ſkilful hands.

But whatever weight there be in this objection againſt the laſt mentioned writers, it certainly cannot be urged againſt Antonius de Haen, one of the moſt eminent phyſicians of the preſent age ; who, after ſix years uninterrupted uſe of it, reckons it among the moſt valuable aſſiſtances of the medical art ; and expreſſly ſays, that though it has often been applied in vain, it has often afforded relief where no other application would have been effectual.   But I ſhall recite in a ſummary manner from his Ratio Medendi, the reſult of all his obſervations on this ſubject.

With reſpect to partial palſies, in particular, he ſays, it never did the leaſt harm ; that one or two perſons who had received no benefit from it in ſix entire months, were yet much relieved by perſevering in the uſe of it.   That ſome perſons diſcontinuing it, after having received ſome benefit, relapſed again ; but afterwards, by recurring to the uſe of electricity, recovered, though more ſlowly than before. Some perſons, he ſays, were relieved who had been paralytic one,

H h h 2   three,

three, six, nine, and twelve years, and some longer; but that in one or two of these cases, the patients had received less relief, and more slowly than was usual in recent cases. In some cases, he says, a most unexpected benefit had been found by those who had been paralytic in their tongues, eyes, fingers, and other particular limbs. A paralysis and trembling of the limbs, from whatever cause it arose, he says, never failed to be relieved by it; and he relates one instance of a perfect cure being performed in a remarkable case of this nature, after receiving ten shocks. *

DE HAEN's custom was to apply the operation for half an hour together at least. He seems to have used gentle shocks, and he joined to electricity, the use of other remedies, which, however, would not have been effectual without it. †

ST. VITUS's dance, he says, never failed to be cured by electricity. ‡ He always observed it to promote a more copious discharge of the menses, and to relieve in cases of obstruction; but, for this reason, he advises that it be not administred to women with child. He found it of use in some cases of deafness, but it entirely failed in its application to a gutta serena, and to a strumous neck. ||

LASTLY, he relates a remarkable case, communicated to him by Mr. Velse at the Hague, of the cure of a mucous apoplexy. §

To the cases which have been mentioned occasionally, in which harm may be apprehended from electrification, may

---

* Ratio Medendi, Vol. 1. p. 234, 199.   † Ib. p. 233   ‡ Ib. p. 389.
      || Ib. Vol. 2. p. 200.   § Ib.

perhaps

perhaps be added the venereal difeafe in which Mr. Veratti advifes, that electrification be by all means avoided. *

I shall conclude this account of medical electricity with obferving that there are two general effects of electricity on the human body, of which, it fhould feem, that phyficians might greatly avail themfelves. Thefe are, that it promotes infenfible perfpiration, and glandular fecretion. The former is effected by fimple electrification, and the latter by taking fparks from the glands, or the parts contiguous to them; on which it acts like a ftimulus. Of the former, inftances have been given in the experiments of the Abbé Nollet, and a few have been given occafionally of the latter.

To thefe I fhall now add, that Linnæus obferved, that when electric fparks have been drawn from the ear, it has inftantly promoted a more copious fecretion of ear-wax; and that it has alfo been obferved, that, when the eye, or the parts about the eye, have been electrified, the tears have flowed copioufly. But the moft remarkable cafe that I have met with, is, of its promoting the fecretion of that matter which forms the hair; whereby hair has been actually re-ftored to a part that had long been bald. †

Hitherto electricity has been generally applied to the human body either in the method of drawing fparks, as it is called, or of giving fhocks. But thefe operations are both violent; and though the ftrong concuffion may fuit fome cafes, it may be of difservice in others, where a moderate fimple electrification might have been of fervice.

* Carmichael Tentamen, p. 34.   † Ib. p. 33.

THE

THE great objection to this method is the tediousness and expence of the application. But an electrical machine might be contrived to go by wind or water, and a convenient room might be annexed to it; in which a floor might be raised upon electrics, and a person might sit down, read, sleep, or even walk about during the electrification. It were to be wished, that some physician of understanding and spirit would provide himself with such a machine and room. No harm could possibly be apprehended from electricity, applied in this gentle and insensible manner, and good effects are, at least, possible, if not highly probable. It would certainly be more for the honour of the faculty, that the practice should be introduced in this manner, than that it be left to some rich Valetudinarian, who may take it into his head, that such an operation may be of service to him.

SECTION

SECTION XV

MISCELLANEOUS EXPERIMENTS AND DISCOVERIES MADE
WITHIN THIS PERIOD.

HAVING diftributed into diftinct fections all the fubjects, under which I had collected materials enow to
form a feparate account; I have referved for the laft place,
thofe fmaller articles, which could neither with propriety be
introduced under the former heads, nor were large enough to
make a fection themfelves.

IT has been a great controverfy among electricians, whether
glafs be permeable to the electric fluid. Mr. Wilfon appeared in favour of the permeability, and, in a paper read at the
Royal Society December the 6th. 1759, produced the following experiments to fupport his opinion; notwithftanding
he, even afterwards, acknowledged, in a paper read at the
Royal Society November the 13th. 1760; that, in the Leyden
experiment, Dr. Franklin had proved that the fluid did not
go through the glafs. *

* Phil. Tranf. Vol. 51. p. 2. p. 896.

HE

He took a very large pane of glaſs, a little warmed; and, holding it upright by one edge, while the oppoſite edge reſted upon wax, he rubbed the middle part of the ſurface with his finger, and found both ſides electrified *plus*. *

Upon this I cannot help obſerving that it ought to be ſo on Dr. Franklin's principles. If one ſide be rubbed by the finger, it acquires from the finger ſome of the electric fluid. This, being ſpread on the glaſs as far as the rubbing extended, repels an equal quantity of that contained in the other ſide of the glaſs, and drives it out on that ſide; where it ſtands as an atmoſphere, ſo that both ſides are found *plus*. If the unrubbed ſide were in contact with a conductor communicating with the earth, the electric fluid would be carried away, and then that ſide would be left *apparently* in the natural ſtate. If the electric fluid found on the unrubbed ſide was really part of that which had been communicated by and from the finger, and ſo had actually *permeated* the glaſs, it might, when conducted away, be continually replaced by freſh permeating fluid communicated in the ſame manner: But if the effect is continually diminiſhing, while the ſuppoſed cauſe, repeated, continues the ſame, there ſeems reaſon to doubt the ſuppoſed relation between that cauſe and the effect. For it appears difficult to conceive how ſome electric fluid, having paſſed through a permeable body, ſhould make it more difficult for other particles of the ſame electric fluid to follow, till, at length, none could paſs at all.

Mr. Wilson alſo ſays, that, holding the ſame pane of

* Phil. Tranſ. Vol. 51. pt. 1. p. 314.

glaſs

glafs within two feet of the prime conductor, which was electrified *plus*, that part of the glafs which was oppofite to the conductor became electrified *minus* on both fides ; but, in a few minutes, the *minus* electricity difappeared, and the *plus* continuing, diffufed itfelf into the place of the other, fo that now the whole was electrified *plus*.

The experiment fo far fucceeding, induced him to make ufe of a lefs piece of glafs, that the whole might be electrified *minus*. Thefe advances, he fays, led him to obferve the power of electrifying that fmall piece of glafs at different diftances.

He expofed the fame fmall piece of glafs to the prime conductor, at the diftance of two feet, and obferved a *minus* electricity at both furfaces.

As he moved the glafs nearer, to a certain diftance, it was more fenfibly electrified *minus* ; and after that, on moving it ftill nearer, the *minus* appearance was lefs and lefs fenfible ; till it came within the diftance of about one inch, and then it was electrified *plus* on both fides.

This *plus* electricity in the glafs, he found, might be changed to a *minus* again, by removing the glafs, and holding it for a time at a greater diftance ; which he thought to be a proof of the repulfive power of that fluid. *

Having by him a pane of glafs, one fide of which was rough and the other fmooth, he rubbed it flightly on one fide ; upon doing which, both fides were electrified *minus*.

On this I muft alfo take the liberty to obferve that, as the

* Ib. Vol. 51. pt 1. p. 328.

I i i   electric

electric fluid contained in glaſs in its natural ſtate, is kept equal in both ſides by the common repulſion; if the quantity in one ſide is diminiſhed, the fluid in the other ſide, being leſs repelled, retires inward, and leaves that ſurface alſo *minus.*

SLIGHT changes, *plus* or *minus,* may be made in either ſurface, that have not ſtrength to act on the other ſide, by repulſion, or by abating repulſion, through the glaſs; and ſo *plus* electricity may be given to one ſurface, and *minus* to the other in ſome degree. Both ſides may alſo be made *plus,* and both *minus,* by rubbing, or by communication, without any neceſſity of ſuppoſing the glaſs *permeable.*

AND yet it is probable that ſome glaſs, from having a greater mixture of non-electric matters in its compoſition, may be permeable, when cold, in ſome ſmall degree, as all glaſs is found to be when warmed.

MR. WILSON treated the other ſide of this pane of glaſs in the ſame manner, after which the *minus* electricity was changed into a *plus* on both ſides.

THOUGH Dr. Franklin was of opinion, that glaſs when cold is not permeable to electricity, he had made no experiments upon it when hot; but Mr. Kinnerſley, a friend of his, made one, which ſeemed to prove, that it was very differently affected in this reſpect, in the different ſtates of hot and cold. He found, that a coated Florence flaſk (made of very thin glaſs, and full of air bubbles) containing boiling water, could not be electrified. The electricity, he ſays, paſſed as readily through it as through metal. The charge of a three pint bottle went freely through it, without injuring the flaſk in

the

the leaft.   When it became cold, he could charge it as before. This effect he attributed to the dilatation of the pores of glafs by heat *

ALL Mr. Wilfon's experiments to prove the permeability of glafs were repeated by Mr. Bergman of Upfal; and, as he fays, with fuccefs. †

ÆPINUS, however, was by no means fatisfied with Mr. Wilfon's experiments concerning the permeability of glafs; and yet he brings no other fact in anfwer to his arguments, but a very common one, which fhows that a glafs tube both receives and lofes its electricity very flowly; fo that he only afferts a *difficulty*, and a *flownefs* in the electric fluid paffing through electric fubftances, as was mentioned before; and confequently Mr. Wilfon feems to have a real advantage in the controverfy: for, as he fays, paffing through, though ever fo flowly, is a real paffing through. ‡

ÆPINUS has fhown, by a curious experiment, that if a metallic conductor and a cork ball be both electrified pofitively, fo as to repel one another; yet, that, if the ball be forcibly brought within two, three, or four lines of the conductor, it will be attracted by it; and be repelled again, if it be forcibly pufhed beyond that limit of attraction.   If the ball be confined to move within the fame fmall diftance, a moderate electrification of the conductor will repel the ball to its utmoft limit; but a ftronger electrification of the conductor will caufe it to be attracted.   He, therefore, limits the general maxim, that bodies poffeffing the fame kind of electricity repel one

* Phil. Tranf. Vol. 53. pt. 1. p. 85.   † Ib. Vol. 52. pt. 2. p. 485.
‡ Ib. Vol. 53. p. 443.

another,

another; and afferts, that this will be the cafe, only when the quantity of electric fluid belonging to them both, as one body, is greater, or lefs than that which is natural to them. * This experiment deferves particular attention.

Signior Beccaria, who has contributed fo largely to feveral of the fections in this period, furnifhes a few articles which well deferve a place in this.

He thought it was evident, that the electric fluid tended to move in a right line, becaufe a longer fpark may be taken in a direct line, from the end of a long conductor, than can be taken from the fame place in any other direction. But he thought it was ftill more evident, from obferving, both in the air, and in vacuo; that, prefenting the finger, or a brafs ball, at a proper diftance, and in a certain angle with the conductor (which experience will foon find) the electric fpark will make an exact curve, to which the conductor produced will be a tangent: as if the electric matter was acted upon by two different forces, one its own acquired velocity, urging it forward in a right line; the other the attraction of the body prefented to it, which throws it out of the right line. †

In his obfervations on pointed bodies, he fays, that two pointed bodies, equally fharp, in their approach to an electrified conductor, will appear luminous only at half the diftance at which one of them would have done.‡

The fame ingenious philofopher reports a curious, but cruel experiment which he made on a live cock. He detached the belly of one of the mufcles from the thigh of the animal, leaving the extremities in their proper infertions, and then

* Æpini Tentamen, p. 146.   † Elettricifmo artificiale, p. 56.   ‡ Ib. p. 67.

discharged

discharged a shock through it. At the instant of the stroke, the leg was violently distended, and the muscle greatly inflated; the motion beginning at the tendon, and the extension of it resembling the opening of a lady's fan. No pricking with a pin could make it act so strongly. *

MR. HAMILTON, professor of Philosophy at the university of Dublin, made a curious experiment with a wire five or six inches long, finely pointed at each end. To the middle of this wire he fitted a brass cap, which rested on the point of a needle communicating with the conductor. Half an inch at each extremity of this wire he bent, in opposite directions, perpendicular to the rest of the wire, and in the plane of the horizon. The consequence of electrifying this apparatus was, that the wire would turn round with very great velocity; moving, as he says, always in a direction contrary to that in which the electric fluid issues from its point, without having any conducting substance near it, except the air. He also observes, that if this wire were made to turn the contrary way, it would stop, and turn as before. †

THE same experiment was also made by Mr. Kinnersley of Boston, with this addition, that he electrified the wire negatively; and observed, to his great surprise, that it still turned the same way. This he endeavoured to account for by supposing that, in the former case, the points, having more e-lectricity than the air, were attracted by it; in the latter case, the air, having more than the points, was attracted by them. ‡

* Lettere dell' ellettricismo, p. 129.   † Phil. Transf. Vol. 51. pt. 2. p. 905.
‡ Ib. Vol. 53. pt. 1. p. 86.

IT

IT may, by some, be thought that this pointed wire turning the same way, whether it be electrified negatively or positively, is a proof that the electric fluid issues out at the points in both cases alike, and by the reaction of the air is, together with the points, driven backwards; contrary to what ought to have been the case if the electric fluid had really issued out of the point in one case, and entered it in the other. But it will be found by experiment, that an eolipile, with its stem bent like the wire above-mentioned, and suspended on its center of gravity by a fine thread, will move in the same direction, whether it be throwing steam out at the orifice; or, after it is exhausted, and cooling, it be drawing the air, or water in.

I SHALL close this section of miscellaneous articles, and the whole history of electricity, with a succinct account of some of the chief particulars in which the analogy between electricity and magnetism consists; very nearly as it was drawn up, in an abridgment of Æpinus, and communicated to me for this purpose, by Mr. Price.

1. As a rod of iron held near a magnet will have several successive poles; so will a glass tube touched by an excited tube have a succession of positive and negative parts.

2. BODIES positively and negatively electrical, when in contact, will unite to one another; as will magnets, when they are laid with their opposite poles to one another.

3. GLASS is a substance of a nature similar to hardened steel. The positive and negative sides of the former answer to the attracting and repelling ends of the latter, when magnetical.

4. As

4. As it is difficult to move the electric fluid in the pores of the former; so, likewise, it is difficult to move the magnetic fluid in the pores of the latter.

5. As there can be no condensation of the electric fluid in the former, without a rarefaction; so, in the latter, if there be a condensation, or positive magnetism in one end of a bar, there must be an evacuation, or negative magnetism in the other end.

6. STEEL corresponds to electric per se, and iron in some measure, to conductors of electricity.

7. STEEL is less susceptible of the magnetic virtue, but when it has acquired it, it retains it more strongly than iron; just as electrics per se will not so easily receive the electric fluid, but when it is forced into them, will retain it more strongly than conductors.

8. ÆPINUS adds, and reckons it one of his discoveries, that an electrified body does not act on other bodies, except they are themselves electrified; just as a magnet will not act on any other substances, except they are themselves possessed of the magnetic virtue. So that an electrified body attracts and repells another body, only in consequence of rendering it first of all electrical; as a magnet attracts iron, only in consequence of, first of all, making it a magnet.

9. MR. CANTON has also found, that if the tourmalin be cut into several pieces, each piece will have a positive and negative side, just as the pieces of a broken magnet would have.

THUS far, says Mr. Price, there is an analogy, and, in some instances, a striking one, between magnetism and electricity, upon the supposition that the cause of magnetism is a fluid.

fluid. But there is no magnetic fubftance which anfwers perfectly to the conductors of electricity. There is no afflux and efflux of the magnetic fluid ever vifible. The equilibrium in a magnet cannot be inftantaneoufly reftored, by forming a communication between the oppofite ends with iron, as it may in charged glafs. Nor are there any fubftances pofitively or negatively magnetical only, as there are bodies which are pofitively or negatively electrical only.

P A R T

PART II.

A

SERIES of PROPOSITIONS,

COMPRISING ALL THE

GENERAL PROPERTIES of ELECTRICITY.

AFTER tracing, at large, the progress of all the discoveries relating to electricity, and giving an historical account of them, in the order in which they were made; it will, probably, be no disagreeable repetition, if I give, at the close of it, a SERIES OF PROPOSITIONS, comprising all the general properties of electricity, drawn up in as succinct a manner as possible. And, notwithstanding the large detail which has been made, it will be found, that a few propositions are sufficient to comprise almost all that we know of the subject.

THIS circumstance may be regarded as a demonstration of the real progress that has been made in this science. And as this progress advances, and the history enlarges, paradoxical

K k k

as

as the affertion may feem, this part may be expected to contract itfelf in the fame proportion.   For the more we know
of any fcience, the greater number of particular propofitions
are we able to refolve into general ones; and, confequently,
within narrower bounds fhall we be able to reduce its principles.

I MIGHT have made this part of my work much fhorter,.
even in the prefent ftate of the fcience, if I would have admitted into it any thing theoretical; but I have carefully a-
voided the principles of any theory, even the moft probable,
and the neareft to being perfectly afcertained, in this feries of
propofitions; in which I propofe to comprehend only *known
facts*; that my younger readers may carefully diftinguifh between *fact* and *theory*; things which are too often confounded,

I HAVE not, in this part of my work, defcended to any
minutiæ, in the defcription of electrical appearances, becaufe
they have been entered into before, and a repetition of them
would have been tedious.   At the fame time, I think it will
be found, upon examination, that I have not omitted to take
notice of any difcovery of importance.   I have alfo introduced into it the definition of all the moft neceffary *technical
terms*; that this part of the work might ferve as a methodical, and fufficient introduction, to thofe who are beginning
the ftudy of electricity; and defire a general knowledge of the
firft elements of the fcience, before they enter upon the detail
of particulars, which will be beft learned afterwards, from
the hiftory.

BY *electricity*, in the following propofitions, I would be
                                                    underftood

underſtood to mean, only thoſe *effects* which will be called electrical; or elſe the *unknown cauſe* of thoſe effects, uſing the term, as we uſe the letters *(x)* and *(y)* in Algebra.

ALL known ſubſtances are diſtributed by electricians into two ſorts. Thoſe of one ſort are termed *electrics*, or *non-conductors*; and thoſe of the other *non-electrics*, or *conductors of electricity*.

METALS of all kinds, together with ſemimetals, and all ſubſtances in a ſtate of fluidity, except air, are conductors. So alſo is charcoal, and other ſubſtances of a ſimilar nature, as will be ſhown at large in the laſt part of this work. All other ſubſtances, whether mineral, vegetable, or animal, are non-conductors. But many of theſe when they are made very hot, as glaſs, roſin, baked wood; and, perhaps, all the reſt on which the experiment can be made in this ſtate, are conductors of electricity.

ALL bodies, however, though in the ſame ſtate of heat and cold, are not equally perfect electrics, or perfect conductors. Vegetable and animal ſubſtances, for inſtance, in their natural ſtate, are ſeldom perfect electrics, on account of the moiſture that is contained in them. And, independent of moiſture, there is a gradation in all ſubſtances, from the moſt perfect conductors to the moſt perfect non-conductors of electricity.

IT is the property of all kinds of electrics, that when they are rubbed by bodies differing from themſelves (in roughneſs or ſmoothneſs chiefly) to attract light bodies of all kinds which are preſented to them; to exhibit an appearance of light (which is very viſible in the dark) attended with a ſnapping noiſe, upon the approach of any conductor; and, if

K k k 2   the

the noftrils be prefented, they are affected with a fmell like that of phofphorus.

An electric fubftance exhibiting thefe appearances, is faid to be *excited*, and fome of them, particularly the tourmalin, are excited by heating and cooling, as well as by rubbing.

It is neceffary, however, to a confiderable excitation of any electric, that the fubftance againft which it is rubbed (ufually called the *rubber*) have a communication with the earth, or bodies abounding with electricity, by means of conductors; for if the rubber be *infulated*, that is, if it be cut off from all communication with the earth by means of electrics, the friction has but little effect.

When infulated bodies have been attracted by, and brought into contact with any excited electric, they begin to be repelled by it, and alfo to repel one another: nor will they be attracted again, till they have been in contact with fome conductor communicating with the earth; but, after this, they will be attracted as at firft.

If conductors be infulated, electric powers may be *communicated* to them by the approach of excited electrics. They will then attract light bodies, and give fparks, attended with a fnapping noife, like the electrics themfelves. But there is this difference between excited and communicated electricity, that a conductor to which electricity has been communicated parts with its whole power at once, on the contact of a conductor communicating with the earth; whereas an excited electric, in the fame circumftances, lofes its electricity only partially; it being difcharged only from the part which was actually touched by the conductor, or thofe in the neighbour-

hood

hood of it; so that the spark of electric fire is not so dense, nor the explosion made by parting with it so loud, from excited, as from communicated electricity.

ELECTRIC substances, brought into contact with excited electrics, will not destroy their electricity; whence it is that they are called non-conductors, because they will not convey or *conduct* away whatever is the cause of electric appearances in bodies.

WHEN electricity is strongly communicated to insulated animal bodies, the pulse is quickened, and perspiration increased; and, if they receive, or part with their electricity on a sudden, a painful sensation is felt at the place of communication.

THE growth of vegetables is quickened by electricity.

No electric can be excited without producing electric appearances in the body with which it is excited, provided that body be insulated. For this insulated rubber will attract light bodies, give sparks, and make a snapping noise upon the approach of a conductor, as well as the excited electric.

IF an insulated conductor be pointed, or if a pointed conductor communicating with the earth be held pretty near it, little or no electric appearance will be exhibited; only a light will appear at each of the points, during the act of excitation, and a current of air will be sensible from off them both.

THESE two electricities, viz. that of the electric itself, and that of the rubber, though similar to, are the reverse of one another. A body attracted by the one will be repelled by the other, and they will attract, and in all respects act upon one another more sensibly than upon other bodies; so that two

pieces

pieces of glafs or filk, poffeffed of contrary electricities, will cohere firmly together, and require a confiderable force to feparate them.

THESE two electricities, having been firft difcovered by producing one of them. from glafs, and the other from amber, fealing wax, fulphur, rofin, &c. firft obtained the names of *vitreous* and *refinous electricity*; and it being afterwards imagined that one of them was a redundancy, and the other a deficiency of a fuppofed electric fluid, the former (viz. that which is produced from the friction of fmooth glafs tubes or globes by the human hand, or a common leathern rubber) obtained the name of *pofitive*; and the latter (viz. that which is produced from the friction of fticks or globes of fulphur, &c. or collected from the rubber of a glafs globe abovementioned) that of *negative* electricity : and thefe terms are now principally in ufe.

IF a conductor, not infulated, be brought within the *atmofphere*, that is, the fphere of action, of any electrified body, it acquires the electricity oppofite to that of the electrified body ; and the nearer it is brought, the ftronger oppofite electricity doth it acquire ; till the one receive a fpark from the other, and then the electricity of both will be difcharged.

THE electric fubftance which feparates the two conductors, poffeffing thefe two oppofite kinds of electricity, is faid to be *charged*. Plates of glafs are the moft convenient for this purpofe, and the thinner the plate, the greater charge it is capable of holding. The conductors contiguous to each fide of the glafs are called their *coating*.

AGREEABLE to the above-mentioned general principle, it

is

is neceſſary, that one ſide of the charged glaſs have a communication with the rubber, while the other receives the electricity from the conductor ; or with the conductor, while the
other receives from the rubber.

It follows alſo, that the two ſides of the plate thus charged
are always poſſeſſed of the two oppoſite electricities ; that ſide
which communicates with the excited electric having the
electricity of the electric, and that which communicates with
the rubber, that of the rubber.

There is, conſequently, a very eager attraction between
theſe two electricities with which the different ſides of the
plate are charged ; and, when a proper communication is
made by means of conductors, a flaſh of electric light, attended with a report (which is greater or leſs in proportion
to the quantity of electricity communicated to them, and the
goodneſs of the conductors) is perceived between them, and
the electricity of both ſides is thereby diſcharged.

The ſubſtance of the glaſs itſelf, in or upon which theſe
electricities exiſt, is impervious to electricity, and does not
permit them to unite ; but if they be very ſtrong, and the
plate of glaſs very thin, they will force a paſſage through the
glaſs. This, however, always breaks the glaſs, and renders
it incapable of another charge.

The flaſh of light, together with the exploſion between
the two oppoſite ſides of a charged electric, is generally called
the *electric ſhock*, on account of the diſagreeable ſenſation it
gives any animal, whoſe body is made uſe of to form the communication between them.

This electric ſhock is always found to perform the circuit
from

from one fide of the charged glafs to the other by the fhorteft paffage, through the beft conductors. Common communicated electricity alfo obferves the fame rule, in its tranfmifion from one body to another.

It has not been found that the electric fhock, takes up the leaft fenfible fpace of time in being tranfmitted to the greateft diftances.

The electric fhock, as alfo the common electric fpark, difplaces the air through which it paffes; and if its paffage from conductor to conductor be interrupted by non-conductors of a moderate thicknefs, it will rend and tear them in its paffage; and in fuch a manner as to exhibit the appearance of a fudden expanfion of the air about the center of the fhock.

If a ftrong fhock be fent through a fmall animal body, it will often deprive it inftantly of life.

When the electric fhock is very ftrong, it will give polarity to magnetic needles, and fometimes it reverfes their poles.

Great fhocks, by which animals are killed, are faid to haften putrefaction.

Electricity and lightning are, in all refpects, the fame thing. Every effect of lightning may be imitated by electricity, and every experiment in electricity may be made with lightning, brought down from the clouds, by means of infulated pointed rods of metal.

PART

PART III.

THEORIES of ELECTRICITY.

SECTION I.

OF PHILOSOPHICAL THEORIES IN GENERAL, AND THE
THEORIES OF ELECTRICITY PRECEDING THAT OF DR.
FRANKLIN.

ONE of the moſt intimate of all aſſociations in the human
mind is that of *cauſe* and *effect*. They ſuggeſt one a-
nother with the utmoſt readineſs upon all occaſions; ſo that
it is almoſt impoſſible to contemplate the one, without having
ſome idea of, or forming ſome conjecture about the other.
In viewing the works of nature, we neceſſarily become firſt

L l l   acquainted

acquainted with appearances, or effects. We naturally attend to the circumstances in which such appearances always arise, and cannot help considering them as the causes of those appearances. Then, considering these circumstances themselves as new appearances, we are desirous of tracing out other circumstances that gave birth to them. Thus, constantly ascending in this chain of causes and effects, we are led, at last, to the first cause of all : and then we consider all secondary, and inferior causes, as nothing more than the various methods in which the supreme cause acts, in order to bring about his great designs.

In all science, we first ascend from particular to general. For nature exhibits nothing but particulars; and all general propositions, as well as general terms, are artificial things; being contrived for the ease of our conception and memory; in order to comprehend things clearly, and to comprise as much knowledge as possible in the smallest compass. It is no wonder then that we take pleasure in this process. Besides, we actually see in nature a vast variety of effects proceeding from the same general principles, operating in different circumstances; so that judging from appearances, that nature is every where uniform with itself, we are led, by analogy, to expect the same in all cases; and think it an argument in favour of any system, if it exhibit a variety of effects springing from a few causes. For such variety in effects, and such simplicity in causes, we generally see in nature.

Having discovered the cause of any appearance, it is the business of philosophy to trace it in all its effects, and to predict

dict

dict other similar appearances from similar previous situations of things. By this means, the true philosopher, knowing what will be the result of putting every thing, which the present system exhibits, into every variety of circumstances, is master of all the powers of nature, and can apply them to all the useful purposes of life. Thus does *knowledge*, as Lord Bacon observes, become *power*; and thus is the philosopher capable of providing, in a more effectual manner, both for his own happiness and for that of others; and thereby of approving himself a good citizen, and an useful member of society.

It is obvious, from this general view of the business of philosophy, that, in order to trace those circumstances in which any appearance in nature is certainly and invariably produced, it is chiefly useful to observe what there is *in common* in the circumstances attending similar appearances: for on those common circumstances, all that is common in the appearances must depend. And the easiest possible method, by which we can trace out the connection of causes and effects in nature, is to begin with comparing those appearances which are most similar, where the difference consists in a single circumstance; the whole effect of which, in different appearances, is thereby perfectly known. And when we have, by this means, noted the whole effect of all the separate circumstances and situations of things, we can easily judge of their effect in all possible combinations.

Hence *analogy* is our best guide in all philosophical investigations; and all discoveries, which were not made by mere

accident,

accident, have been made by the help of it. We obferve a complex appearance, attended with a proportionable variety of circumftances and fituations. We alfo fee another appearance, in fome refpects fimilar, in others diffimilar; the circumftances, likewife, fimilar and diffimilar in the fame proportion; or we purpofely vary the circumftances of the former appearance, and obferve what difference it occafions. But, unlefs there be a very great analogy, or fimilarity, between them, fo that the influence of a fingle circumftance, or of a few circumftances, can be traced feparately, no probable judgment can be formed of their real operation.

But in all this procefs, a man who acts from defign, and not abfolutely at random, would never think of trying the influence of any circumftance in an appearance, unlefs, from fome other analogies in nature, more or lefs-perfect, he had formed fome idea what its influence would probably be; at leaft, he muft, from analogies in nature previoufly obferved, have formed an idea of feveral poffible confequences, and try which of them will really follow. That is, in other words, every experiment, in which there is any defign, is made to afcertain fome *hypothefis*. For an hypothefis is nothing more than a preconceived idea of an event, as fuppofed to arife from certain circumftances, which muft have been imagined to have produced the fame, or a fimilar effect, upon other occafions. An hypothefis abfolutely verified ceafes to be termed fuch, and is confidered as a fact; though, when it has long been in an hypothetical ftate, it may continue to be called, occafionally, by the fame name.

The

THE only danger in the use of hypotheses arises from making this transition too soon. And when an hypothesis is no longer considered as a mere probable supposition, but a real fact; a philosopher not only acquiesces in it, and thereby mistakes the cause of one particular appearance; but, by its analogies, he mistakes the cause of other appearances too, and is led into a whole system of error. A philosopher who has been long attached to a favourite hypothesis, and especially if he have distinguished himself by his ingenuity in discovering or pursuing it, will not, sometimes, be convinced of its falsity by the plainest evidence of fact. Thus both himself, and all his followers, are put upon false pursuits, and seem determined to warp the whole course of nature, to suit their manner of conceiving of its operations.

BUT, provided philosophers can be upon their guard against this species of vanity (which must be owned to be very tempting) and against the obstinacy which is the consequence of it; hypotheses, and even a great variety of them, are certainly very promising circumstances to philosophical discoveries. Hypotheses, while they are considered merely as such, lead persons to try a variety of experiments, in order to ascertain them. In these experiments, new facts generally arise. These new facts serve to correct the hypothesis which gave occasion to them. The theory, thus corrected, serves to discover more new facts; which, as before, bring the theory still nearer to the truth. In this progressive state, or method of approximation, things continue; till, by degrees, we may hope that we shall have discovered all the facts, and have formed a perfect theory of them. By this perfect theory, I mean a system

of

of propofitions, accurately defining all the circumftances of every appearance, the feparate effect of each circumftance, and the manner of its operation.

I have dwelt fo long upon this fubject, becaufe I apprehend, that electricians have generally been too much attached to their feveral theories, fo as to have retarded the progrefs of real difcoveries. Indeed, no other part of the whole compafs of philofophy affords fo fine a fcene for ingenious fpeculation. Here the imagination may have full play, in conceiving of the manner in which an invifible agent produces an almoft infinite variety of vifible effects. As the agent is invifible, every philofopher is at liberty to make it whatever he pleafes, and afcribe to it fuch properties and powers as are moft convenient for his purpofe. And, indeed, if he can frame his theory fo as really to fuit all the facts, it has all the evidence of truth that the nature of things can admit.

With the firft electricians, electrical attraction was performed by means of unctuous effluvia emitted by the excited electric. Thefe were fuppofed to faften upon all bodies in their way, and to carry back with them all that were not too heavy. For, in that age of philofophy, all effluvia were fuppofed to return to the bodies from which they were emitted; fince no perfon could, otherwife, account for the fubftance not being fenfibly wafted by the conftant emiffion. When thefe light bodies, on which the unctuous effluvia had faftened, were arrived at the excited electric, a frefh emiffion of the effluvia was fuppofed to carry them back again. But this effect of the effluvia was not thought of, till electrical repulfion had been fufficiently obferved.

When

When the Newtonian philofophy had made fome progrefs, and the extreme fubtilty of light, and other effluvia of bodies, was demonftrated ; fo that philofophers were under no apprehenfion of bodies being wafted by continual emiffion, the doctrine of *the return of effluvia* was univerfally given up, as no longer neceffary ; and they were obliged to acquiefce in the unknown principles of attraction and repulfion, as fuppofed to be properties of certain bodies, communicated to them by the Divine Being, the mechanical caufe of which they fcarce attempted to explain.

When Mr. Du Faye difcovered the two oppofite fpecies of electricity, which he termed the *vitreous* and the *refinous* electricity, he neceffarily formed the idea of *two diftinct electric fluids*, repulfive with refpect to themfelves, and attractive of one another. But he had no idea of both fpecies being actually concerned in every electrical operation, and that glafs or rofin alone always produced them both. This theory, therefore, was as fimple in its application as the other.

While nothing more was known of electricity but attraction and repulfion, this general theory was fufficient. The general attraction of all bodies to all bodies was called (and by fome abfurdly enough fuppofed to be accounted for by) gravitation, and many fuperficial philofophers thought they had given a very good account of electricity, cohefion, and magnetifm, by calling them particular fpecies of attraction peculiar to certain bodies.

But when electricity began to fhow itfelf in a greater variety of appearances, and to make itfelf fenfible to the fmell, the fight, the touch, and the hearing : when bodies were not

only

only attracted and repelled, but made to emit strong sparks
of fire, attended with a considerable noise, a painful sensation,
and a strong phosphoreal smell; electricians were obliged to
make their systems more complex, in proportion as the facts
were so. It was then generally supposed, that the matter of
the electric fluid was the same with the chymical principle of
fire ; though some thought it was a fluid *sui generis*, which
very much resembled that of fire ; and others, with Mr. Boulanger at their head, thought that the electric fluid was
nothing more than the finer parts of the atmosphere, which
crouded upon the surfaces of electric bodies, when the grosser parts had been driven away by the friction of the rubber.

THE great difficulty common to all these theories was to
ascertain the direction of the electric matter. It is no wonder that, when electrical appearances were first observed, all
electric powers were supposed to reside in, and therefore to
proceed from the excited electric. Consequently, the electric spark was first imagined to be darted from the electrified
body towards any conductor that was presented to it. It was
never imagined there could be any difference in this respect
whether it was amber, glass, sealing wax, or any thing else
that was excited. Nothing was thought to be more evident
to the senses than this progress of the electric matter: What
then must have been the astonishment of all electricians, when
they first observed electric appearances at an insulated rubber;
at the same time that it was demonstrated, that the action of
the rubber did not produce, but only collect the electric
fluid.

IN this case, the current could not have been supposed to
flow

flow both from the conductor and the rubber; and yet the first appearances were the same. To provide a supply of the electric matter, they were obliged to suppose that, notwithstanding appearances were nearly the same, the electric fluid was really received by the electrified body in the one case, and emitted by it in the other. But now, being obliged to give up the argument for the manner of its progress from the evidence of sight, they were at a loss whether, in the usual method of electrifying by excited glass, the fluid proceeded from the rubber to the conductor, or from the conductor to the rubber; and nothing was found to obviate these difficulties, till an excellent theory of positive and negative electricity was suggested by Dr. Watson, and digested and illustrated by Dr. Franklin.

It was soon found, that the electricity, at the rubber was the reverse of that at the conductor, and in all respects the same with that which had before been produced by the friction of sealing wax, sulphur, rosin, &c. Seeing, therefore, that both the electricities, as they had heretofore been called, were produced at the same time, by one and the same electric, and by the same friction, all electricians and among the rest Mr. Du Faye himself concluded, that they were both modifications of one and the same fluid; and the old doctrine of the different electricities was universally discarded.

The accidental discovery of the Leyden phial most clearly demonstrated the imperfection of all the theories preceding that of positive and negative electricity, by exhibiting an astonishing appearance, which no electricians, with the help of

M m m   any

any theory, could have foreseen, and of which they could have formed no idea, *a priori.*

Upon this great event, new theories of electricity multiplied apace, so that it would be to no purpose to enumerate them all. Indeed many of them were no more than the beings of a day. For no sooner were they started, but the authors themselves, upon the appearance of some new fact, saw reason to new model, or entirely reject them. I shall, therefore, content myself with giving the outlines of some of the principal theories of electricity, which have their adherents at present, without considering whether they took their rise before or after this discovery.

With some, and particularly Mr. Wilson, the chief agent in all electrical operations is Sir Isaac Newton's ether; which is more or less dense in all bodies in proportion to the smallness of their pores, except that it is much denser in sulphureous and unctuous bodies.* To this ether are ascribed the principal phenomena of attraction and repulsion, whereas the light, the smell, and other sensible qualities of the electric fluid are referred to the grosser particles of bodies, driven from them by the forcible action of this ether. Many phenomena in electricity are also attempted to be explained by means of a subtile medium, at the surface of all bodies, which is the cause of the refraction and reflection of the rays of light, and also resists the entrance and exit of this ether. † This medium, he says, extends to a small distance from the body, and is of the same nature with what is called the electric fluid.

* Wilson's Differtation, p. 5.   † Hoadley and Wilson, p. 55.

On

On the furface of conductors this medium is rare, and eafily admits the paffage of the electric fluid ; whereas on the furface of electrics it is denfe, and refifts it. This medium is rarified by heat, which converts non-conductors into conductors.* On this theory I fhall make no particular remarks, becaufe I cannot fay that I clearly comprehend it.

But the far greater number of philofophers fuppofe, and with the greateft probability, that there is a fluid *fui generis* principally concerned in the bufinefs of electricity. They feem, however, though perhaps without reafon, entirely to overlook Sir Ifaac Newton's ether ; or if they do not fuppofe it to be wholly unconcerned, they allow it only a fecondary and fubordinate part to act in this drama. And among thofe who fuppofe a fluid *fui generis*, there is a great diverfity of opinions about the mode of its exiftence, and the manner of its operation.

The ingenious Abbé Nollet whofe theory has been more the fubject of debate than all the other theories before Dr. Franklin's, fuppofes that, in all electrical operations, the fluid is thrown into two oppofite motions ; that the *affluence* of this matter drives all light bodies before it, by impulfe, upon the electrified body, and its *effluence* carries them back again. But he feems very much embarraffed in accounting for facts where both thefe currents muft be confidered, at the fame time that he is obliged to find expedients to prevent their impeding the effects of each other. To obviate this great difficulty, he fuppofes, that every excited electric, and likewife

* Ib. p. 78.

M m m 2   every

every body to which electricity is communicated, has two orders of pores, one for the emission of the effluvia, and the other for the reception of them. A man of less ingenuity than the Abbé could not have maintained himself in such a theory as this; but, with his fund of invention, he was never at a loss for resources upon all emergencies, and in his last publication appears to be as zealous for this strange hypothesis as at the first.

He more than once requested a deputation of the members of the Academy of Sciences, to be witnesses of some experiments, in which, he thought, there was a visible effluence of the electrical effluvia from the conductor, both to the globe at one of its extremities, and to any non-electric presented to it at the other; and their testimony was signed and registered in proper form. * But it does not seem to the honour of Mr. Nollet, or those gentlemen of the Academy, to be so very positive in a matter which does not admit of the evidence of sense.

The Abbé's confidence upon this subject is very remarkable. These effects, says he, well considered, and reviewed a thousand times, in the course of thirty years, in which I have applied to electricity, make me say with confidence, that those pencils of rays are currents of electric matter, which fly from the conductor towards the excited globe. This is so evident, that I would freely appeal to the ocular testimony of any unprejudiced person, who should see the experiments which I have recited. But, says he, the fact in question is contrary

* Leçons de Physique, p. 368, 395.

to

to a fyftem of electricity, which fome perfons perfift in main-
taining.   They have the affurance to tell me, that the matter
of the luminous pencil, in my experiment, moves in a direction quite oppofite to that which I fuppofe; that it proceeds
from the excited globe, and is from thence thrown upon any
non-electric within its reach. *   In another place, he fays,
that the principle of fimultaneous effluences and affluences is
by no means a *fyftem,* but a *fact* well proved. †

THE Abbé Nollet propofes an hypothefis to explain the
difference between common electricity and the electric
fhock.   All the effects of common electricity, he fays, plainly fhow, that the electric matter is animated with a progreffive motion, which really carries it forwards; whereas the
remarkable cafe of the electric fhock appears to be an inftantaneous percuffion, which the contiguous parts of the fame
matter communicate to one another, without being difplaced.
Sound and wind, he fays, are motions of the air; but would
a philofopher, be permitted to take the one for the other,
in meafuring their velocity or extent.   But this comparifon
is by no means juft. ‡

IT muft be acknowledged, that far the greater part of the
Abbé Nollet's arguments in favour of his doctrine of effluences and affluences are very unfatisfactory, and that his method
of accounting for electrical attraction and repulfion, with o-
ther phenomena in electricity, by means of it, is more ingenious than folid.   It is a great pity that this truly excellent philofopher had not fpent more time in diverfifying facts, and

* Ib. p. 363.   † Lettres fur l' electricité, p. 98.   ‡ Lecons de Phyfique, 293.

less

leſs in refining upon theory. But it is in ſome meaſure the natural fault of a diſpoſition to philoſophize.

MR. DU TOUR improves upon this hypotheſis of the Abbé
Nollet, by ſuppoſing that there is a difference between the
affluent and effluent current; and that the particles of the
fluid are thrown into vibrations of different qualities, which
makes one of theſe currents more copious than the other,
according as ſulphur or glaſs is uſed. Difficult as it is to form
any idea of this hypotheſis, the author appears very much
attached to it, and has no doubt of its accounting for all
electrical appearances.

SECTION

SECTION II.

THE THEORY OF POSITIVE AND NEGATIVE
ELECTRICITY.

THE English philosophers, and perhaps the greater part of foreigners too, have now generally adopted the theory of *positive* and *negative* electricity. As this theory has been extended to almost all the phenomena, and is the most probable of any that have been hitherto proposed to the world, I shall give a pretty full account of it, and show how it agrees with all the propositions of the last part, to which it has hitherto been applied.

THIS theory generally goes by the name of Dr. Franklin, and there is no doubt of his right to it; but justice requires that I distinctly mention the equal, and, perhaps, prior claim of Dr. Watson, to whom I have before said it had occurred. Dr. Watson showed a series of experiments to confirm the doctrine of *plus* and *minus* electricity to Martin Folkes Esq. then president, and to a great number of fellows of the Royal Society, so early as the beginning of the year 1747, before it was known in England that Dr. Franklin had discovered
the

the fame thing in America. See the Philofophical Tranfactions, Vol. 44, p. 739; and Vol. 45, p. 93———101. Dr. Franklin's paper, containing the fame difcovery, was dated at Philadelphia, June the 1ft. 1747.

ACCORDING to this theory, all the operations of electricity depend upon one fluid *fui generis*, extremely fubtile and elaftic, difperfed through the pores of all bodies; by which the particles of it are as ftrongly attracted, as they are repelled by one another.

WHEN the equilibrium of this fluid in any body is not difturbed; that is, when there is in any body neither more nor lefs of it than its natural fhare, or than that quantity which it is capable of retaining by its own attraction, it does not difcover itfelf to our fenfes by any effect. The action of the rubber upon an electric difturbs this equilibrium, occafioning a deficiency of the fluid in one place, and a redundancy of it in another.

THIS equilibrium being forcibly difturbed, the mutual repulfion of the particles of the fluid is necefiarily exerted to reftore it. If two bodies be both of them overcharged, the electric atmofpheres (to adopt the ideas of all the patrons of this hypothefis before Æpinus) repel each other, and both the bodies recede from one another to places where the fluid is lefs denfe. For, as there is fuppofed to be a mutual attraction between all bodies and the electric fluid, electrified bodies go along with their atmofpheres. If both the bodies be exhaufted of their natural fhare of this fluid, they are both attracted by the denfer fluid, exifting, either in the atmofphere contiguous to them, or in other neighbouring bodies;

which

which occasions them still to recede from one another, as much as when they were overcharged.

Some of the patrons of the hypothesis of positive and negative electricity conceive otherwise of the immediate cause of this repulsion. They say that, as the denser electric fluid, surrounding two bodies negatively electrified, acts equally on all sides of those bodies, it cannot occasion their repulsion. Is not the repulsion, say they, owing rather to an accumulation of the electric fluid on the surfaces of the two bodies; which accumulation is produced by the attraction of the bodies, and the difficulty the fluid finds in entering them? This difficulty in entering is supposed to be owing, chiefly, to the *air* on the surface of bodies, which is probably a little condensed there; as may appear from Mr. Canton's experiment above-mentioned on the double barometer.

Lastly, if one of the bodies have an overplus of the fluid, and the other a deficiency of it, the equilibrium is restored with great violence, and all electrical appearances between them are more striking.

The influence of points in drawing or throwing off the electric fluid has not been quite satisfactorily accounted for upon any hypothesis, but it is as agreeable to this as any other. As it is evident that every electric atmosphere meets with some resistance, both in entering and quitting any body, whatever be the cause of that resistance, it is natural to suppose, that it must be least at the points of bodies where there are fewer particles of the body (on which the resistance depends) opposed to its passage, than at the flat parts of the

N n n   surface,

surface, where the refisting power of a greater number of particles is united.

The *light* which is visible in electrical appearances is generally fuppofed to be part of the compofition of the electric fluid, which appears when it is properly agitated. But this fuppofition concerning electric light is not neceffary to the general hypothefis. It may be fuppofed, upon this as well as Mr. Wilfon's theory, that the light, and the phofphoreal fmell, in electrical experiments arife from particles of matter much groffer than the proper electric fluid, but which may be driven from bodies by its powerful action.

The *found* of an electrical explofion is certainly produced by the air being difplaced by the electric fluid, and then fuddenly collapfing, fo as to occafion a vibration, which diffufes itfelf every way from the place where the explofion was made. For in fuch vibrations found is known to confift.

But the chief excellence of this theory of pofitive and negative electricity, and that which gave it the greateft reputation, is the eafy explication which it fuggefts of all the phenomena of the Leyden phial. This fluid is fuppofed to move with the greateft eafe in bodies which are conductors, but with extreme difficulty in electrics per fe; infomuch that glafs is abfolutely impermeable to it. It is moreover fuppofed, that all electrics (and particularly glafs) on account of the fmallnefs of their pores, do at all times contain an exceeding great, and always an equal quantity of this fluid; fo that no more can be thrown into one part of any electric fubftance, except the fame quantity go out at another, and the gain be exactly equal to the lofs. Thefe things being

previoufly

previously supposed, the phenomena of charging and discharging a plate of glass admit of an easy solution.

In the usual manner of electrifying, by a smooth glass
globe, all the electric matter is supplied by the rubber from
all the bodies which communicate with it. If it be made to
communicate with nothing but one of the coatings of a plate
of glass, while the conductor communicates with the other,
that side of the glass which communicates with the rubber
must necessarily be exhausted, in order to supply the conductor, which must convey the whole of it to the side with which
it communicates. By this operation, therefore, the electric
fluid becomes almost entirely exhausted on one side of
the plate, while it is as much accumulated on the other;
and the discharge is made by the electric fluid rushing,
as soon as an opportunity is given it, by means of proper
conductors, from the side which was overloaded to that which
was exhausted.

It is not necessary, however, to this theory, that the very
same individual particles of electric matter which were thrown
upon one side of the plate, should make the whole circuit of
the intervening conductors, especially in very great distances,
so as actually to arrive at the exhausted side. It may be sufficient to suppose, as was observed before, that the additional
quantity of fluid displaces and occupies the space of an equal
portion of the natural quantity of fluid belonging to those
conductors in the circuit, which lay contiguous to the charged
side of the glass. This displaced fluid may drive forwards an
equal quantity of the same matter in the next conductors;

N n n 2   and

and thus the progreſs may continue, till the exhauſted ſide of the glaſs is ſupplied by the fluid naturally exiſting in the con- ductors contiguous to it. In this caſe the motion of the elec- tric fluid in an exploſion will rather reſemble the vibration of the air in ſounds, than a current of it in winds.

It will eaſily be acknowledged, that while the ſubſtance of the glaſs is ſuppoſed to contain as much as it can poſſibly hold of the electric fluid, no part of it can be forced into one of the ſides, without obliging an equal quantity to quit the other ſide; but it may be thought a difficulty upon this hy- potheſis, that one of the ſides of a glaſs plate cannot be *ex- hauſted*, without the other receiving more than its natural ſhare, particularly as the particles of this fluid are ſuppoſed to be repulſive of one another. But it muſt be conſidered, that the attraction of the glaſs is ſufficient to retain even the large quantity of the electric fluid which is natural to it, a- gainſt all attempts to withdraw it, unleſs that eager attrac- tion can be ſatisfied by the admiſſion of an equal quantity from ſome other quarter. When this opportunity of a ſup- ply is given, by connecting one of the coatings with the rubber, and the other with the conductor, the two attempts to introduce more of the fluid into one of the ſides, and to take from it on the other, are made, in a manner, at the ſame inſtant. The action of the rubber tends to diſturb the equilibrium of the fluid in the glaſs, and no ſooner has a ſpark quitted one of the ſides, to go to the rubber, but it is ſupplied by the conductor on the other; and the difficulty with which theſe additional particles move in the ſubſtance of

the

the glafs effectually prevents its reaching the oppofite exhaufted fide; near as the two fides are to one another, and eager as is the endeavour of the fluid to go whither it is fo ftrongly attracted.

It is not faid, however, but that either fide of the glafs may give or receive *a fmall quantity* of the electric fluid, without altering the quantity on the oppofite fide. It is only a very confiderable part of the charge that is meant, when one fide is faid to be filled, while the other is exhaufted.

It is a little remarkable, that the electric fluid, in this, and in every other hypothefis, fhould fo much refemble the ether of Sir Ifaac Newton in fome refpects, and yet differ from it fo effentially in others. The electric fluid is fuppofed to be, like ether, extremely fubtile and elaftic, that is, repulfive of itfelf; but, inftead of being, like the ether, repelled by all other matter, it is ftrongly attracted by it; fo that, far from being, like the ether, rarer in the fmall than in the large pores of bodies, rarer within the bodies than at their furfaces, and rarer at their furfaces than at any diftance from them; it muft be denfer in fmall than in large pores, denfer within the fubftance of bodies than at their furfaces, and denfer at their furfaces than at a diftance from them. But no other property can account for the extraordinary quantity of this fluid contained within the fubftance of electrics per fe, or for the common atmofpheres of all excited and electrified bodies.

To account for the attraction of light bodies, and other electrical appearances, in air of the fame denfity with the common atmofphere, when glafs (which is fuppofed to be

impermeable

impermeable to electricity) is interposed; it is conceived, that the addition or substraction of the electric fluid, by the action of the excited electric, on one side of the glass, occasions (as in the experiment of the Leyden phial) a substraction or addition of the fluid on the opposite side. The state of the fluid, therefore, on the opposite side being altered, all light bodies within the sphere of its action must be affected, in the very same manner as if the effluvia of the excited electric had actually penetrated the glass, according to the opinion of all electricians before Dr. Franklin.

THE manner in which clouds acquire their positive or negative electricity is not determined, according to this, or any other theory, with sufficient certainty. Mr. Canton's conjecture is, that the air resembles the tourmalin, and, consequently, acquires its electricity by heating or cooling; but whether it gains or loses the electric fluid in either state must be determined by experiment. Signior Beccaria's theory of the electricity of the clouds has been related at large.

THIS hypothesis of positive and negative electricity has been adopted, and, in some measure, rendered more systematical by Æpinus, in his elaborate treatise entitled, *Tentamen Theoriæ Electricitatis et Magnetismi.*

HE has extended the property of impermeability to air and all electrics as well as glass, and defined it in a better manner. He supposes impermeability to consist in the great difficulty with which electric substances admit the electric fluid into their pores, and the slowness with which it moves in them. Moreover, in consequence of this impermeability of air to the electric fluid, he denies the reality of electric atmospheres,

and

and thinks, as was obferved before, that Dr. Franklin's theory will do much better without them.

He thinks that all the particles of matter muft repel one another: for that, otherwife (fince all fubftances have in them a certain quantity of the electric fluid, the particles of which repel one another, and are attracted by all other matter) it could not happen, that bodies in their natural ftate, with refpect to electricity, fhould neither attract nor repel one another.

He that reads the firft chapter, as well as many other parts of his elaborate treatife above-mentioned, may fave a good deal of time and trouble by confidering, that the refult of many of his reafonings and mathematical calculations cannot be depended upon ; becaufe he fuppofes the repulfion or elafticity the electric fluid to be in proportion to its condenfation ; which is not true, unlefs the particles repel one another in the fimple reciprocal ratio of their diftances, as Sir Ifaac Newton has demonftrated, in the fecond book of his Principia.

Mr. WILKE, as well as Æpinus, adopts all the general principles of Dr. Franklin's theory of pofitive and negative electricity, but thinks that no experiments which have hitherto been made fhow which of the electricities is pofitive and which negative. Suppofing, however, what is called pofitive to be really fo, and that fmooth glafs, for inftance, rubbed upon fulphur attracts the electric fluid from it, he would account for it upon the fame principles whereby water ftands in drops on rough furfaces, but is diffufed on fmooth ones. The electric fluid, he would fuppofe, is more ftrongly attracted by the

the smooth surface of the glass, and therefore diffuses itself over it, while it retreats from electrics of rougher surfaces. * But this explanation, I imagine, will give little satisfaction to sceptical electricians.

MR. WILKE acknowledges there is great difficulty in accounting for the repulsive power of bodies electrified negatively, and thinks that it obliges us to suppose the mutual repulsion of all homogeneous matter. Mr. Waitz, he says, was of the same opinion. Mr. Wilke observes upon this subject, that the attraction of light bodies to negative electrics cannot be owing to the repulsive power of the electric fluid in the neighbouring air, driving them, or the electric matter in them to the place where there is a want of it ; because the velocity ought to decrease as it recedes from the impulsive power : whereas it is accelerated, as if it were attracted by the negative electric. †

BUT to this it may be replied, that a succession of impulses, though every subsequent one should be weaker than the preceding, will produce an accelerated motion. Besides, the nearer the light body is to the negative electric, the nearer it is to the point where the equilibrium of the fluid is most destroyed ; or the less force there is on the side of the electric to balance the force that drives the light body towards it, and therefore the impulses themselves must increase.

MR. WILKE, whose treatise on the two electricities is admirable, both for its materials, and methodical arrangement of them, distinguishes three causes of excitation, viz. *warm-*

* Wilke, p. 65.   † Ib. p. 15.

ing,

ing, *liquefaction*, and *friction* ; and he advifes, that we care-fully diftinguifh between *fpontaneous* and *communicated* electri-city.   By the former he means that which is the refult of the appofition, or mutual action of two bodies ; in confequence of which, one of them is left pofitively electric and the other negative.   Whereas *communicated* electricity is that which is fuperinduced upon a body, or part of a body, electric or non-electric, without its being previoufly heated, melted, or rub-bed ; or without any mutual action between it and any other body.   This diftinction is, in general, very obvious ; but Mr. Wilke defines it more accurately than it had been done before, and mentions feveral cafes in which they are often confounded.

Signior Beccaria admits the theory of pofitive and ne-gative electricity, though he explains fome electrical pheno-mena in a manner different from the other patrons of that fyftem.

He fuppofes that electrified bodies move to one another only in the act of giving and receiving the electric fluid : *
this effect being produced by the electric matter making a vacuum in its paffage, and the contiguous air afterwards col-lapfing, and thereby pufhing the bodies together. †   This vacuum, he fays, is very obfervable upon great explofions of thunder, when animals have been ftruck dead without being touched with the lightning ; a vacuum being only fuddenly made near them, and the air immediately rufhing out of their lungs to fill it, whereby they are left flaccid and empty ;

* Lettere dell' elettricifmo, p. 36.   † Ib. p. 41.

O o o   whereas

whereas when persons are properly killed by lightning, their lungs are found distended. *

IN confirmation of this hypothesis, he says, that less motion is given to bodies by electricity as the air is excluded from them, and that in vacuo no motion at all can be given to them. † He also says, that no electric light is visible in a barometer in which there is a perfect vacuum : whence he infers, that electric light is visible only by means of some vibrations which it excites in the air. ‡

THIS hypothesis seems to be unworthy of so eminent an electrician, and it is certainly not fact, that there is no electric light, if there be no electrical attraction in vacuo.

To account for the collection or dissipation of electricity by points, he says that the electric fluid appears, from experiments, to move with the greatest violence in the smallest bodies. All electrical appearances will, therefore, be most sensible at the points of bodies; and, consequently, it will be soonest dissipated there. But this does not seem to touch the real difficulty.

DR. FRANKLIN the author of this excellent theory of positive and negative electricity, with a truly philosophical greatness of mind, to which few persons have ever attained, always mentions it with the utmost diffidence. Every appearance, says he, which I have yet seen, in which glass and electricity are concerned, are, I think, explained with ease by this hypothesis. Yet, perhaps, it may not be a true one, and I shall be obliged to him who affords me a better. ||

* Lettere dell' elettricismo, p. 42.   † Ib. p. 48.
‡ Ib. p. 50. || Franklin's Letters, p. 78.

IT

IT is no wonder, indeed, that this excellent philofopher fhould treat even his own hypothefis with fuch indifference, when he had fo juft a fenfe of the nature, ufe, and importance of all hypothefes. Nor is it, fays he, of much importance to us, to know the manner in which nature executes her laws. It is enough if we know the laws themfelves. It is of real ufe to us to know that china left in the air, unfupported, will fall and break; but how it comes to fall, and why it breaks, are matters of fpeculation. It is a pleafure indeed to know them, but we can preferve our china without it. *

THE great merit of this writer as an electrician ftands independent of all hypothefes, upon the firm bafis of the difcovery of many new and important facts. and, what is more, applied to the greateft ufes. Suppofing him, for inftance, to have been miftaken in his account how the clouds come to be poffeffed of electricity, muft not all the world acknowledge themfelves indebted to him for the difcovery of the famenefs of the electric fluid and the matter of lightning; and efpecially for a certain method of preferving their buildings and perfons from the fatal effects of thunder ftorms.

* Franklin's Letters, p. 59.

O o o 2   SECTION

SECTION III.

OF THE THEORY OF TWO ELECTRIC FLUIDS.

CONVINCED, as the reader may have perceived that I am, of the ufefulnefs of various theories, as fuggefting a variety of experiments, which lead to the difcovery of new facts; he will excufe me, if I recall his attention to the old theory of vitreous and refinous electricity, as it was firft fuggefted by Mr. Du Faye, upon his difcovery of the different properties of excited glafs, and excited amber, fulphur, rofin, &c. and as it has been new modelled by Mr. Symmer. To fhow my abfolute impartiality, I fhall, notwithftanding the preference I have given to Dr. Franklin's theory, endeavour to reprefent this to as much advantage as poffible, and to do it more juftice than has yet been done to it, even by Mr. Symmer himfelf; who, as I obferved before, has fallen into fome miftakes in his application of it. Indeed, hitherto very little pains has been taken with this theory, nor has it been extended to any great variety of phenomena,

LET us fuppofe then, that there are two electric fluids, which have a ftrong chymical affinity with each other, at the fame time that the particles of each are as ftrongly repulfive of

one

one another.   Let us suppose these two fluids, in some measure, equally attracted by all bodies, and existing in intimate
union in their pores, and while they continue in this union to
exhibit no mark of their existence.   Let us suppose that the
friction of any electric produces a separation of these two fluids, causing (in the usual method of electrifying) the vitreous
electricity of the rubber to be conveyed to the conductor, and
the resinous electricity of the conductor to be conveyed to the
rubber.   The rubber will then have a double share of the resinous electricity, and the conductor a double share of the
vitreous; so that, upon this hypothesis, no substance whatever can have a greater or less quantity of electric fluid at
different times; the quality of it only can be changed.

THE two electric fluids, being thus separated, will begin
to show their respective powers, and their eagerness to rush
into reunion with one another.   With whichsoever of these
fluids a number of bodies are charged, they will repel one a-
nother, they will be attracted by all bodies which have a less
share of that particular fluid with which they are loaded, but
will be much more strongly attracted by bodies which are
wholly destitute of it, and loaded with the other.   In this
case they will rush together with great violence.

UPON this theory, every electric spark consists of both the
fluids rushing contrary ways, and making a double current.
When, for instance, I present my finger to a conductor loaded with vitreous electricity, I discharge it of part of the vitreous, and return as much of the resinous, which is supplied to
my body from the earth.   Thus both the bodies are unelectrified, the balance of the two powers being perfectly restored.

WHEN

WHEN I present the Leyden phial to be charged, and, consequently, connect the coating of one of its sides with the rubber, and that of the other with the conductor, the vitreous electricity of that side which is connected with the conductor is transmitted to that which is connected with the rubber, which returns an equal quantity of its resinous electricity; so that all the vitreous electricity is conveyed to one of the sides, and all the resinous to the other. These two fluids, being thus separated, attract one another very strongly through the thin substance of the intervening glass, and rush together with great violence, whenever an opportunity is presented, by means of proper conductors. Sometimes they will force a passage through the substance of the glass itself; and, in the mean time, their mutual attraction is stronger than any force that can be applied to draw away either of the fluids separately.

HAVING stated the general principles of this hypothesis of two fluids, I shall now enter into a brief comparison of it with that of a single fluid, as explained by the mode of positive and negative electricity; that we may see which of them will account for the same facts in the easier manner, and more agreeable to the analogy of nature in other respects. For, allowing that no fact can be shown to be absolutely inconsistent with either of them; yet, certainly, that will be judged preferable, which is attended with the least difficulty in conceiving of its mode of operation.

IN the first place, the supposition itself, of two fluids, is not quite so easy as that of one, though it is far from being disagreeable to the analogy of nature, which abounds with affinities,

and

and in which we fee innumerable cafes of fubftances formed, as it were to unite with and counteract one another.   Here, likewife, agreeable to the theory of two electric fluids, while thofe fubftances are in union, we fee nothing of their feparate and peculiar powers, though they be ever fo remarkable. What, for inftance, do we fee of the ftriking properties of the *acid* and *alkali* while they are united in a neutral falt? What powers in nature are more formidable than the vitriolic acid, and phlogifton (which confifts principally, if not wholly, of mephitic air) and what more innocent than common fulphur, which is a compofition of them both, and from which the action of fire feparates them.

THE two fluids being fuppofed, the double current from the rubber to the conductor and from the conductor to the rubber is an eafy and neceffary confequence.   For if, upon the common fuppofition, the action of the rubber puts a fingle fluid into motion in one direction, we might expect that, if there were two fluids, which counteracted each other, they would, by the fame operation, be made to move in contrary directions.   And a perfon who has been ufed to conceive that a fingle fluid may be made to move either way, viz. from the rubber to the conductor, or from the conductor to the rubber, at pleafure, according as a rough or a fmooth globe is ufed, can have much lefs objection to this part of the hypothefis.

ADMITTING then this different action of the rubber and the electric upon the two different fluids, the manner of conveying electric atmofpheres, or powers to bodies is the fame on this as on any other theory; and it is apprehended, that
the

the phenomena of negative electricity are more easily conceived by the help of a real fluid, than by no fluid at all.
Indeed Dr. Franklin himself ingenuously acknowledges, that
he was a long time puzzled to account for bodies that were
negatively electrified repelling one another; whereas Mr. Du
Faye, who observed the same fact, had no difficulty about
it, supposing that he had discovered another electricity, similar, with respect to the properties of elasticity and repulsion,
to the former.

By this double action of the rubber, the method of charging a plate of glass is exceeding easy to conceive. Upon this
hypothesis, all the vitreous electricity quits its union with the
resinous on the side communicating with the conductor, and
is brought over to the side communicating with the rubber;
which, by the same operation, had been made to part with
its resinous electricity in return.

All the vitreous electricity being thus brought to one side
of the plate of glass, and all the resinous to the other, the
phenomena of the plate while standing charged, or when
discharged, are, perhaps, more free from all difficulty than
upon any other hypothesis. When one of the sides of the glass
is conceived to be loaded with one kind of electricity, and
the other side with the other kind; the strong affinity between them, whereby they attract each other with a force
proportioned to their nearness, immediately supplies a satisfactory reason, why so little of either of the fluids can be drawn
from one of the sides without communicating as much to the
other. Upon this supposition, that consequence is perhaps
more obvious than upon the supposition of one half of the

glass

glaſs being crowded with the electric matter, and the other half exhauſted. In the former caſe, every attempt to withdraw the fluid from one of the ſides is oppoſed by the more powerful attraction of the other fluid on the oppoſite ſide. On the other hypotheſis, it is only oppoſed by the attraction of the empty pores of the glaſs.

LASTLY, the exploſion upon the diſcharge of the glaſs has as much the appearance of two fluids ruſhing into union, in two oppoſite directions, as of one fluid, proceeding only in one direction. The ſame may be ſaid of the appearance of every common electric ſpark, in which, upon this hypotheſis, there is always ſuppoſed to be two currents, one from the electric, or the electrified body, and the other to it.

I DO not ſay that the bur which is uſually ſeen on both ſides of a quire of paper pierced by an electric exploſion, and the current of air flowing from the points of all bodies electrified negatively as well as poſitively, are material objections to the doctrine of a ſingle fluid. I have even ſhown how they may be explained in a manner conſiſtent with it; but upon the ſuppoſition of two fluids, and two currents, the difficulty of accounting for theſe facts would hardly have occurred.

THE phenomena of diſcharging a plate of glaſs, upon the hypotheſis of two fluids, are indeed, injudiciouſly explained by Mr. Symmer; who ſuppoſes that the two fluids do not always make the whole circuit of the intervening conductors, but enter them more or leſs, from each ſide of the plate, according to the ſtrength of the charge. But upon this ſuppoſition, the fire of the ſmalleſt charge performs the

P p p   whole

whole circuit, as well as the fire of the greateſt, in order to reſtore the equilibrium of the two fluids on each ſide of the glaſs.

IT is almoſt needleſs to obſerve, that the influence of points is attended with exactly the ſame difficulty upon this theory as upon the other. It is equally eaſy, or equally difficult, to ſuppoſe one fluid to enter and go out at the point of an e-lectrified conductor, at different times, as to ſuppoſe that, of two fluids, one goes out, and the other goes in, at the ſame time.

THAT bodies immerged in electric atmoſpheres muſt acquire the contrary electricity, is quite as eaſy to conceive upon this, as upon any other hypotheſis. For, in this caſe, ſuppoſe the electrified body to be poſſeſſed of the vitreous electricity, all the vitreous electricity of the body which is brought near it will be driven backwards, to the more diſtant parts, and all the reſinous electricity will be drawn forwards. And when the attraction between the two electricities, in theſe different bodies, is ſo great, as to overcome the oppoſition to their union, occaſioned by the attraction of the bodies that contained them, the form of their ſurfaces, and the reſiſtance of the interpoſing medium, they will ruſh together; an electric ſpark will be viſible between them, and the electricity of both will appear to be diſcharged; the prevailing electricity of each being ſaturated with an equal quantity of the oppoſite kind, from the other body.

THIS hypotheſis will likewiſe eaſily account for the difficulty of charging a very-thick plate of glaſs, and the impoſſibility of charging it beyond a certain thickneſs: for theſe

fluids

fluids at a greater diſtance, will attract one another leſs forcibly; and at a certain ſtill greater diſtance will not attract
at all.

HAVING given the moſt favourable view that I can of this
hypotheſis of two electric fluids, I ſhall, with the ſame fairneſs,
make the beſt anſwer I am able to the principal objection
that will probably be made to it.

IF it be aſked, why the two fluids, meeting on the ſurface
of the globe, or in the electric exploſion, do not unite, by
means of their ſtrong affinity, and make no further progreſs;
it may be anſwered, that the attraction between all other
bodies and the particles of both theſe fluids may be ſuppoſed
to be, at leaſt, as ſtrong, as the affinity between the fluids
themſelves; ſo that the moment that any body is diſpoſſeſſed
of one, it may recruit itſelf, to its uſual point of ſaturation,
from the other.

BESIDES, in whatever manner it be that one of the electric fluids is diſlodged from any body (ſince, upon every theory, the two electricities are always produced at the ſame
time) the oppoſite electricity will, by the ſame action, be
diſlodged from the other ſubſtance. And (as upon the common theory) whatever it be that diſlodges the fluid from
any ſubſtance, it will be ſufficient to prevent its return;
conſequently, ſuppoſing both the ſubſtances neceſſarily to
have a certain proportion of electric matter, each may be
immediately ſupplied from that which was diſlodged from
the other.

THE rubber, therefore, at the time of excitation, gives its
vitreous electricity to that part of the ſmooth glaſs againſt
which

which it has been preffed, and takes an equal quantity of the refinous in return. The glafs, being a non-conductor, does not allow this additional quantity of vitreous electricity to enter its fubftance. It is therefore diffufed upon the furface; and, in the revolution of the globe, is carried to the prime conductor. There (as in the experiments begun by Mr. Canton, and profecuted by Mr. Wilke, &c.) it repels the vitreous, and violently attracts the refinous electricity; and (the points of the conductor favouring the mutual tranfition) the vitreous, which abounds upon the globe, paffes to the conductor; and the refinous, which abounds on the neareft parts of the conductor, rufhes upon the globe. There it mixes with, and faturates what remained of the vitreous electricity, on the part on which it flows, and thereby reduces it to the fame ftate in which it was before it was firft excited. Every part of the furface of the globe performs the fame office, firft exchanging electricities with the rubber, and then with the conductor.

THE folution of this difficulty will likewife folve that of the electric explofion, in which there is a collifion, as it were, of the two fluids, while yet they completely pafs one another. For ftill each furface of the glafs may be fuppofed to require its certain portion of electric matter, and therefore cannot part with one fort without receiving an equal quantity of the other. It muft be confidered alfo, that the air, through which thefe fluids pafs, has already its natural quantity of electricity, fo that being fully faturated, it can contain no more; and that the two fluids only rufh to the places from whence they had been forcibly diflodged, and where the greater body of the oppofite fluid waits to embrace them.

MR.

Mr. Symmer's hypothefis of a double current differed in fome refpects from that of the Abbe Nollet. The Abbe, however, according to his ufual candour, fpeaks of him with the higheft refpect; at the fame time, he ftill appears an advocate for his old favourite hypothefis.

Johannes Franciscus Cigna, who purfued the experiments above recited of Mr. Symmer, obferves, with refpect to his theory; that it is not contradicted by any phenomena that are yet known, and that it fuits fome of them in a peculiarly clear and elegant manner; particularly every thing relating to charging and difcharging a plate of glafs; all the experiments which feem to fhow a mutual attraction between the two electricities, when they are kept afunder; and that curious experiment above-mentioned of Signior Beccaria, of difcharging a plate of glafs fufpended by a filken ftring, without either touching or moving the plate. Yet, upon the whole, he declares in favour of Dr. Franklin's theory of pofitive and negative electricity, on account of its admirable fimplicity, and becaufe philofophers ought not to multiply caufes without neceffity.

Dr. Franklin's theory, he fays, completely folves all the cafes of the two electricities deftroying one another when they are mixed; but doth not fo clearly account for their attracting, and counteracting one another when they are feparate. He concludes with faying, he doth not chufe to fay much on fo very obfcure a queftion, which has divided the opinions of very great men; and that any hypothefis of the two electricities, which will account for the deftruction of all

the

the signs of electricity when they are united, and their mutual attraction when they are separate, will equally suit all the phenomena that have yet been discovered.

I HAVE taken a little pains with this theory, because I thought, it had been hitherto, too much overlooked, and that sufficient justice had not been done to it, even by those who proposed it. For the future, I hope it will be seen to more advantage, and appear a little more respectable among its sister hypotheses; and then, *valeat quantum valere potest.* If any electrician will favour me with the communication of any other theory, not obviously contradicted by facts, I shall think myself obliged to him, and shall think I do a piece of real service to the science in the publication of it. If more persons favour me with more different theories, I should think my book, as far as theories are of any use, so much the more valuable.

SECTION

PART IV.

DESIDERATA in the science of electricity, and hints for the further extension of it.

SECTION I.

GENERAL OBSERVATIONS ON THE PRESENT STATE OF ELECTRICITY.

THAT real progrefs has been made in electricity, has, I prefume, been fufficiently demonftrated in the courfe of the preceding hiftory; that a great deal ftill remains to be done, will, I think, be evident from this part of the work. Thofe perfons who think that nothing has been done to any purpofe in Natural Philofophy, or that the advances have been made very flowly, fince the time of Sir

Ifaac

Iſaac Newton, need only read the preceding hiſtory, to be convinced, both that a great deal has been done, and that the progreſs in this kind of knowledge, inſtead of being ſlow, has been amazingly rapid. To quicken the ſpeed of philo-ſophers in purſuing this progreſs, and at the ſame time, in ſome meaſure, to facilitate it, is the intention of this treatiſe, and more eſpecially of this part of it. When a traveller imagines he is near his journey's end, he is little ſolicitous about making diſpatch, thinking that, without any haſte, the labour of the day will quickly be over; whereas, if he find that, whatever progreſs he may have made, he has a great deal ſtill to make, he continues, or quickens his ſpeed.

THE principal reaſon why many ingenious perſons have ſo ſoon got to their *ne plus ultra* in philoſophical diſcoveries, has evidently been their attachment to favourite theories; which they imagined both accounted for all the phenomena that had been obſerved, and would likewiſe account for all that ſhould be obſerved. Having therefore attained to the great object of a ſcience, and diſcovered the ultimate and moſt general principles of it, there was nothing more that was worth their notice; it being beneath men of genius to ſpend their time in diverſifying effects, when there were no new cauſes to be found. I hope that what has hitherto been ſaid concerning the nature and uſe of hypotheſes, and about the progreſs and preſent imperfect ſtate of thoſe which reſpect electricity, will convince thoſe electricians who may not yet have been con-vinced of it, that our buſineſs is ſtill chiefly with *facts*, and the *analogy of facts*; that far too few of theſe have been diſ-covered to aſcertain a perfect general theory, and that all that

the

the prefent hypothefes can do for us muft confift in fuggefting further experiments.

If we look back upon the hiftory of electricity, and confider the ftate of facts and of hypothefes at any particular period of time paft, we fhall fee that there was always the fame apparent reafon for acquiefcing in what had been done, as at prefent. The theories of the firft electricians, lame and imperfect as they were, were yet fufficient to account for all the facts they were acquainted with; and as for other facts, they could have no idea or apprehenfion of them, and therefore could not be folicitous about them.

Mr. Boyle, no doubt, was as fully fatisfied with his fimple hypothefis of the unctuous effluvia, as Mr. Nollet with his theory of affluences and effluences; or the greateft part of the prefent race of electricians with that of pofitive and negative electricity. Mr. Hawkefbee, when he made his furprifing difcoveries concerning the properties of electric light, and many curious circumftances concerning electric attraction and repulfion, might very naturally think that little more was to be done. Indeed, who could have thought otherwife, when the fcience was actually at a ftand for feveral years after him? All that the indefatigable Mr. Grey (who made the great difcovery of the communication of electric powers to bodies not electric per fe) imagined to remain undone, were mere chimeras and illufions. Mr. Du Faye, who made the difcovery of vitreous and refinous electricity, had no idea of the electric fhock; and the German philofophers, who accidentally obferved it, knew nothing of its moft remarkable properties. Notwithftanding a great number of

Q q q

treatifes

treatifes on the fubject of electricity appeared prefently after this difcovery, and fome of them very fyftematical, comprehending, no doubt, what the authors of them thought to be the whole of the fcience, yet none of them had the leaft idea of the amazing difcoveries of Dr. Franklin, relating either to the Leyden phial, or to the nature of lightning. And though numbers of Dr. Franklin's admirers thought that he had exhaufted the whole fubject, he himfelf was far from thinking fo; and the hiftory of electricity, fince the date of his capital difcoveries, demonftrates that his fufpicion was true.

IT may be faid, that there is a *ne plus ultra* in every thing, and therefore in electricity. It is true: but what reafon is there to think that we have arrived at it. Mr. Grey might have ufed the fame language above twenty years ago; but every body will, now, acknowledge, that it would have been above twenty years too foon: and yet, I think, it is evident, that Mr. Grey had really more reafon to think he had arrived at the *ne plus ultra* of electricity, than we have to think that we are arrived at it. Time has brought to light a great number of *incomplete*, as well as complete experiments, and perhaps more of the former than of the latter; concerning all which, as he could have no knowledge, fo he could have no doubts; fo that, though we know much more than he did; we, at the fame time, know how much more is unknown better than he could. Hitherto the acquifition of electrical knowledge has been like the acquifition of riches. The more we poffefs, the more we wifh to poffefs; and, I hope, the

the more indefatigable we fhall be to acquire the poffeffion of it.

ONE thing extremely ufeful to the progrefs of further dif-coveries, is to know what has really been done by others, and where the fcience ftands at prefent. For want of this knowledge, many a perfon has loft his time upon experiments which he might have known had either failed or fucceeded with others; and which it was, therefore, not worth his while to repeat. But the fources of this kind of information are too much fcattered, and too diftant for moft perfons to have accefs to them. This was the firft motive of the prefent undertaking, intended to exhibit a diftinct view of all that has been done in electricity to the prefent time, and likewife the order and manner in which every thing has been done; that electricians, having a diftinct idea of what the progrefs of electrical knowledge has been, might fee more clearly what remains to be done, and what purfuits beft promife to reward their labour.

INDEED it is almoft impoffible for any perfon to read the hiftory of electricity without gathering many hints for new experiments. When he has the whole before him at one view, he can better bring the diftant parts together; and from the comparifon of them, new lights may arife. When he fees what experiments have failed, and what have fucceeded; what branches of the fcience have been moft attended to, and what things feem to have been overlooked; what has been difcovered by accident, and what by theory; when he fees both the true lights which directed fome happy difcove-

rers

rers, and the falſe lights which miſled others, he will have the beſt preparation for purſuing his own inquiries.

To point out many of the *deſiderata* in the ſcience of electricity, I am ſenſible, will, for this reaſon, be ſuperfluous to many perſons, and probably to moſt who will have read thus far of this treatiſe: for ſufficient hints of them muſt have been ſuggeſted by the peruſal of the hiſtory. But if I have been anticipated in this part of my work by ſome of my readers, it will not diſpleaſe them to find it; and to others the contents of this chapter will be peculiarly uſeful.

IF, indeed, I had conſulted my reputation as a writer, or a philoſopher, I ſhould not have attempted this chapter at all. For not only will many of the articles which I ſhall now put down as *deſiderata* in the ſcience be ſoon no longer ſo, and even young electricians be able to give ſatisfactory anſwers to ſome difficult queries I am going to propoſe; but many of them will probably appear idle, frivolous, or extravagant ones; and, in a more advanced ſtate of the ſcience, it will hardly be imagined why I put them down at all. But if this chapter be a means of haſtening ſo deſirable an event, and of accelerating the progreſs of electrical knowledge, I am very willing that it ſhould, ever after, ſtand as a monument of my preſent ignorance.

" THESE thoughts," to adopt the words of Dr. Franklin, with much more propriety than he himſelf firſt uſed them, " are many of them crude and haſty, and if I were merely " ambitious of acquiring ſome reputation in philoſophy, I " ought to keep them by me, till corrected and improved by " time and farther experience. But ſince even ſhort hints and imperfect experiments, in any new branch of ſcience,

being

" being communicated, have oftentimes a good effect, in
" exciting the attention of the ingenious to the subject,
" and so become the occasion of more exact disquisitions, and
" more complete discoveries; you are at liberty," says he,
to Mr. Collinson, " to communicate this paper to whom you
" please, it being of more importance that knowledge should
" increase, than that your friend should be thought an ac-
" curate philosopher."

I would not even propose to draw up the following *queries*
upon the plan of those of Sir Isaac Newton, at the end of his
treatise on Optics. Many of them are such, that I have hardly
the most distant expectation of their being verified; but the
attempt to verify them may possibly lead to some other discoveries of more importance. They are such random thoughts
as led to the new experiments I have made; and not having
any more leisure to pursue them myself, I freely impart them
to my reader, that he may make as much advantage of them
as he can: being determined, upon taking leave of the subject, *to write myself fairly out*, as Mr. Addison says; or, as
the Spanish writers say, *to leave nothing in my inkhorn*.

Happy would it be for science, if all philosophers who
are engaged in the same pursuits, would make one common
chapter of all their hints and queries: and greatly honoured
should I think myself if the present chapter in this treatise
might be made use of for that purpose, and if, in future editions of the work, it should be looked into as the common
receptacle of the *present desiderata* among the whole body of
electricians, and of their imperfect hints for new discoveries.
With pleasure should I see each of them distinguished by the
name

name of some generous and illustrious contributor. A few, the reader will find, have been added to my own, and are distinguished in this manner.

MANY persons can throw out hints, who either have not leisure, or a proper apparatus for pursuing them: others have leisure, and a proper apparatus for making experiments, but are content with amusing themselves and their friends in diversifying the old appearances, for want of hints and views for finding new ones. By this means, therefore, every man might make the best use of his abilities for the common good. Some might strike out lights, and others pursue them; and philosophers might not only enjoy the pleasure of reflecting upon their own discoveries, but also upon the share they had contributed to the discoveries of others.

SECTION

SECTION II.

QUERIES AND HINTS CALCULATED TO PROMOTE
FURTHER DISCOVERIES IN ELECTRICITY.

I.

QUERIES AND HINTS CONCERNING THE ELECTRIC
FLUID.

WHAT is the proportion of the feveral colours in electric light, in different cafes, and in different appearances of it ?

Is not the electric light a real vapour ignited, fimilar to
that of phofphorus ; and may not experiments be, hereafter,
made, where we fhall have the explofion, the fhock, and the
other effects of electricity, without the light ? Is the electric light ever vifible except in vacuo ?   In the open air the
electric fluid makes itfelf a vacuum in order to its paffage.

COLLECT the electric fluid, not from the general mafs of
the earth but from bodies of particular kinds, and obferve if
it have any different properties, with refpect to light, &c.

Is

Is it exactly the same at sea, as on land; below the surface of the earth, as above it, &c. &c. &c. ?

Dr. Franklin observed, that iron was corroded by being exposed to repeated electric sparks. Must not this have been effected by some acid? What other marks are there of an acid in the electric matter? May not its phosphoreal smell be reckoned one? Is it not possible to change blue vegetable juices into red by some application of electricity? This, I think I have been told has been done at Edinburgh.

Is there only one electric fluid, or are there two? Or is there any electric fluid *sui generis* at all, distinct from the ether of Sir Isaac Newton? If there be, in what respect does it differ from the ether?

Are the particles which affect the organ of smelling, as well as the particles of light, parts of the proper electric fluid, or are they merely adventitious, being, some way or other, brought into action by electricity?

Does not some particular order of the particles, which Sir Isaac Newton supposes to be continually flying from the surfaces of all bodies, constitute the electric fluid; as others, he imagined, constituted the air, and others the ether, &c. ?

Is it probable that there is even any temporary, or growing addition to, or diminution of the whole stock of electricity?

Whence arises the elasticity of the electric fluid, and according to what law do its particles repel one another? *Mr. Price.*

II.

II.

QUERIES and HINTS concerning ELECTRICS and CONDUCTORS.

In what does the difference between electrics and conductors confist? In other words, what is it that makes fome bodies permeable to the electric fluid, and others impermeable to it?

Are the pores of electric bodies fmaller than thofe of conductors, and do they contain very much, or very little of the electric fluid?

What is it in the internal ftructure of bodies that makes them break with a polifh? Perhaps all folid electrics do fo.

Has elafticity any connection with electricity, fome electrics being extremely elaftic?

What is the reafon why, in fome of Mr. Hawkefbee's experiments, the electric light was vifible through a confiderable thicknefs of very opaque electrics, as rofin, fulphur, pitch, &c; but not through the thinneft metallic conductors?

What fimilarity is there in the procefses of calcination, vegetation, animalization, and in fome meafure chryftalization; fince all bodies which have gone through any of thofe procefses, and perhaps no others, are found to be electrics?

R r r   Are

ARE not both electrics and conductors more perfect in their kind in proportion to their specific gravity?

WILL not water conduct electricity the best in its state of greatest condensation; and metals the least in their greatest expansion, as shown by a pyrometer?

TRY the conducting power of different metals, by sending a large shock through wires of the same size, and observing the different lengths that may be melted of the different wires. *Dr. Franklin.*

COMPARE the invisible effluvia of water with the invisible effluvia of a burning candle, and also those proceeding from other bodies, with respect to their power of conducting electricity.

OBSERVE what degree of heat will discharge any given degree of electricity, in order to find in what degree heat makes air a conductor.

III.

QUERIES AND HINTS CONCERNING EXCITATION.

WHAT is the difference, in the internal structure of electrics, that makes some of them excitable by friction, and others by heating and cooling?

WHAT have friction, heating, cooling and the separation after close contact in common to them all? How do any of
them

them contribute to excitation? And in what manner is one or the other kind of electricity produced by rubbers and electrics of different furfaces?

Is not Æpinus's experiment of preſſing two flat pieces of glaſs together, when one of them contracts a poſitive and the other a negative electricity ſimilar to the experiments of Mr. Wilke, concerning the production of electricity by the liquefaction of various ſubſtances in others; when the ſubſtance which melts and contracts is in one ſtate, and that which contains it is in the oppoſite? And are not both theſe caſes ſimilar to the excitation of the tourmalin, &c. by heating and cooling? In this caſe may not the tourmalin and the air act upon one another and be in oppoſite ſtates?

Is not the circumſtance common to all theſe caſes, ſome affection of that ſpace near the ſurface of the bodies in which the refractive power lies? When bodies which have been preſſed together within that ſpace recede from one another, more ſurface, and conſequently more of that ſpace is made, doth not the electric fluid flow into it from that body which has the leaſt power of retaining it, and which it can permeate with the moſt eaſe; when not being able to enter the ſubſtance of the other it reſts upon its ſurface?

Are not the particles of the electric and rubber thrown into a vibration in the act of excitation, which makes frequent recedings of the parts from one another, and thereby promotes the effect above-mentioned?

What is the real effect of putting moiſture or amalgam upon the rubber? Do not thoſe ſubſtances increaſe the power of excitation, as conductors more diſtant from

the

the fmooth glafs, in the gradation of electrics, than the furface of the leather? Or do they only make the rubber touch in more points, or alter the furface of the rubber?

HAS that difference of furface on which colour depends any influence upon the power of excitation?

THE tourmalin and a veffel of charged glafs hermetically fealed are both excited by heating and cooling.   What other properties have they in common?

IV.

QUERIES AND HINTS CONCERNING ELECTRIFICA-TION.

DOES electrification increafe the exhalation of vapours, either from cold or from boiiing water?   If it do, is the in-creafed exhalation the fame in all ftates of the atmofphere?

DOES not the electric matter pafs chiefly on the furfaces of bodies?

Is the action of electrified bodies upon one another more properly an attraction, or a repulfion?

WOULD not continued electrification promote putrefaction?

IN what manner is the mutual repulfion of two bodies electrified negatively performed?   Is it by the attraction of the denfer electric fluid in the neighbourhood, or by the quan-
                                                                    tity

tity of it which may be fuppofed to be accumulated on the furfaces of fuch bodies in the manner defcribed? P. 457.

V.

QUERIES and HINTS concerning the power of CHARGING electrics.

WHAT is the real operation of conductors in coating electric fubftances?

WHY may not one phial be charged by connecting it with another (while it is charging) as high as if it were charged at the prime conductor? Or by what rule muft the force of thofe different chargings be eftimated? To all appearance, two phials charged together, fo as that one of them receives the fire from the other, do not give fo large a fhock, as only one of them charged in the ufual way.

WHAT is the *maximum* of charging a glafs jar, with refpect to the quantity of its furface, covered by the coating? It is evident that fome jars will difcharge themfelves, when only a fmall part at the bottom of them is coated, and when the explofion is very inconfiderable.

ENDEAVOUR to charge a plate of glafs with the coating preffed into actual contact with its furface, by means of heavy weights. Alfo endeavour to excite a plate of glafs in the fame manner. It is pretty certain that, in the ufual method of exciting

citing and charging, the real fubftance of the glafs is not touched; and though water be attracted by glafs, it may only be to a certain diftance from it.

VI.

QUERIES CONCERNING THE ELECTRICITY OF GLASS.

THROUGH what thicknefs of glafs will an excited electric, of any given ftrength, attract and repel light bodies? Is not the fame thicknefs the limit of charging glafs with the electric fluid?

Is not a plate of glafs contracted in its dimenfions by charging, the two electricities ftrongly compreffing it, fo as to increafe its fpecific gravity?

Is the tone of a glafs veffel, made in the form of a bell, the fame when it is charged as when it is uncharged? Or would the ringing of it make it more liable to break in thofe circumftances?

DOES the electric fluid with which glafs is charged refide in the pores of the glafs, or only on its furface; or rather within the fpace that is occupied by the power of refraction, i.e. a fmall fpace within, and likewife without the furface?

Is the refractive power of glafs the fame when it is charged or excited?

How

How does the different refractive power of glafs, or its denfity (which is probably in the fame proportion with its refractive power) affect its property of being excited or charged?

Is there not a confiderable difference in glafs when it is new made, and when it has been kept a month or two, both with refpect to excitation and charging?

Let glafs of every different compofition be tried both with refpect to excitation, and charging. Would it not be found that differences with refpect to metallic ingredients, hardnefs, annealing, continuance in fufion, &c. would influence both the properties; and that, in feveral cafes, the fame circum-ftance that was favourable to one would be unfavourable to the other?

Glass has hitherto been fuppofed to be full of the electric fluid, and its impermeability has been accounted for upon the difficulty with which the electric fluid moves in its pores. But may we not fuppofe the fubftance of glafs to be abfolutely impermeable to electricity, that no foreign electric matter ever fo much as enters a fingle pore of it, but lodges wholly on its furface; for inftance, between the point of contact and the real furface, or within the limits of the refractive power that is a little way on both fides the furface. This place is, I think, on many accounts, extremely convenient to difpofe of the electric matter, whether we make it to confift of two fluids, or of one. Their being kept afunder, if there be two, or its being prevented from getting through, if there be but one, will be much eafier to conceive in this cafe, than upon the fuppofition that the electric fluid *can* enter and move

move in the fubftance of the glafs, though it can only enter and move with difficulty, as Æpinus expreffes it. For, let the motion be ever fo difficult, one would think, that this circumftance could only make it move fo much the flower, and that, give the electricity in the charged plate of glafs time enough, and it would at length, without any external communication, perform the journey to the other fide, whither it has fo ftrong a tendency to go.

MOREOVER, one would think, that, upon the hypothefis of the admiffion of the electric fluid within the pores of the glafs, when the difcharge of a phial was actually made through the fubftance of the glafs, it might be in a filent manner, without breaking the glafs; whereas when the furfaces of the glafs are fuppofed to be violently preffed, and the pores of it not in the leaft entered by any particle of the fluid, or fluids, the impoffibility of the electric charge getting through the glafs is evident, as well as the neceffity of its breaking the glafs, if it do force a paffage.

VII.

QUERIES AND HINTS CONCERNING THE EFFECT OF ELECTRICITY ON ANIMAL BODIES.

Is the fluid on which electricity depends at all concerned in any of the functions of an animal body? In what manner

is

is the pulſe of a perſon electrified quickened, and his perſpiration increaſed?

Does not the air, by being heated in the lungs, communicate an electric virtue to the blood? What connection has
this circumſtance with the mephitic air which is exhaled
from the lungs in great quantities, as well as contained in all
the other excrements of the animal body?

May not the increaſed perſpiration of an animal body be
greater in a moiſt atmoſphere than in a dry one, there being
then more conducting particles in the atmoſphere, to act and
react upon the effluvia in the pores of the body; on which
the copious perſpiration does, probably, in a great meaſure,
depend?

VIII.

QUERIES and HINTS CONCERNING THE ELECTRICITY
OF THE ATMOSPHERE.

In what manner do the clouds become poſſeſſed of electricity?

Does the wind in any meaſure contribute to it?

Is it effected by the gradual heating and cooling of the air?
If ſo, whether is it the heating or the cooling that produces
poſitive electricity? Which ever it be, the contrary will

S ſ ſ   probably

probably produce negative electricity. Let the experiment be made by an electrical kite. *Mr. Canton.*

As thunder generally happens in a fultry ftate of the air, when it feems replenifhed with fome fulphureous vapours; may not the electric matter then in the clouds be generated by the fermentation of fulphureous vapours with mineral or acid vapours in the air. *Mr. Price.*

LET rain, fnow, and hail be received in infulated veffels, in different ftates of the atmofphere, to obferve whether they contain any electricity, and in what degree.

MAY not the void fpace above the clouds be always occupied with an electricity oppofite to that of the earth ? And may not thunder, earthquakes, &c. be occafioned by the rufhing of the electric fluid between them whenever the redundancy in either is exceffive ? Is not the aurora borealis, and other electrical meteors, which are remarkably bright and frequent before earthquakes, fome evidence of this ?

Is not the earth in a conftant ftate of moderate electrification, and is not this the caufe of vegetation exhalation and other the moft important proceffes in nature ? Thefe are promoted by increafed electrification. And it is probable that earthquakes, hurricanes, &c. as well as lightning, are the confequence of too powerful an electricity in the earth.

SECTION

SECTION III.

BRANCHES OF KNOWLEDGE PECULIARLY USEFUL TO AN ELECTRICIAN.

IN the hiftorical part of this work I have fhown what has been done on the fubject of electricity, and under the preceding *defiderata* I have endeavoured to give fome idea of what yet remains to be done, with a few hints concerning further experiments.   In the clofe of this part, I would willingly do fomething more towards enabling my reader to make farther advances in electrical inquiries.   However, all that can be done in this way muft, in its own nature, be more imperfect than even the account of the defiderata : for it is evident, that he who is able to teach others to make difcoveries might make them himfelf.   Notwithftanding this, it is poffible that fome general obfervations may be of ufe to this purpofe ; fuch for inftance as Lord Bacon makes, in his *Novum Organon* ; a book which, though it contain few or no philofophical difcoveries itfelf, has contributed not a little to the difcoveries contained in others.   A few fuch general obferva-

S f f 2   tions,

tions, confined to the fubject of electricity, I fhall endeavour to fuggeft in this place.

It is an obfervation which the progrefs of fcience daily confirms, that all truths are not only confiftent, but alfo connected with one another. The obfervation has, with no fmall appearance of juftice, been extended even to the arts; there being no two of them fo remote, but that fome of the methods and proceffes ufed in the one have fome analogy to fome that are ufed in the other. Hence the knowledge of one art or fcience is fubfervient to the knowledge of others; and no perfon can prefume that he is perfectly mafter of any one, till he has received all the affiftance he can from, at leaft, all its fifter arts or fciences.

Indeed the very exiftence of the various arts and fciences is almoft a demonftration of their relation to each other. For it were highly unreafonable to fuppofe, that the elements of any new art or fcience were difcovered by means independent of the ftudy or practice of thofe already known. As it is by eafy tranfitions that we pafs from one part of any particular fcience to another, fo it is by tranfitions equally eafy that mankind have paffed from one diftinct fcience to another. Confequently, to thofe previoufly difcovered arts and fciences muft we have recourfe, in order to underftand the full evidence, on which the firft principles of any new art or fcience reft.

Electricity is by no means an exception to this general rule. It has its fifter fciences as well as others. In purfuance of them were its own principles firft difcovered; and the further profecution of electrical experiments has fhown

its

its connection with more sciences than it was at first apprehended to have any relation to. Now the study of all these cannot but, reciprocally, contribute to perfect and extend the knowledge of electricity.

GILBERT, the first of modern electricians, was led to make his electrical experiments by their relation to those of magnetism, into which he was professedly inquiring. The study of chymistry seems to have led Mr. Boyle to attend to electricity, as well as to other occult qualities of particular bodies. Electric light was considered by all those who first observed it as a species of phosphorus; and with this view was Mr. Hawkesbee conducted in all the experiments he made upon it.

THESE, and other discoveries in electricity, having been made thus indirectly, excited the attention of philosophers to the subject, and induced them to sit down to the study of it in a direct and professed manner. Upon this it soon appeared, that electricity was no secondary, or occasional, but a principal, and constant agent in the works of nature, even in some of its grandest scenes; and that its agency, far from being confined to bodies of a particular class, extended its influence to all without exception; that the mineral, vegetable, and animal world, with the human frame in a particular manner, were all subject to its power; and that electrical experiments and principles enter into the most interesting arts and sciences which have them for their object. We also see every day, that electricity is extending itself still more into the subjects of other sciences, both by means of the analogy of their operations, and also by their reciprocal influences.

On

ON these accounts, to be an electrician at present, requires a much more extensive fund of various knowledge than it did but ten years ago; and a man must have a very comprehensive knowledge of nature in all its known operations, before he can reasonably expect to make any further discoveries. For it can only be by applying electricity to various parts of nature, and by combining its operations with other operations, both of nature and art, that any thing new can be found out. Almost all that can be done by the common electrical machines, and the usual apparatus of them, has been done already; so that we must look farther in quest of new discoveries. I hope, therefore, that I shall be excused, if I endeavour to give a hint of that kind of knowledge which, I apprehend, may be peculiarly subservient to improvements in electricity, and furnish views and materials for new experiments.

NATURAL PHILOSOPHY, cannot but be of the greatest use for this purpose; but, of all its branches, none promises to be of more use to the electrician than CHYMISTRY. Here seems to be the great field for the extension of electrical knowledge: for chymistry and electricity are both conversant about the latent and less obvious properties of bodies; and yet their relation to each other has been but little considered, and their operations hardly ever combined; few of our modern electricians having been either speculative or practical chymists.

AMONG other branches of Natural Philosophy, let the doctrine of LIGHT AND COLOURS be also particularly attended to.

It

It was this that Newton thought would be the key to other, at prefent, occult properties of bodies.

Let particular attention be alfo given to every thing that the imperfect ftate of Natural Philofophy furnifhes refpecting the Atmosphere, its compofition and affections. The phenomena of lightning fhow the connection of this fubject with electricity; and probably, electricity may be our key to a much more extenfive knowledge of meteorology than we are yet poffeffed of.

The fhock of the Leyden phial, the difcovery of the famenefs of lightning and electricity, together with the cure of feveral difeafes by electrical operations, are fufficient to convince us of the fingular importance of the ftudy of Anatomy, and every thing relating to the animal economy to an electrician. And had phyficians more generally attended to electricity, as an article of the materia medica, many more important and ufeful difcoveries might, no doubt, have been made. Enow, however, have been made to excite us to farther inquiries.

Æpinus has lately given us an excellent fpecimen of what ufe Mathematics, and efpecially algebraical calculations may be of to an electrician, and their ufe, will probably, in time, be found ftill more extenfive.

As electricity has much to expect from feveral branches of Natural Philofophy, fo it will be ready, in its turn, to lend its affiftance to them. It already fupplies arguments and proofs of fome principles in Natural Philofophy, which ftrengthen thofe that are drawn from other quarters. By e-
electricity,

lectricity, as well as by the principles of light and colours, we can demonstrate, that it requires a considerable force to bring bodies which are contiguous to one another, and even lie upon one another, into actual contact; and the moisture of the air may perhaps be shown to more exactness by Mr. Canton's electrical balls than by any other hygrometer whatever. But I do not mean to pursue this subject, and only mention these cases by way of example.

Upon the subject of the proper furniture for an electrician, I think it may be justly added, that a knowledge of MECHANICS will be useful to him; by which I mean, upon this occasion, not only the theory, but in some measure the practice too. For without some mechanical knowledge of his own, his electrical machinery will be very often out of order, and but ill answer his purpose.

If, indeed, a person mean nothing more than to amuse himself and his friends with the experiments that have been made by others (and this is a method of amusement which I am far from discouraging) the machines he may purchase, ready constructed to his hands, will answer his purpose very well; and the directions which are usually given along with the machines will enable him to perform the common experiments with tolerable certainty: and if any damage should happen to his apparatus, any mathematical instrument maker (if he happen to live in or near a large town) can readily repair it for him. But if a man propose to study the subject of electricity as a philosopher, with a view to extend the knowledge of it, the assistance of others will not be sufficient for him.

THE

THE common electrical machines, and the usual electrical apparatus, will enable a person to do little more than exhibit the common experiments. If he propose to go further, he must diversify his apparatus ; he must often alter the construction of his machines, and will find that common workmen cannot execute any thing out of their usual way, without more than general directions. Besides, unless a person be fortunately situated, workmen of every kind cannot always be at hand, to do every little thing he may want in the mechanical way, whenever he may happen to get a hint of a new experiment that requires it.

AN electrician, therefore, ought never to be without the common tools of a *cabinet maker*, *clock*, and *watch maker*, at least, and know, in some measure, how to use them. With respect to glass, he ought, by all means, to learn the use of a *blow pipe*, the method of drawing out and bending glass tubes, and performing, with some degree of dexterity, other operations upon glass, which he will want to use in a great variety of forms. An electrician, thus furnished, will be able, upon any occasion, to serve himself : and the slowness and blunders of mechanicks do but ill suit with the ardour of persons engaged in philosophical inquiries.

IT were much to be wished, that philosophers would attend more than they do to the construction of their own machines. We might then expect to see some real and capital improvements in them ; whereas little can be expected from mere mathematical instrument makers ; who are seldom men of any science, and whose sole aim is to make their goods elegant and portable.

T t t   FORMERLY

FORMERLY, indeed, philofophers were obliged to conftruct their own machines. Mr. Boyle, Mr. Hawkefbee, and Dr. Defaguliers would have done nothing by giving tradefmen orders for what they wanted. There were no fuch things to be had. Neceffity therefore drove them to the ftudy and practice of mechanics, and from their contrivances are derived almoft all the philofophical inftruments which are now in ufe.

EVERY original genius, like them, muft, in this refpect, follow their fteps. He will extend his views beyond the power of the prefent machinery, which can only be adapted to the prefent ftate of fcience. And, I think, one principal reafon of the imperfect ftate of feveral branches of electrical knowledge with us, may be evidently traced to fome general imperfections in the ftructure of all our common machines in England; which render feveral kinds of experiments very difficult, or almoft impoffible to be made; as may be fhown in the next part of the work; in which I fhall treat at large of the conftruction of machines, and give the beft directions I am able for ufing them.

LASTLY, if an electrician intend that the public fhould be benefited by his labours, he fhould, by all means, qualify himfelf to draw according to the rules of PERSPECTIVE; without which he will often be unable to give an adequate idea of his experiments to others. There is fo much beauty in the rules of this ingenious art, and fo much pleafure in the application of them, that I cannot help wondering, that all gentlemen of a liberal education do not take the fmall degree of pains, that is neceffary to make themfelves mafters of it. All
the

the mechanical methods of drawing, efpecially where a great number of right lines are ufed, as in drawing machines, &c. are exceedingly imperfect, and infufficient. They admit not of half the variety of perfpective drawings. They can hardly ever be near fo correct : befides that, I know by experience, they take up much more time, and the operation is exceedingly flavifh and troublefome.

Ttt2   PART

PART V.

OF THE CONSTRUCTION OF ELECTRICAL MACHINES, AND THE PRINCIPAL PARTS OF AN ELECTRICAL APPARATUS.

SECTION I.

GENERAL OBSERVATIONS ON THE CONSTRUCTION OF AN ELECTRICAL APPARATUS.

IMPROVEMENTS in electrical machines have, as might well be expected, kept pace with improvements in the science of electricity. While nothing more than electrical attraction and repulsion were known, nothing that we should now call an *electrical apparatus* was neceffary. Every thing that was known might be exhibited by means of a piece of amber,

sealing

fealing wax, or glafs; which the philofopher rubbed againft his coat, and prefented to bits of paper, feathers, and other light bodies that came in his way, and coft him nothing.

To give a greater degree of friction to electric fubftances Otto Guericke and Mr. Hawkefbee contrived to whirl fulphur and glafs in a fpherical form; but their limited knowledge of electricity did not fuggeft, or require the more complex ftructure of a modern electrical machine. Mr. Hawkefbee's contrivances, indeed, were excellent, and the apparatus for many of his experiments well adapted to the purpofes for which they were intended.

When no further ufe could be made of globes, philofophers had recourfe to the eafier and cheaper apparatus of glafs tubes, and fticks of fulphur or fealing wax; and the firft conductors they made ufe of were nothing more than hempen cords fupported by filken lines. To thefe, bars of metal were foon fubftituted. After that, recourfe was again had to the globe, as much more convenient to give an uniform fupply of electric matter to thefe infulated conductors; and, in due time, a rubber was ufed to fupply the place of a human hand.

The difcovery of the Leyden phial occafioned ftill more additions to our electrical apparatus; and the more modern difcoveries of Dr. Franklin and others have likewife made proportionable additions highly requifite. No philofopher, for inftance, can now be fatisfied, if he be not able to fupply a conductor from the clouds, as well as from the friction of his glafs globes or tubes. But having already marked the
progrefs

progreſs of improvements in electrical machines, as well as in electrical ſcience, I ſhall content myſelf with this brief recapitulation, and proceed to deſcribe what experience (in many caſes dear bought) has taught me to think the beſt me-thod of conſtructing machines, and to lay down the beſt rules for conducting electrical operations.

NOTWITHSTANDING globes or cylinders are now of the moſt extenſive uſe in electrical experiments, GLASS TUBES are, nevertheleſs, moſt convenient for ſeveral purpoſes, and no electrician ought to be without them. They ſhould be made as long as a perſon can well draw through his hand at one ſtroke, which is about three feet, or ſomething more; and as wide as he can conveniently graſp. The thickneſs of the metal is not material, perhaps the thinner they are the better, if they will bear ſufficient friction; which, however, needs only to be very gentle, when the tube is in good order. It is moſt convenient to have the tube cloſed at one end: for, be-ſides that the electric matter is thereby retained beſt on its ſurface, the air may more eaſily be drawn out of it, or condenſed in it, by means of a braſs cap fitted to the open end. A tube thus furniſhed is requiſite for various experi-ments. [a. Pl. II.]

THE beſt rubber that has yet been found for a ſmooth glaſs tube is the rough ſide of black oiled ſilk, eſpecially when a little amalgam, of mercury and any metal, is put upon it.

AN electrician ſhould be furniſhed with rough glaſs tubes, i. e. tubes with their poliſh taken off, as well as with ſmooth ones; but a cylinder of baked wood will do nearly as well. The beſt rubber for a rough glaſs tube, or a cylinder of baked

wood,

wood, as well as for a ftick of fulphur or fealing wax, is foft new flannel; or rather fkins, fuch as hare fkins, or cat fkins, tanned with the hair on, being fmoother, and having a more exquifite polifh.

ELECTRICIANS are not quite agreed whether the preference is, upon the whole, to be given to GLOBES, or CYLINDERS. In favour of cylinders it is faid, that more of their furface may be touched by the rubber. On the other hand, in favour of globes, it is faid, that they can more eafily be blown true, fo as to prefs the rubber equally : they may alfo be made larger in diameter, and by this means, the axis (if they have any) may be further from the excited furface : for when the *axes* are near the furface, the electric fire will feem to ftrike to them, fo that they will fometimes appear luminous in the dark, and, if they be infulated, the extremities of the axes will give fparks; which is certainly a diminution of the electric fire at the conductor.

FOR this reafon, I would advife, that all *axes* be avoided as much as poffible, having found by experience, that they are in no cafe whatever neceffary, the largeft globes being whirled horizontally, with the greateft eafe, and in every refpect to more advantage, with one neck than two. This method of fitting up globes alfo makes electrical machines much lefs complex, expenfive and troublefome.

LET every globe intended to be thus fitted up have its neck inclofed in a pretty deep brafs cap, ending in a dilated brim, of about half an inch broad, if the globe be a large one. To this neck let there be fitted a fhort iron axis, and on that a PULLEY; and let a fpace of about three quarters of an inch

of

of the axis be left between the pulley and the cap. In this place the axis is to be fupported by a ftrong BRASS ARM, [c. Pl. vii.] proceeding from the pillar into which the extremity of the axis is put, and in which it turns. This brafs arm may be made to receive globes of any fize whatever, room being left in it for pullies of any fize, that may be wanted for them.

In this manner globes may be fixed much more truly than they can with two necks, and they are mounted with much more eafe, and lefs expence. The weight of large globes is no objection to this method. The largeft need not to weigh above eight or ten pounds, and thefe have been found to turn with great eafe in this manner. The rubber, if it be placed under the globe, will contribute to fupport the weight of it.

LET there be a hole made in the brafs cap above-mention-ed, in order to preferve a communication between the exter-nal air, and the air within the globe: for if the air within the globe be either rarer or denfer than that without, the excita-tion is found to be leffened in proportion; and, judging from experience, nothing is to be apprehended from any moifture which might be fuppofed to infinuate itfelf into the globe by fuch a communication.

IT will be found convenient to have the axis project about an inch beyond the pillar in which it turns, [as at d. Pl. vii.] that a handle may be fitted to it, and that it may thereby be turned without a wheel, for the greater variety of experiments.

IF an axis be ufed, let both the extremities of it be care-fully turned in a lathe; otherwife it will not turn without a very difagreeable rattling; and let the part within the globe

be

be made round, and smooth, or covered with some electric substance, to prevent its taking off much of the electric virtue of the globe.

ONE of the pillars, in which these globes or cylinders with two axes are turned, should be moveable; for then a globe or cylinder of any size may be used, and they should be made high enough, and have holes at small distances quite to the top, to take two globes upon occasion, one above the other. [Plate vii.]

IT has not yet been determined by electricians what kind of glass is the fittest for electrical purposes, but the best flint is commonly used. I have not made so many experiments, as I could wish, to ascertain this circumstance; as they are both very uncertain, and expensive; but I have some reason to think that common bottle metal is fittest for the purpose of excitation: at least, the best globe I have yet seen is one that I have of that metal. Its virtue is certainly exceeding great, and I attribute it in part to the great hardness of the metal, and in part to its exquisite polish. The blowing of any thing spherical in this metal, and especially the making large globes smooth is very precarious; and they can hardly be made with two necks.

THE globe above-mentioned is about ten inches in diameter, but nothing has been determined about the best size. have used almost every size from three inches to near eightee in diameter, without knowing what advice to give. Perhaps *ceteris paribus,* twelve or thirteen inches may be, upon the whole, as convenient as any; but much larger, if they could be whirled with the same ease, would probably do better.

U u u

IF

IF a perſon chuſe to have no aſſiſtant, but would turn the globe, and manage the apparatus himſelf (which is, on many accounts, very deſirable) it will be moſt convenient to have the axis of the WHEEL level with the table at which he ſits. But if he chuſe rather to ſtand all the time he is making his experiments, it ſhould be raiſed proportionably higher. It will, perhaps, be moſt convenient to make the diameter of the wheel about eighteen inches; and the diameter of the pullies ſhould be ſuch as will give them, at leaſt, four or five revolutions for one of the wheel. For the globe ſhould generally revolve at leaſt four or five times in a ſecond, which is much ſwifter than it can be well turned without a wheel.

THE wheel ſhould be made moveable with reſpect to the frame in which the globes are hung, or the frame ſhould be moveable with reſpect to the wheel, to ſuit the alterations which the weather will make in the length of the *ſtring*, particularly if it be made of hemp; but worſted is ſaid to make an excellent ſtring, and is not ſo apt to alter with the weather. If the diſtance between the wheel and the pulley cannot be altered, the operator muſt occaſionally moiſten his hempen ſtring, in order to make it tighter, which is, on many accounts, very inconvenient. Several grooves in the ſame wheel are very uſeful, and almoſt neceſſary, if more than one globe be uſed at the ſame time. They ſhould be cut ſharp at the bottom; as ſhould alſo the grooves in the pulleys, that the ſtring may lay faſter hold of them, and that ſtrings of different ſizes may be uſed.

THE beſt RUBBERS for globes or cylinders are made of red baſil ſkins, particularly the neck part of them, where the

grain

grain is more open, and the surface rather rough. That the rubber may press the globe equally, it should be put upon a plate of metal bent to the shape of the globe, and be stuffed with any thing that is pretty soft. Bran is good; and if the stuffing be a conductor, as flax, it will be better than if it be a non-conductor, as hair, or wool. It should rest upon a spring, to favour any inequality there may be in the form of the globe or cylinder. The best position of the rubber, for a variety of purposes, is an horizontal one, but it should be capable of being placed in every variety of horizontal position; and the spring which supports the rubber should be made to press more or less at pleasure. The rubber should be made nearly as large one way as the other. If it be made very narrow, some parts of the globe will pass it without a sufficient friction. To remedy that inconvenience, the hand (if it be dry) may be held to the globe, just before the rubber, to add to its breadth; but that posture is very inconvenient.

It is advisable that there be no sharp edges or angles about the rubber, for that would make the *insulation* of it (which is a matter of great consequence) ineffectual. By the insulation of the rubber every electrical experiment may be performed with the twofold variety of positive and negative, and a conductor be made to give or take fire at pleasure. This insulation is best made by means of baked wood, in the form of a plate, five or six inches in diameter, [g Pl. vii.] interposed between the metallic part of the rubber and the steel spring that supports it. When positive electricity is intended to be produced, a chain [n Pl. vii.] must connect the rubber with the floor; but when negative electricity is wanted, the chain must be removed, and hung upon the common conductor, while another prime conductor must be connected with the rubber; which will therefore be electrified negatively.

THE best method of collecting the electric fire from the globe seems to be by three or four pointed wires, [*m* Pl. vii.] two or three inches long, hanging lightly upon the globe; and neither so light as to be thrown off the globe by electrical repulsion (which would occasion a loss of the electric matter) nor so heavy as to prevent their separating to a proper distance, and being drawn backwards or forwards, as the most effectual discharge of the fire, accumulated on different parts of the globe, may require. For this purpose they are best suspended on an open metallic ring. Needles with fine points do admirably well.

IT is requisite, for a variety of uses, that the PRIME CONDUCTOR be fixed very steady. It ought not, therefore, to hang in silken strings, but have a solid support; and, of all others, the preference is to be given to baked wood, being found by experience to make the most perfect insulation, and being the cheapest, and best in all other respects.

FOR common purposes a small conductor is most convenient, but where a strong spark is wanted, it is proper to have a larger conductor at hand, which may be occasionally placed in contact with the smaller, and be removed from it at pleasure. But whatever be the size of a prime conductor, the extremity of it, or that part which is most remote from the globe, should be made much larger and rounder than the

rest:

reſt : [*k* Pl. vii.] for the effort of the electric matter to fly off is always the greateſt at the greateſt diſtance from the globe.

As the electrician will have frequent occaſion to inſulate various bodies, I would adviſe that he make all the ſtands and ſtools which he uſes for that purpoſe of baked wood. It may eaſily be turned into any form, it inſulates better than glaſs, and is not ſo brittle. But care ſhould be taken that the wood be thoroughly baked, even till it be quite brown. It will not then be found very apt to collect moiſture from the air. If it ſhould, a little warming and rubbing will be ſufficient to expel the moiſture again. At moſt it can only be neceſſary to boil it in linſeed oil, or give it a ſlight coating of varniſh after it has come hot out of the oven. If this preparation be uſed, it muſt be well heated once more, immediately after the boiling.

THE electrician, having thus conſtructed his machine, will want METALLIC RODS, [*s* Pl. ii.] to take ſparks from his conductor for various uſes. Theſe ſhould have knobs, larger or ſmaller in proportion to the curvature of the conductor. If the knob be too ſmall, it will not diſcharge the conductor at once, but by degrees, and with a leſs ſenſible effect; whereas the ſpark between broad ſurfaces is thick and ſtrong.

THE more formidable part of an electrical apparatus conſiſts in the COATED GLASS, that is uſed for the Leyden experiment. The form of the plate is immaterial with reſpect to the ſhock; and, for different experiments, both plates of glaſs, and jars, of various forms and ſizes, muſt be uſed. For common uſes, the moſt commodious form is that of a jar, as wide as a perſon can conveniently hold by graſping,

and

and as tall as it will ſtand without any danger of falling; per-haps about three inches and a half in diameter, and eight inches in height. The mouth ſhould be pretty open, that it may be the more conveniently coated on the inſide, as well as the outſide, with tinfoil : but it will generally be moſt convenient to have the mouth narrower than the belly; for then it may be more eaſily kept clean and dry, and the cork, when one is wanted, will be eaſier to manage. But no elec-trician would chuſe to be without a great number of jars of various ſizes and forms. A conſiderable variety may be ſeen in plate ii. fig. *c, d, e, f, g, h, i, j, k.*

THE method of coating is much preferable to that of put-ting water or braſs ſhavings into the jars, which both makes them very heavy, and likewiſe incapable of being inverted, which is requiſite in many experiments. Braſs duſt, how-ever, or leaden ſhot is very convenient for ſmall phials. Theſe ſerve very well where it is neceſſary to remove the coating as ſoon as the jar is charged, but, for this purpoſe, quickſilver will generally anſwer the beſt. The tinfoil may be put on either with paſte, gum water, or bees wax. To coat the inſides of veſſels, which have narrow mouths, moiſten the inſide with gum-water, and then pour ſome braſs duſt upon it. Enough will ſtick on to make an exceeding good coat-ing; and if nothing very hard rub againſt it, it will not eaſily come off. This braſs duſt, which is extremely uſeful in a great variety of electrical experiments, may be had at the pin-makers.

IN the conſtruction of an ELECTRICAL BATTERY I would not recommend very large jars. A number of ſmaller are

preferable

Plate III.

p. 519.

J.Mynde Sc.

Prieflley del.

preferable on several accounts. If one of these break by an explosion, or be cracked by any accident, the loss is less considerable; besides, by means of narrow jars, a greater force (that is a greater quantity of coated surface) may be contained in less room; and, as narrow jars may be made thinner, they will be capable of being charged higher in proportion to their surface than large jars, which must necessarily be made thick. The largest jars that the glass-men can conveniently make are about seventeen inches in height; and they should not be more than three in diameter, and of the same width throughout. Thus they may be easily coated both within and without, and a box of a moderate size will contain a prodigious force: for the jars being coated within two inches of the top, they will contain a square foot of coated glass a piece.

The first battery that I constructed for my own use consisted of forty one jars of this size; but a great number of them bursting by spontaneous discharges, I constructed another, which I much prefer to it, and of which a drawing is given Plate III. It consists of sixty four jars, each eight inches long, and two inches and a half in diameter, coated within an inch and a half of the top. The coated part of each is half a square foot; so that the whole battery contains thirty two square feet. The wire of each jar has a piece of very small wire twisted about the lower end of it, to touch the inside coating in several places; and it is put through a pretty large piece of cork, within the jar, to prevent any part of it touching the side, which would tend to promote a spontaneous discharge. Each wire is turned round, so as to make a
hole

hole at the upper end; and through thefe holes a pretty thick brafs rod with knobs is put, one rod ferving for one row of the jars.

THE communication between thefe rods is made by laying over them all a chain, which is not drawn in the plate, left the figure fhould appear too confufed. If I chufe to ufe only part of the battery, I lay the chain over as many rods as I want rows of jars. The bottom of the box, in which all the jars ftand, is covered with tinfoil and brafs duft; and a *bent wire*, touching this tinfoil, is put through the box, and appears on the outfide, as in the plate. To this wire is faftened whatever is intended to communicate with the outfide of the battery; as the piece of fmall wire in the figure; and the difcharge is made by bringing the brafs knob to any of the knobs of the battery.

THIS is the battery which I have generally ufed in the experiments related in the laft part of this work; though, when I have wanted a very great force, I have joined both the batteries, and even feveral large jars to them. And it will perhaps be allowed to be fome evidence of the goodnefs of this conftruction, that after ufing it fo much, I fee no reafon to wifh the leaft alteration in any part of it. Were I to conftruct another battery, I fhould take jars of the very fame fize, and difpofe them in the very fame manner.

THE glafs of which this battery is made is what the workmen call flint-green; which I think is much better for this purpofe than the beft flint; as I find jars made of it not fo apt to difcharge themfelves. Befides it is much cheaper than flint.

To

To difcover the kind and degree of electricity, many forms of ELECTROMETERS have been thought of, as the reader may have perceived in the courfe of the hiftory, but this bufinefs is ftill imperfect. Mr. Canton s balls are of excellent ufe both to difcover fmall degrees of electricity, to obferve the changes of it from pofitive to negative, and *vice verfa* ; and to eftimate the force of a fhock before the difcharge, fo that the operator fhall always be able to tell, very nearly, how high he has charged his jars, and what the explofion will be whenever he chufes to make it.

MR. CANTON'S BALLS (reprefented on a glafs, ftanding on the ftool [*c* Plate ii.] are only two pieces of cork, or pith of elder, nicely turned in a lathe, to about the fize of a fmall pea, and fufpended on fine linen threads. It is convenient to have thefe in fmall boxes for the pocket; the box being the full length of the ftrings, that they may lie there without being bent.

MR. KINNERSLEY's electrometer defcribed p. 216, is ufeful to afcertain how great fhocks have been, and for many curious experiments in electricity. A drawing is given of it [*r* Plate ii.] but the glafs tube is reprefented as much fhorter than it was made by Mr. Kinnerfley. I think it in general more convenient; as the bore of the fmall tube may eafily be proportioned to it. But if a perfon get one long tube, of the fame fize throughout, it may be cut into different lengths, and the fame brafs caps will fit any part of it.

AT the top of the ftand of baked wood which fupports Mr. Kinnerfley's electrometer, I have fixed another, contrived by Mr. Lane, to give a number of fhocks, all of precifely the

X x x   fame same degree of strength ; but the reader must wait Mr. Lane's own leisure for a particular account of it, and of its various uses.

To the account of these articles of an electrical apparatus, which must be used within doors, it will not be wholly insignificant to add, that a strong firm table is highly requisite. For if the TABLE on which the apparatus is disposed be apt to shake, a great number of experiments cannot be performed to advantage.

In order to repeat the noble experiment of the sameness of the electric fluid with the matter of lightning, and to make further observations on the electricity of the atmosphere, the electrician must be provided with A MACHINE FOR DRAWING ELECTRICITY FROM THE CLOUDS. For the best construction of such a machine, take the following directions. On the top of any building (which will be the more convenient if it stand upon an eminence) erect a pole, [a fig. 2. Pl. 1.] as tall as a man can well manage, having on the top of it a solid piece of glass or baked wood, a foot in length. Let this be covered with a tin or copper vessel [b] in the form of a funnel, to prevent its ever being wetted. Above this let there rise a long slender rod [c] terminating in a pointed wire, and having a small wire twisted round its whole length, to conduct the electricity the better to the funnel. From the funnel make a wire [d] descend along the building, about a foot distance from it, and be conducted through an open sash, into any room which shall be most convenient for managing the experiments. In this room let a proper conductor be insulated, and connected with the wire coming in at

the

the window. This wire and conductor, being completely insulated, will be electrified whenever there is a considerable quantity of electricity in the air; and notice will be given when it is properly charged, either by Mr. Canton's balls, hung to it, or by such a set of bells as will be described hereafter.

If the electrician be desirous of making experiments upon the electricity of the atmosphere to greater exactness, he must raise a kite, by means of a string in which a small wire is twisted. The extremity of this line must be silk, and the wire must terminate in some metallic conductor, of such a form as shall be thought most convenient. Mr. Romas's experiment will perhaps convince my reader, that it may be dangerous to raise this kite at the approach of a thunder storm; and upon this occasion the common apparatus above described for drawing electricity from the clouds will, probably, answer his purpose well enough.

But, with the following apparatus, I should apprehend no great danger in any thunder storm. Let the string of the kite [a fig. 3. Pl. i.] be wound upon a reel [b] going through a slit in a flat board, fastened at the top of it; by which more or less of the string may be let out at pleasure. Let the reel be fixed to the top of a tin or copper funnel [c] such as was described above; and from the funnel let a metallic rod [d] with a large knob be projected, to serve for a conductor. This funnel and reel must be supported by a staff [e] the upper end of which, at least, must be well baked; and the lower end may be made sharp, to thrust into the ground, when the kite is well raised.

The

THE safety of this apparatus depends upon the chain [*f*] fastened to the staff, by a hook a little below the funnel, and dragging on the ground: for the redundant lightning will strike from the funnel to the chain, and so be conducted as far as any one chuses, without touching the person who holds the staff.

SPARKS may be taken from the conductor belonging to this apparatus with all safety, by means of a small rod of baked wood [*a* fig. 4.] furnished with a small funnel [*b*] and a brass rod [*c*] and a chain connected with it: for the lightning which strikes the rod, will pass by the funnel and the chain, without touching the person who holds the rod.

SECTION

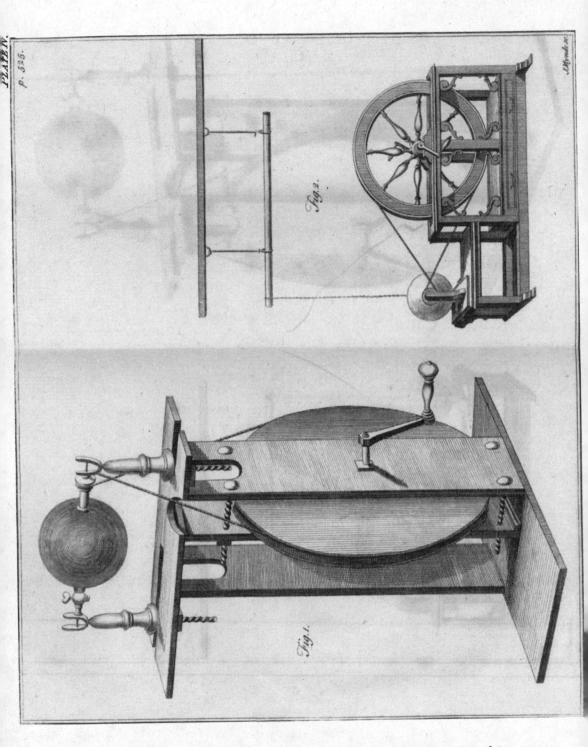

Fig.2.

Fig.1.

SECTION II.

A DESCRIPTION OF SOME PARTICULAR ELECTRICAL MACHINES, WITH OBSERVATIONS ON THEIR PRINCIPAL ADVANTAGES AND DEFECTS.

AFTER this general account of the conftruction of electrical machines, and the principal parts of an electrical apparatus, my reader may perhaps expect a more particular account of fome of the principal varieties with which they are ufually made. And though it might be prefumed, from what has been advanced upon that head, that any perfon might judge for himfelf, I fhall endeavour to gratify thofe who are willing to provide themfelves with an electrical machine, by giving drawings and defcriptions of fome of the beft conftructions that have fallen under my notice, obferving what I apprehend to be their feveral advantages and defects.

I SHALL begin with Mr. Hawkefbee's machine [Plate 4. fig. 1.] which is an excellent conftruction confidering the ftate of electricity in his time. The drawings annexed will

render

render a very particular defcription of this, or the other machine, unneceffary. This has no rubber, no prime conductor, or field for making experiments; for no fuch things were wanted in his time: but it may be eafily accommodated with them all. A conductor may hang from the cieling, a rubber may be fupported by a fpring fixed under the globe; and a table placed near the machine, may receive the apparatus neceffary for making experiments. The inconveniences of this conftruction are, that the operator cannot well turn the wheel himfelf. A fervant is therefore neceffary, who muft fit to his work. The machine admits only of one globe, or cylinder, which muft have two necks; though is admits of a confiderable variety of fuch, and it is by no means portable.

The Abbé Nollet's machine [Plate iv. fig. 2.] refembles the greateft number of the electrical machines that were ufed about the time that the Leyden phial was difcovered. Thefe were the machines, heavy and unwieldy as they feem, that were generally carried from place to place, when electrical experiments made a gainful bufinefs, and would bear the expence of the conveyance.

In thofe early times, electricians had no idea, that it was poffible to make the globe revolve too fwiftly. They, therefore, made their wheels exceeding large, and the frame of the machine proportionably ftrong. The globe was generally rubbed by the hand, the conductor was a bar of iron, or generally a gun barrel, fufpended in filken lines from the top of the room, and the apparatus was difpofed on an adjoining table.

THESE

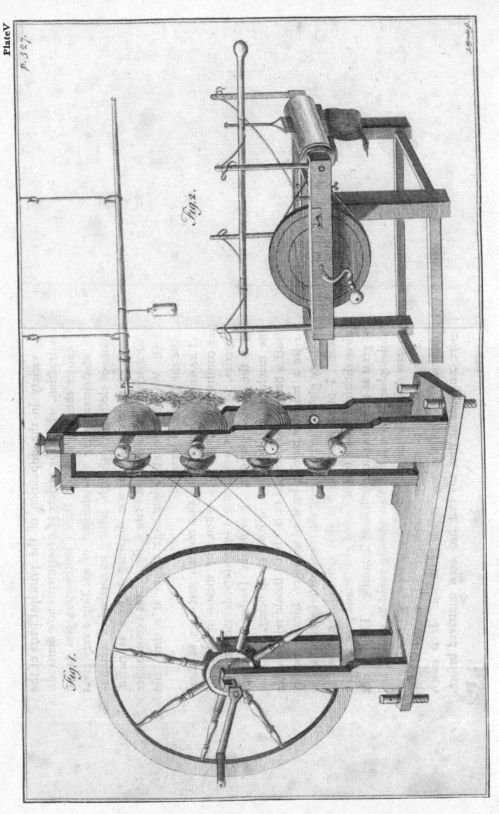

Fig. 2.

Fig. 1.

THESE machines are now univerſally laid aſide, being more fit for a large laboratory than a private ſtudy. Beſides they neceſſarily require an aſſiſtant, and do not admit of half the variety in the diſpoſition of the principal parts of the conſtruction, which the variety of experiments now demands.

ABOUT the time that Mr. Boze's beatification was talked of, electricians were very deſirous of exciting a very great power of electricity; though, having no method of accumulating, or preſerving it, it was diſperſed as ſoon as raiſed. The machine repreſented in Plate v. fig. 1. was a contrivance of Dr. Watſon's, to whirl four large globes at a time, and unite the power of them all.

I CANNOT help regretting that no ſuch machines as theſe are conſtructed at this day, when, by means of electrical batteries, ſo great a power might be preſerved, and employed to the greateſt purpoſes. I wiſh the Doctor would refit the machine here deſcribed, if it be yet in being, and conſtruct a battery proportioned to it. But I ſhould rejoice more to ſee a machine moving by wind or water, turning twenty or thirty globes, and charging electrical batteries adequate to them. I make no doubt but that a full charge of two or three thouſand ſquare feet of coated glaſs would give a ſhock as great as a ſingle common flaſh of lightning. They are not philoſophers who will ſay, that nothing could be gained, and no new diſcoveries made by ſuch a power.

PLATE v. fig. 2. exhibits a machine which Mr. Wilſon conſtructed, about the time above-mentioned. It is much more commodious than any that had been contrived before,
as

as all the parts are brought within a moderate compaſs; ſo that the ſame perſon may turn the wheel, and conduct the experiments. One of the moſt complete machines, and one that had the greateſt power of any that I have ſeen, except my own, is of this conſtruction.

ITS inconveniences are, that it admits but little variety of globes or cylinders, and both theſe and the rubber are not ſufficiently diſtant from other bodies. The rubber is not inſulated, and the conductor is unſteady. This machine has a frame ſtanding upon the ground, but the general conſtruction may be preſerved, and the machine be made to ſcrew to a table. Some I have ſeen which, by this means, were made very portable; and a box was contrived in the inſide, to contain the apparatus.

OF the more modern conſtructions (of which there is an endleſs variety) the more elegant are thoſe in which the globe is turned by tooth and pinion. This reduces the wheel work, contained in the box [a Pl. vi. fig. 1.] to an exceeding ſmall compaſs, and gives the workmen an opportunity of making the machine all in braſs, very elegant, and portable. But I object to them, as liable to accidents, which electricians in general cannot eaſily repair; and I would wiſh philoſophers to be as independent as poſſible of all workmen. Of this more elegant kind of machines, is that made by Mr. Nairne, a drawing of which may be ſeen in this plate. The conductor belonging to machines of this conſtruction is generally hung in ſilk, ſupported, either by wooden pillars, in a frame, as in the figure annexed, or by two braſs arms extending from the machine.

THESE

PLATE VI.
p. 529.

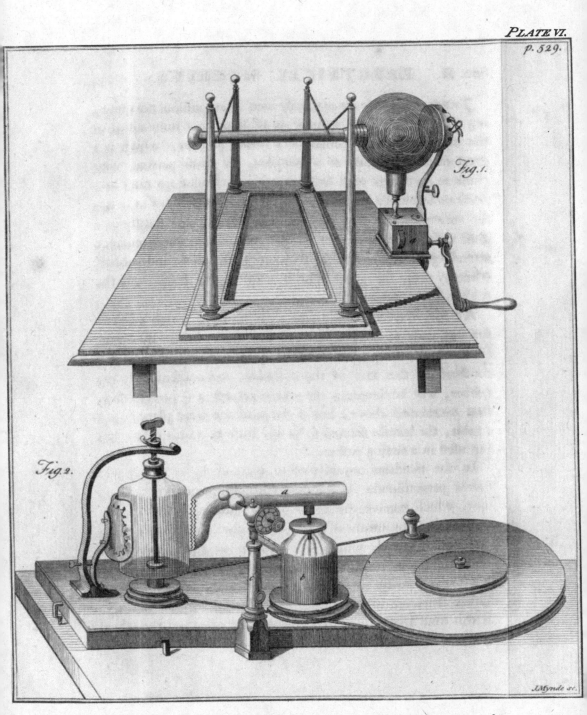

Fig. 1.

Fig. 2.

J. Mynde sc.

THESE machines are certainly very commodious for screwing to a table. They require no assistant, and they admit of the experiments being made in a sitting posture; which is a great recommendation of a machine, to those persons who chuse to do things with little trouble, and who are fond of a studious sedentary life. This construction admits of very little variety in the size or number of globes, and hardly of a glass vessel of any other form. But the greatest inconvenience attending it, is the upright position of the globe and rubber, whereby every thing put upon it is apt to slip down; and the rubber is not insulated.

AT Mr. Martin's I have seen a machine, in which a cylinder was made to revolve between two brass pillars, by means of a wheel and pinion, concealed in a neat box. In this construction the axis of the cylinder, and consequently the rubber, was horizontal; in which respect it is preferable to that mentioned above; but if the machine were placed upon a table, the handle seemed to be too high to admit of its being used in a sitting posture.

IN the machine represented in plate vi. fig. 2. a cylinder stands perpendicular to the horizon, supported by a brass bow, which receives the upper end of the axis; and motion is given to it by means of a pulley, at the lower end of the axis, and a wheel, which lies parallel to the table. The conductor [a] is furnished with points to collect the fire, and it is screwed to the wire of a coated jar, [b] standing in a socket, between the cylinder and the wheel. One machine of this kind I have seen, in which the cylinder and the wheel were not separated by the conductor.

Y y y

THIS

THIS conftruction is peculiarly ufeful to phyficians and apothecaries; and, with Mr. Lane's electrometer [c] annexed to it, (the figure of which he has given me leave to infert in the drawing annexed, taken from his own machine) as many fhocks as are requifite may be given, of precifely the fame, and any degree of force, without any change of pofture, either in the patient, or the operator; who has nothing to do but turn the wheel, without fo much as touching any other part of the apparatus.

WHEN this machine is ufed for fimple electrification, and other purpofes where the fhock is not required; the coated jar muft be taken away, and another jar, without any coating, put in its place. By this means the conductor is fixed, which is a very great advantage, and which few machines are pof-feffed of. But thefe machines, befides that they admit no variety of globes or cylinders, and no infulation of the rub-ber, require a motion of the arm, which I fhould think not quite eafy.

MR. WESLEY's people, I believe, generally ufe a machine in which two cylinders are turned by the fame wheel: but one that I faw, in the poffeffion of a very intelligent perfon of that perfuafion, had the cylinders and rubbers fo confined in a cheft, that, though it might do very well for medical ufes, it was very ill adapted to the purpofes of philofophy.

BUT the machine which I would advife a philofopher to conftruct for his own ufe, is that of which a drawing is given plate vii. This conftruction is the refult of my beft attention to this fubject. I have ufed it above fix months (how much I leave the reader to imagine) without feeing the leaft reafon

to

PLATE VII.

p. 531.

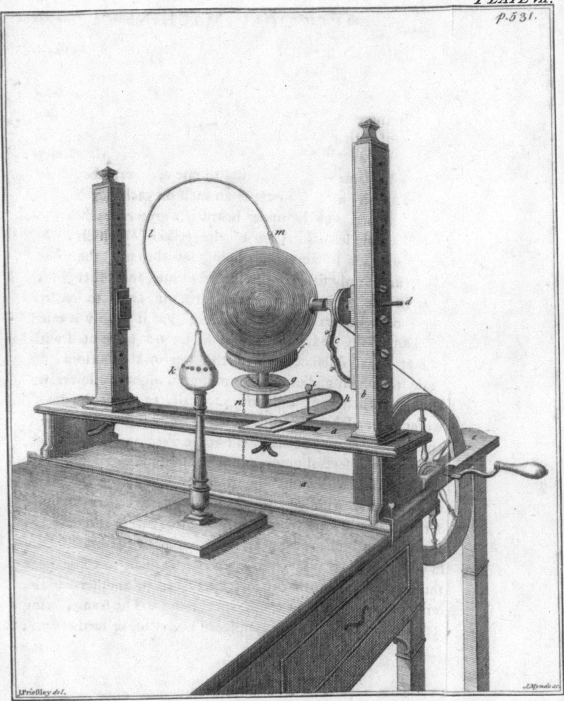

J.Priestley del.

J.Mynde sc.

to make any alteration of confequence in it; and believe it to have almoft all the advantages, which an electrical machine, defigned for the clofet, can have. The reader will, therefore, allow me to be a little longer in the defcription of it than I have been on the reft.

THE FRAME confifts of two ftrong boards of Mahogena, [a a] of the fame length, parallel to one another, about four inches afunder; and the lower is an inch on each fide broader than the upper. In the upper board is a *groove*, reaching almoft its whole length. One of the *pillars* [b] which are of baked wood, is immoveable, being let through the upper board, and firmly fixed in the lower, while the other pillar flides in the groove above-mentioned, in order to receive globes or cylinders of different fizes; but it is only wanted when an axis is ufed. Both the pillars are perforated with *holes* at equal diftances, from the top to the bottom; by means of which globes may be mounted higher or lower according to their fize; and they are tall, to admit the ufe of two or more globes at a time, one above the other. Four of a moderate fize may be ufed, if two be fixed on one axis: and the wheel has feveral grooves for that purpofe.

IF a globe with only one neck be ufed, as in the plate, a BRASS ARM with an open focket [c] is neceffary to fupport the axis beyond the pulley; and this part is alfo contrived to be put higher or lower, together with the brafs focket in which the axis turns. The axis [d] is made to come quite through the pillar, that it may be turned by another handle, without the wheel, if the operator chufes. The frame, being fcrewed to the table, may be placed nearer to, or farther from

Y y y 2   the

the wheel, as the length of the string requires, in different states of the weather. The WHEEL is fixed in a frame by itself, [*e*] by which it may have any situation with respect to the pulley, and be turned to one side, so as to prevent the string from cutting itself. The hinder part of this frame is supported by a foot of its own.

THE RUBBER [*f*] consists of a hollow piece of copper, filled with horse hair, and covered with a basil skin. It is supported by a socket, which receives the cylindrical axis of *a round and flat piece of baked wood,* [*g*] the opposite part of which is inserted into the socket of a bent steel spring. [*h*] These parts are easily separated; so that the rubber, or the piece of wood that serves to insulate it, may be changed at pleasure. The SPRING admits of a twofold alteration of position. It may be either slipped along the groove, or moved in the contrary direction; so as to give it every desirable position with respect to the globe or cylinder; and it is, besides, furnished with a *screw,* [*i*] which makes it press harder or lighter, as the operator chuses.

THE PRIME CONDUCTOR [*k*] is a hollow vessel of polished copper, in the form of a pear, supported by a pillar, and a firm basis of baked wood, and it receives its fire by means of a *long arched wire, or rod of very soft brass,* [*l*] easily bent into any shape, and raised higher or lower, as the globe requires; and it is terminated by an *open ring,* in which are hung *some sharp pointed wires* [*m*] playing lightly on the globe when it is in motion. The body of the conductor is furnished with *holes, and sockets,* for the insertion of metallic rods, to convey

the

the fire wherever it is wanted, and for many other purpofes convenient in a courfe of electrical experiments. The conductor is, by this means, fteady, and yet may be eafily put into any fituation. It collects the fire perfectly well, and (what is of the greateft confequence, though but little attended to) retains it equally every where.

WHEN pofitive electricity is wanted, a wire, or chain, as is reprefented in the plate [*n*] connects the rubber with the table, or the floor. When negative electricity is wanted, that wire is connected with another conductor fuch as that reprefented [*t* Pl. ii.] while the conductor in plate vii. is connected, by another wire or chain, with the table. If the rubber be made tolerably free from points, the negative power will be as ftrong as the pofitive. In this machine I do not know which is the ftronger of the two.

IN fhort, the capital advantages of this machine are, that glafs veffels, or any other electric body, of any fize or form, may be ufed, with one neck, or two necks, at pleafure; and even feveral of them at the fame time, if required. All the effential parts of the machine, the *globe*, the *frame*, the *wheel*, the *rubber*, and *conductor*, are quite feparate; and the pofition of them to one another may be varied in every manner poffible. The rubber has a complete infulation, by which means the operator may command either the negative or the pofitive power, and may change them in an inftant. The conductor is fteady, and eafily enlarged, by rods inferted into the holes, with which it is furnifhed, or by the conjunction of other conductors, in order to give larger fparks, &c. The wheel

may

may be ufed or not at pleafure; fo that the operator may either fit, or ftand to his work, as he pleafes; and he may, with the utmoft eafe, both manage the wheel and his apparatus.

IF a perfon would be fatisfied with ufing globes with one neck only, which I greatly prefer, one pillar, and the greateft part of the frame of this machine may be fpared; and if tooth and pinion be preferred, the wheel may be thrown out: and fo an elegant inftrument may be made in brafs, upon the fame general plan, which may be brought into a fmall compafs, and be greatly fuperior to any inftrument I have yet feen fold at any of the fhops. But to philofophers I would, for the reafons above-mentioned, recommend the prefent conftruction in plain wood.

PART

PART VI.

PRACTICAL MAXIMS FOR THE USE OF YOUNG ELECTRICIANS.

AS the chapter I am now entering upon is profeſſedly defigned for the ufe of young electricians, it is hoped that the proficient will excuſe my inferting a few plain and trite maxims; which, though they be fuperfluous with re-fpect to him, may not be fo to all my readers. The greateſt electricians (who are generally thoſe who have had the feweſt inſtructions) may remember the time when the knowledge of a rule or maxim, which they would perhaps ſmile to fee in a book, would have faved them a great deal of trouble and expence; and it is hoped they will not envy others acquiring wifdom cheaper than they did. In a general treatiſe every man has an equal right to expect to find what he wants; and it is for the intereſt of the ſcience in general, that every thing be made as eafy and inviting as poſſible to beginners. It is this circumſtance only that can increaſe the number of elec-

tricians,

tricians, and it is from the increafe of this number that we may moft reafonably expect improvements in the fcience.

WHEN the air is dry, particularly when the weather is frofty, and when the wind is North, or Eaft, there is hardly any electrical machine but will work very well. If the air be damp, let the room in which the machine is ufed be well aired with a fire, and let the globe and every thing about it be made very dry, and it may be made to work almoft as well as in the beft ftate of the air.

WHEN a tube is ufed, the hand fhould be kept two or three inches below the upper part of the rubber; otherwife the electricity will difcharge itfelf upon the hand, and nothing will remain upon the tube for electrical purpofes.

A LITTLE bees wax drawn over the furface of a tube will greatly increafe its power. When the tube is in very good order, and highly excited, it will, at every ftroke, throw off many pencils of rays from its furface, without the approach of any conductor, except what may float in the common atmofphere.

To increafe the quantity of electric fire from a globe, moiften the rubber a little from time to time; or rather moiften the under fide of a loofe piece of leather, which may occafionally be put upon the rubber. But the moft powerful exciter of electricity is a little amalgam, which may be made by rubbing together mercury and thin pieces of lead or tinfoil in the palm of the hand. If the rubber fhould be placed perpendicular to the horizon, it will be neceffary to ufe a little tallow to make it ftick. With this excellent refource, almoft all ftates of the weather are equal to an electrician.

A

A LITTLE time after fresh amalgam has been put upon the globe, and often at other times, if there be any foulness upon the cushion; and sometimes when there is none, there will be formed upon the globe small black spots, of a hard rough substance; which grow continually larger, till a considerable quantity of that matter be accumulated upon the surface. This must be carefully picked off, or it will obstruct the excitation, and in a great measure defeat the electrical operations.

WHEN the amalgam has been used for some time, there will be formed upon the rubber a thick incrustation of the same kind of black substance which is apt to adhere to the globe. This incrustation is a very great improvement of the rubber. For when once a considerable body of this matter is formed, and it is a little moistened, or scraped, as much fire will be produced, as if fresh amalgam were used : so that it seems to supersede the further use of the amalgam.

As the electric matter is only collected at the rubber, it is necessary that it have a communication with the common mass of the earth, by means of good conductors. If, therefore, the table on which the machine stands, or the floor of the room in which it is used, be very dry, little or no fire will be got, be the machine ever so good. In this case it will be necessary to connect the rubber, by means of chains or wires, with the floor, the ground, or even the next water, if the neighbouring ground be dry. This Dr. Franklin informs me he was frequently obliged to do in Philadelphia.

WHEN the electricity of a globe is very vigorous, the electric fire will seem to dart from the cushion towards the wires

Z z z   of

of the conductor. I have seen those lucid rays (which are in part visible in day light) make the circuit of half the globe, and reach the wires: and they will frequently come in a confiderable number, at the same time, from different parts of the cushion, and reach within an inch or two of the wires. The noise attending this beautiful phenomenon exactly resembles the crackling of bay leaves in the fire. Frequently these lucid arches have *radiant points*, often four or five in different parts of the same arch. These radiant points are intensely bright, and appear very beautiful. It is peculiarly pleasing to observe these circles of fire rise from those parts of the cushion where the amalgam or moisture has been put, or which have been lately scraped. Single points on the rubber will then seem intensely bright, and for a long time together will seem to pour out continual torrents of flame. If one part of the rubber be pressed closer than another, the circles will issue in that place more frequently than in any other.

WHEN the conductor is taken quite away, circles of fire will appear on both sides the rubber, which will sometimes meet, and completely encircle the globe. If a finger be brought within half an inch of the globe, in that state, it is sure to be struck very smartly, and there will often be a complete arch of fire from it to the rubber, though it be almost quite round the globe.

THE smaller the conductor is made, the more fire may be collected from it : for there is less surface from which the fire may escape. But in charging a phial, if the wire be
placed

placed close to the conductor, the difference will be inconsiderable, whether a smaller or larger conductor be used, till it begin to be charged pretty high : for, till that time, the conductor will not have acquired any considerable atmosphere.

If the conductor be made perfectly well, and the air be dry, there will never be any loss of fire from any part of it. For when the whole surface has received as high a charge as the machine can give to it, it will, in all places alike, perfectly resist all further efforts to throw more upon it, and the circulation of the fluid by the rubber will be stopped, being balanced, as it were, by equal forces. Or if it lose in all places alike, the dissipation must be invisible. This maxim almost admits of ocular demonstration. For when the rubber is perfectly insulated, and the conductor has an opportunity of discharging itself, the rubber will take sparks from a wire placed near it very fast; but when the conductor has little opportunity of emptying itself, it will take fewer of those sparks.

To form a just estimate of the electrical power of any machine, and to compare different machines in this respect, take two wires with knobs of any size, and fix one of them at the conductor, and the other at some certain distance from it, about an inch, or an inch and a half; and when the wheel is turned, count the number of sparks that pass between them during any given time. Fix the same wires to any other conductor, belonging to any other machine (but the same conductor would be more exact) and the difference between the number of strokes in any given time will ascertain the difference between the strength of the two machines.

Z z z 2

THE

THE larger the conductor is made, the stronger spark it will give: for the more extended the electrified surface is, the greater quantity of the electric atmosphere it contains, and the more sensible will be its effect when it is all discharged at once. The conductor, however, may be made so large, that the necessary dissipation of the electric matter from its surface into the air will be equal to the supply from the machine, which will constitute the MAXIMUM of the power of that machine, and will be different in different states of the air.

A CERTAIN degree of friction is necessary to give a globe its greatest power. A number of globes increases the power, but the increase of friction will make it more difficult for a man to excite their power. A few trials with a number of globes would enable any man to judge of the *maximum* of his strength in exciting electricity. I should imagine, from my own experience, that no person could excite much more electricity from any number of globes, than he could from one; supposing him to continue the operation an hour, or even only half an hour together.

WHEN a long conductor is used, the longest and the strongest spark may be drawn from the extremity of it, or from that part which is the most remote from the globe.

THE largest and most pungent sparks are drawn from any conductor along an electric substance. Thus if the conductor be supported by pillars of glass or of baked wood, the longest sparks will be taken close to the pillar.

IF the conductor bend inwards in any place, so as to make the surface concave, a peculiarly large, strong, and undivided

spark

spark may be drawn from that place. Where the surface is convex, the spark is more apt to be divided and weakened.

IF a smooth cork ball be hung in a long silken string, and electrified positively, it will always be repelled by positive, and attracted by negative electricity. But the strongest repulsion will be changed into attraction at a certain distance.

IF two pith balls hung by linen threads, and diverging with positive electricity, be insulated, though in connection with conductors of considerable length, the approach of a body electrified positively will first make them separate, and then (if the electricity of the balls be small, and that of the approaching body great) it will, at a certain distance, make them approach, and at length come into contact with it. Sometimes the divergence previous to the convergence is very slight, and, without great attention, is apt to be overlooked.

IF the balls have a free communication with the earth, the approach of positive electricity will make them converge, and negative electricity will make them diverge: the electric matter of the approaching body, in the former case, repelling that of the balls, and thereby, as it were, unelectrifying them; whereas, in the latter case, the negative electricity of an approaching body draws it more powerfully into the threads, and makes them diverge more. This method of judging is, therefore, excellently adapted to ascertain the kind of electricity in the atmosphere, or of a charged jar or battery, the balls being held in the hand of a person standing on the earth or the floor.

To discover smaller degrees of electricity than the balls can show to advantage, use a very fine thread, or two of them.

If

If insulation be necessary, fasten it to a stick of baked wood. But the most accurate measure of electricity I have yet hit upon, is a single thread of silk as it comes from the worm. When the end of this has received a small degree of electricity, it will retain it a considerable time, and the slightest electric force will give it motion. Before any experiments be made, let it be carefully observed how long, in any particular situation, it will retain the degree of electricity that is intended to be given to it; and let allowance be made for that in the course of the experiments. It will retain electricity much longer, if a small piece of down from a feather be fastened to it, but it will not acquire the virtue so soon. And it will be most easy to manage, if two or three threads of silk be used, and the piece of down be so adjusted to them, that it shall but just prefer a perpendicular situation, and not absolutely float in the air at random. This electrometer is not liable to the inaccuracies of those that have a sensible weight: for as there is always a sphere of attraction within a sphere of repulsion, the weight of the electrified body will allow another to pass the boundary of those two spheres, without a sensible obstruction; but the body I am describing immediately retires, with all its spheres of attraction and repulsion about it.

THE force of the ELECTRIC SHOCK is in proportion to the quantity of surface coated, the thinness of the glass, and the power of the machine. That this last circumstance ought to be taken into consideration is evident; for different machines will charge the same jar very differently. With one

machine,

machine, for inftance, it may be made to difcharge itfelf, when it cannot with another.

THE globe which charges the fafteft, or whofe electrical power is greateft, charges alfo higheft. Therefore a number of globes will not only charge fafter, but alfo, probably, higher than a fingle globe.

THE moft effectual method of charging a jar is to connect the outfide by means of wires, with the rubber, while the wire proceeding from the infide is in contact with the conductor. In this manner the infide of the jar will be fupplied with the very fame fire that left the outfide. In this cafe alfo the jar will receive as high a charge as it is capable of receiving, though the rubber be infulated, and have no communication but with the outfide coating; fo that, in the cafe of charging, there can be no occafion for the directions given above, when the table, the floor of the room, or the ground are very dry.

THE greateft quantity of fire that a jar will hold is not always the quantity it will contain when it is coated juft fo low as not to difcharge itfelf. In this cafe, indeed, the part that is coated is charged as high as it can be, but then a confiderable part of the furface is not charged at all, or very imperfectly. On the other hand, if the jar be coated very high, it may be made to difcharge itfelf with as fmall an explofion as one chufes. The exact *maximum* of the charge of any jar is not eafy to afcertain.

THE greateft effort in a jar to make a difcharge feems to be about half a minute, or a minute, after it is removed from the conductor; owing, perhaps, to non-electric duft or

moifture

moisture attracted by and adhering to the glass, between the outside and inside coating; so that if there be any apprehension of its discharging itself, it is advisable to discharge it before it has stood charged at all.

When a thin jar is discharged, it is advisable not to do it by placing the discharging rod opposite to the thinnest part. It will endanger the bursting of the jar in that place.

The more persons join hands to take a shock the weaker it is.

If two jars, of the same thickness, be used together, the stronger of them will receive no higher a charge than the weaker. If one of them, for instance, be coated so high as that it will discharge itself, either with or without bursting, after a few turns of the wheel; the other will always be discharged along with it, though it was capable of being charged ever so high by itself. The method, therefore, of estimating the force of a number of jars, is to consider each of them as capable of containing no more fire than the weakest in the company. It follows from hence, that if a single jar in a large battery have the smallest crack in the coated part of it, not one of them is capable of being charged in conjunction with it.

In large batteries, it is advisable to coat the jars pretty high, the dissipation of the electric matter from so great a surface when the charge is high being very considerable. The battery might be made so large, as that, after a very moderate charge, the machine would be able to throw no more fire in than was exhaled, as we may say, from the surface. This would be the MAXIMUM OF THE POWER OF THAT MACHINE IN CHARGING.

In

In order to judge of the strength of a charge (which, in large batteries, is a thing of considerable consequence) present Mr. Canton's balls to the wires, from time to time. A comparison of the degree of their divergence, compared with the actual explosion, will soon enable the operator to tell how high his battery is charged, and what will be the force of the explosion.

In comparing different explosions by their power to melt wires, let it be observed, that, in wires of the same thickness, the forces that melt them will be in the proportion of the lengths; and in wires of the same length, in the proportion of the squares of their diameters.

Do not expect that the explosion of a battery will pierce a number of leaves of paper in proportion to its force in other respects. That depends upon the height of the charge much more than the quantity of coated surface. I have known an explosion which would have melted a pretty thick wire not able to pierce the cover of a book, which a small common jar would have done with ease. If it had been pierced with the explosion of the battery, the hole would have been larger in proportion.

Let no person imagine that, because he can handle the wires of a large battery without feeling any thing, that therefore he may safely touch the outside coating with one hand, while the other is upon them. I have more than once received shocks that I should not like to receive again, when the wires showed no sign of a charge; even two days after the discharge, and when papers, books, my hat and many other things had lain upon them the greatest part of the time. If

4 A

the

the box be tolerably dry, the refiduum of the charge will not diſperſe very ſoon. I have known even the *refiduum of a refiduum* in my batteries to remain in them ſeveral days. For preſently after an exploſion, I ſeldom fail to diſcharge the reſiduum, which, in ſome caſes, is very conſiderable, for fear of a diſagreeable accident.

A SMALL ſhock paſſing through the body gives a ſenſation much more acute and pungent than a large one. I cannot boaſt, like Dr. Franklin, of being twice ſtruck ſenſeleſs by the electric ſhock; but I once, inadvertently, received the full charge of two jars, each containing three ſquare feet of coated glaſs. The ſtroke could not be called painful, but, though it paſſed through my arms and breaſt only, it ſeemed to affect every part of my body alike. The only inconvenience I felt from it was a laſſitude, which went off in about two hours.

PART VII.

A DESCRIPTION OF THE MOST ENTERTAINING EX-
PERIMENTS PERFORMED BY ELECTRICITY.

ELECTRICITY has one confiderable advantage over
most other branches of fcience, as it both furnifhes
matter of fpeculation for philofophers, and of entertainment
for all perfons promifcuoufly. Neither the air pump, nor the
orrery; neither experiments in hydroftatics, optics, or magnetifm; nor thofe in all other branches of Natural Philofophy
ever brought together fo many, or fo great concourfes of people, as thofe of electricity have done fingly. Electrical experiments have, in almoft every country in Europe, occafionally furnifhed the means of fubfiftance to numbers of ingenious and induftrious perfons, whofe circumftances have not
been affluent, and who have had the addrefs to turn to their
own advantage that paffion for the marvellous, which they
faw to be fo ftrong in all their fellow creatures. An empe-
ror need not defire a greater revenue than the fums which,

4 A 2   have

have been received in shillings, sixpences, threepences, and twopences, for exhibiting the Leyden experiment.

AND if we only confider what it is in objects that makes them capable of exciting that pleafing aftonifhment, which has fuch charms for all mankind; we fhall not wonder at the eagernefs with which perfons of both fexes, and of every age and condition, run to fee electrical experiments. Here we fee the courfe of nature, to all appearance, entirely re-verfed, in its moft fundamental laws, and by caufes feemingly the flighteft imaginable. And not only are the greateft effects produced by caufes which feem to be inconfiderable, but by thofe with which they feem to have no connection. Here, contrary to the principles of gravitation, we fee bodies attract-ed, repelled, and held fufpended by others, which are feen to have acquired that power by nothing but a very flight fric-tion; while another body, with the very fame friction, re-verfes all its effects. Here we fee a piece of cold metal, or even water, or ice, emitting ftrong fparks of fire, fo as to kindle many inflammable fubftances; and in *vacuo* its light is prodigioufly diffufed, and copious; fo as exactly to refemble, what it really is, the lightning of heaven. Again, what can feem more miraculous than to find, that a common glafs phi-al or jar, fhould, after a little preparation (which, however, leaves no vifible effect, whereby it could be diftinguifhed from other phials or jars) be capable of giving a perfon fuch a violent fenfation, as nothing elfe in nature can give, and even of deftroying animal life; and this fhock attended with an explofion like thunder, and a flafh like that of lightning? Laftly, what would the antient philofophers, what would

Newton

Newton himself have said, to see the present race of electricians imitating in miniature all the known effects of that tremendous power, nay disarming the thunder of its power of doing mischief, and, without any apprehension of danger to themselves, drawing lightning from the clouds into a private room, and amusing themselves at their leisure, by performing with it all the experiments that are exhibited by electrical machines.

So far are philosophers from laughing to see the astonishment of the vulgar at these experiments, that they cannot help viewing them with equal, if' not greater astonishment themselves. Indeed, all the electricians of the present age can well remember the time, when, with respect to these things, they themselves would have ranked among the same ignorant staring vulgar.

Besides, so imperfectly are these strange appearances understood, that philosophers themselves cannot be too well acquainted with them; and therefore should not avoid frequent opportunities of seeing the same things, and viewing them in every light. It is possible that, in the most common appearances, some circumstance or other, which had not been attended to, may strike them; and that from thence light may be reflected upon many other electrical appearances.

Whether philosophers may think this consideration worth attending to or not, I shall, for the sake of those electricians who are young enough, and, as it may be thought, childish enough, to divert themselves and their friends with electrical experiments, describe a number of the most beautiful and surprising appearances in electricity; that the young operator

may

may not be at a loſs what to exhibit when a company of gentlemen or ladies wait upon him, and that he may be able to
perform the experiments to the moſt advantage, without diſ-
appointing his friends, or fretting himſelf.

To make this buſineſs the eaſier to the young operator, I
ſhall conſult his convenience in the order in which I ſhall
relate the experiments, beginning with thoſe which only require ſimple electrification, then proceeding to thoſe in which
the Leyden experiment is uſed, and concluding with thoſe in
which recourſe muſt be had to other philoſophical inſtruments
in conjunction with the electrical machine.

SECTION

SECTION I.

ENTERTAINING EXPERIMENTS IN WHICH THE LEYDEN
PHIAL IS NOT USED.

THE phenomena of electrical attraction are shown in as
pleasing a manner by the tube, as they can be by any
methods that have been found out since the later improvements in electricity. It is really surprising to see a feather, or
a piece of leaf gold first attracted by a glass tube excited by
a slight friction, then repelled by it, and held suspended in
the air, or driven about the room wherever the operator pleases; and the surprise is increased by seeing the feather, which
was repelled by the smooth glass tube, attracted by an excited
rough tube, or a stick of sealing wax, &c. and jumping from
the one to the other, till the electricity of both be discharged.
Nor is the observation of Otto Guericke the least pleasing
circumstance, viz. that in turning the tube round the feather, the same side of the feather is always presented towards
it.

BUT

But since electrical substances part with their electricity but slowly, the more rapid alternate attractions and repulsions are shown to the best advantage at the prime conductor. Thus present a number of seeds of any kind, grains of sand, a quantity of brass dust, or other light substances in a metal dish (or rather in a glass cylindrical vessel standing on a metal plate) to another plate hanging from the conductor [as at *n* and *o* Pl. ii.] and the light substances will be attracted and repelled with inconceivable rapidity, so as to exhibit a perfect shower, which, in the dark will be all luminous.

Suspend one plate of metal to the conductor, and place a metal stand, of the same size, at the distance of a few inches exactly under it, and upon the stand put the figures of men, animals, or whatever else shall be imagined, cut in paper or leaf gold, and pretty sharply pointed at both extremities; and then, upon electrifying the upper plate, they will perform a dance, with amazing rapidity of motion, and to the great diversion of the spectators.

If a downy feather, or a piece of thistle down be used in this manner, it will be attracted and repelled with such astonishing celerity, that both its form and motion will disappear; all that is to be discerned being its colour only, which will uniformly fill the whole space in which it vibrates. *

If a piece of leaf gold be cut with a pretty large angle at one extremity, and a very acute one at the other, it will need no lower plate, but will hang by its larger angle at a small distance from the conductor, and by the continual waving

Lovet, p. 28.

motion

motion of its lower extremities, will have the appearance of something animated, biting or nibling at the conductor. It is therefore called by Dr. Franklin the *golden fish*.

To the dancing figures above-mentioned, it is very amusing to add a set of ELECTRICAL BELLS. These consist of three small bells, the two outermost of which are suspended from the conductor by chains, and that in the middle by a silken string, while a chain connects it with the floor; and two small knobs of brass, to serve instead of clappers, hang by silken strings, one between each two bells. In consequence of this disposition. when the two outermost bells, communicating with the conductor, are electrified, they will attract the clappers, and be struck by them. The clappers, being thus loaded with electricity, will be repelled, and fly to discharge themselves upon the middle bell. After this, they will be again attracted by the outermost bells; and thus by striking the bells alternately, a continual ringing may be kept up as long as the operator pleases. In the dark, a continual flashing of light will be seen between the clappers, and the bells; and when the electrification is very strong, these flashes of light will be so large, that they will be transmitted by the clapper from one bell to the other, without its ever coming into actual contact with either of them, and the ringing will, consequently, cease. When these two experiments of the bells and the figures are exhibited at the same time, they have the appearance of men or animals dancing to the music of the bells; which, if well conducted, may be very diverting.

IF a piece of burnt cork, about the bigness of a pea, cut into the form of a spider, with legs of linen thread, and a

4 B   grain

grain or two of lead put in it, to give it more weight, be
fufpended by a fine filken thread, it will, like a clapper between the two bells, jump from an electrified to an unelectrified body and back again, or between two bodies poffeffed
of different electricities, moving its legs as if it were alive,
to the great furprize of perfons unacquainted with the conftruction of it. This is an American invention, and is defcribed by Dr. Franklin. *

SEVERAL very beautiful experiments, which depend on
electrical repulfion, may be fhown to great advantage by bun-
dles of thread, or of hair, fufpended from the conductor, or
prefented to it. They will fuddenly ftart up, and feparate
upon being electrified, and inftantly collapfe when the electricity is taken off. If the operator can manage this experiment with any degree of dexterity, the hair will feem to the
company to rife and fall at the word of command.

IF a large plumy feather be fixed upright on an electrified
ftand, or held in the hand of a perfon electrified, it is very
pleafing to obferve how it becomes turgid, its fibres extending themfelves in all directions from the rib; and how it
fhrinks, like the fenfative plant, when any unelectrified body
touches it, the point of a pin or needle is prefented to it, or
the prime conductor with which it is connected.

BUT the effects of electrical repulfion are fhown in a more
furprifing manner by means of water iffuing out of a capillary
tube. If a veffel of water be fufpended from the conductor,
and a capillary fyphon be put into it, the water will iffue

* Letters, p. 17.

flowly,

flowly, and in the form of large drops from the lower leg of the fyphon; but, upon electrifying this little apparatus, inftead of drops, there will be one continued ftream of water; and if the electrification be ftrong, a number of ftreams, in the form of a cone, the apex of which will be at the extremity of the tube; and this beautiful fhower will be luminous in the dark.

LASTLY, Mr. Rackftrow's experiment (as it is generally called) is a ftriking inftance of electrical attraction and repulfion, and, at the fame time, exhibits a very pleafing fpectacle. Electrify a hoop of metal, fufpended from the prime conductor (or fupported with fmall pieces of fealing wax, &c.) about half an inch above a plate of metal, and parallel to it. Then place a round glafs bubble, blown very light, upon the plate, near the hoop, and it will be immediately attracted to it. In confequence of this, the part of the bubble which touched the hoop will acquire fome electric virtue, and be repelled; and, the electricity not being diffufed over the whole furface of the glafs, another part of the furface will be attracted, while the former goes to difcharge its electricity upon the place. This will produce a revolution of the bubble quite round the hoop, as long as the electrification is continued, and will be either way, juft as it happens to fet out, or as it is driven by the operator. If the room be darkened the glafs ball will be beautifully illuminated. Two bubbles may be made to revolve about the fame hoop, one on the infide, and the other on the outfide; and either in the fame, or contrary directions. If more hoops be ufed, a greater number of bubbles may be made to revolve, and thus a kind

4 B 2   of

of *planetarium* or *orrery* might be conftructed, and a ball hung over the center of all the hoops would ferve to reprefent the fun in the center of the fyftem. Or the hoops might be made elliptical, and the fun be placed in one of the *foci*. N. B. A bell or any metallic veffel inverted would ferve inftead of a fingle hoop.

ALL the motions above-mentioned are the immediate effect of electrical attraction and repulfion. The following amufing experiments are performed by giving motion to bodies through the medium of air, i. e. by firft putting the air in motion. Let the electrician provide himfelf with a fet of vanes, made of gilt paper or tinfel, each about two inches in length and one in breadth. Let thefe be ftuck in a cork, which may be fufpended from a magnet by means of a needle; and then, if they be held at a fmall diftance on one fide of the end of a pointed wire proceeding from the conductor, they will be turned round with great rapidity by the current of air which flows from the point. If the vanes be removed to the other fide of the point, the motion will prefently ftop, and begin again with the fame rapidity in a contrary direction; and thus the motion may be changed at pleafure. This experiment may be diverfified by vanes cut in the form of thofe of a fmoke jack; when, being held over the end of a pointed wire, turned upwards, and electrified, they will be turned round very fwifty, by the current of air flowing upwards. If they be held under a point projecting downwards, they will be turned the contrary way.

ON the top of a finely pointed wire, rifing perpendicularly from the conductor, let another wire fharpened at each end

be

be made to move freely as on a center. If it be well balanced, and the points be bent horizontally, in opposite directions, it will, when electrified, turn very swiftly round, by the re-action of the air against the current which flows off the points. These points may be nearly concealed, and horses or other figures placed upon the wires, so as to turn round with them, and look as if the one pursued the other. This experiment Mr. Kinnersley calls the ELECTRICAL HORSE RACE. If the number of wires proceeding from the same center be increas-ed, and different figures be put upon them, the race will be more complicated and diverting. If this wire which sup-ports the figures have another wire finely pointed rising from its center, another set of wires, furnished with other figures, may be made to revolve above the former, and either in the same, or in a contrary direction, as the operator pleases.

IF such a wire, pointed at each end, and the ends bent in opposite directions, be furnished like a dipping needle with a small axis fixed in its middle, at right angles with the bending of the points, and the same be placed between two insulated wire strings, near and parallel to each other, so that it may turn on its axis freely upon and between them; it will when electrified have a progressive as well as circular motion, from one end of the wires that support it to the other, and this even up a considerable ascent.

A VARIETY of beautiful appearances may be exhibited by means of electrical LIGHT, even in the open air, if the room be dark. Brushes of light from points electrified positively, and not made very sharp, or from the edges of metallic plates, diverge in a very beautiful manner, and may be excited to a

great

great length, by prefenting to them a finger, or the palm of of the hand, to which they feel like foft lambent flames, which have not the leaft pungency, nor give a difagreeable fenfation of any kind. It is alfo amufing to obferve the difference there is between brufhes of light from pointed bodies electrified pofitively, or negatively.

In the electrical horfe race above-mentioned, a fmall flame will be feen in the dark at every point of the bent wires ; fo that, if the operator can contrive to make the wire terminate in the horfe's tail, it will feem to be all on fire. And if a circular plate of metal be cut into the fhape of a ftar, fo that every point may be at the fame diftance from the center, and the center be made to turn freely on a point, like the wires in the preceding experiment, a fmall flame will be feen at e-very point ; and if the ftar be turned round, it will exhibit the appearance of a lucid circle, without any difcontinuance of the light.

As the motion of electric matter is, to the fenfes, inftanta-neous, a variety of beautiful appearances may be exhibited by a number of fmall electric fparks, difpofed in various forms. This may be done by means of a board and a number of wires, in the following manner. Let two holes be made through the board, about a quarter of an inch on each fide of the fpot where a fpark is defired. Let the extremities of the wires neatly rounded, come through thefe holes, and be brought near together, exactly over the place ; and let the wires on the back fide of the board be fo difpofed, as that an electric fpark muft take them all in the fame circuit. When they are thus prepared, all the points will appear luminous at once.

whenever

whenever a spark is taken by them at the prime conductor. In this manner may beautiful representations be made of any of the constellations, as of the great bear, Orion &c; and in this manner, also, may the outlines of any drawing, as of figures in tapestry, be exhibited.

The force of an electric spark in setting fire to various substances was one of the first experiments that gave an *eclat* to electricity, and it is still repeated with pleasure. Spirits of wine a little warmed are commonly made use of for this purpose. The experiment will not fail to succeed, if a pretty strong spark be drawn, in any manner, or direction whatever, through any part of it: and this may easily be done many ways, if it be contained in a metal spoon with a pretty wide mouth. A candle newly blown out may be lighted again by the electric spark passing through the gross part of the smoke, within half an inch of the snuff; though it is perhaps blown in again by the motion given to the air by the force of the explosion. Also air produced by the effervescence of steel filings with oil of vitrol diluted with water, and many other substances, which throw out an inflammable vapour, may be kindled by it.

The strong phosphoreal or sulphureous smell, which may be perceived by presenting the nostrils within an inch or two of any electrified point, makes a curious experiment, but it does not give a pleasing sensation.

Lastly, the most entertaining experiment that can be performed by simple electrification, is when one or more of the company stand upon an insulated stool, holding a chain from the prime conductor. In this case, the whole body is,

in

in reality, a part of the prime conductor, and will exhibit all the same appearances, emitting sparks wherever it is touched by any person standing on the floor. If the prime conductor be very large, the sparks may be too painful to be agreeable, but if the conducter be small, the electrification moderate, and none of the company present touch the eyes, or the more tender parts of the face of the person electrified, the experiment is diverting enough to all parties.

MOST of the experiments above-mentioned may also be performed to the most advantage by the person standing upon the stool, if he hold in his hand whatever was directed to be fastened to the prime conductor. Spirits of wine may be fired by a spark from a person's finger as effectually as in any other way. Care must be taken that the floor on which the stool is placed be free from dust, but it is most advisable to have a large smooth board for the purpose.

SECTION

SECTION II.

ENTERTAINING EXPERIMENTS PERFORMED BY MEANS OF THE LEYDEN PHIAL.

NO electrical experiments answer the joint purpose of pleasure and surprise in any manner comparable to those that are made by means of the Leyden phial. All the varieties of electrical attraction and repulsion may be exhibited, either by the wire, or the coating of it; and if the knobs of two wires, one communicating with the inside, and the other with the outside of the phial, be brought within four or five inches of one another, the electrical spider above-mentioned will dart from the one to the other in a very surprising manner, till the phial be discharged. But the peculiar advantage of the Leyden experiment is, that, by this means, the electrical flash, report, and sensation, with all their effects, may be increased to almost any degree that is desired.

When the phial, or the jar, is charged, the shock is given through a person's arms and breast, by directing him to hold a chain communicating with the outſide in one hand, and to touch the wire of the phial, or any conductor communicating with it, with the other hand. Or the ſhock may be made to paſs through any particular part of the body without much affecting the reſt, if that part, and no other, be brought into the circuit through which the fire muſt paſs from one ſide of the phial to the other.

A GREAT deal of diverſion is often occaſioned by giving a perſon a ſhock when he does not expect it; which may be done by concealing the wire that comes from the outſide of the phial under the carpet, and placing the wire which comes from the inſide in ſuch a manner in a perſon's way; that he can ſuſpect no harm from putting his hand upon it, at the ſame time that his feet are upon the other wire. This, and many other methods of giving a ſhock by ſurpriſe, may eaſily be executed by a little contrivance; but great care ſhould be taken that theſe ſhocks be not ſtrong, and that they be not given to all perſons promiſcuouſly.

WHEN a ſingle perſon receives the ſhock, the company is diverted at his ſole expence; but all contribute their ſhare to the entertainment, and all partake of it alike, when the whole company forms a circuit, by joining their hands; and when the operator directs the perſon who is at one extremity of the circuit to hold a chain which communicates with the coating, while the perſon who is at the other extremity of the circuit touches the wire. As all the perſons who form this circuit are ſtruck at the ſame time, and with the ſame degree of force, it is often very pleaſant to ſee them ſtart at the ſame

moment,

moment, to hear them compare their fenfations, and obferve the very different accounts they give of it.

This experiment may be agreeably varied, if the operator, inftead of making the company join hands, direct them to tread upon each others toes, or lay their hands upon each others heads; and if, in the latter cafe, the whole company fhould be ftruck to the ground, as it happened when Dr. Franklin once gave the fhock to a number of very ftout robuft men, the inconvenience arifing from it will be very inconfiderable. The company which the Doctor ftruck in this manner neither heard nor felt the ftroke, and immediately got up again, without knowing what had happened.

The moft pleafing of all the furprifes that are given by the Leyden phial is that which Dr. Franklin calls the MAGIC PICTURE, which he defcribes in the following manner. Having a large metzotinto, with a frame and glafs (fuppofe of the king) take out the print, and cut a pannel out of it, near two inches diftant from the frame all round. If the cut be through the picture, it is not the worfe. With thin pafte, or gum water, fix the board that is cut off on the infide of the glafs, preffing it fmooth and clofe, then fill up the vacancy, by gilding the glafs well with leaf gold, or brafs. Gild likewife the inner edge of the back of the frame all round, except the top part, and form a communication between that gilding and the gilding behind the glafs; then put in the board, and that fide is finifhed. Turn up the glafs, and gild the forefide exactly over the back gilding; and when it is dry, cover it, by pafting on the pannel of the picture that has been cut out, obferving to bring the correfpondent parts

4 C 2   of

of the board and picture together, by which the picture will appear of a piece as at firft, only part is behind the glafs, and part before. Laftly, hold the picture horizontally by the top, and place a little moveable gilt crown on the king's head.

If now the picture be moderately electrified, and another perfon take hold of the frame with one hand, fo that his fingers touch its infide gilding, and with the other hand endeavour to take off the crown, he will receive a terrible blow, and fail in the attempt. The operator who holds the picture by the upper end, where the infide of the frame is not gilt, to prevent its falling, feels nothing of the fhock, and may touch the face of the picture without danger, which he pretends to be a teft of his loyalty. If a ring of perfons take a fhock among them, the experiment is called the CONSPIRA-TORS.

As the electric fire may be made to take whatever circuit the operator fhall pleafe to direct, it may be thrown into a great variety of beautiful forms. Thus, if a charged phial be placed at one extremity of the gilding of a book, and the difcharge be made by a wire which touches the other extremity, the whole gilding will be rendered luminous. But if feveral pretty ftrong fhocks be fent through the fame gilding, they will foon render it incapable of tranfmitting any more, by breaking and feparating the parts too far afunder. Alfo the electric conftellations and figures, mentioned above, may be lighted up much more ftrongly by a charged phial than by fparks from the conductor; only, they cannot be lighted up fo often in this way.

On

On the same principle that the wires of phials charged differently will attract and repel differently, is made an ELEC-
TRICAL WHEEL, which Dr. Franklin says, turns with considerable strength, and of which he gives the following description.   A small upright shaft of wood passes at right angles through a thin round board, of about twelve inches diameter, and turns on a sharp point of iron, fixed in the lower
end; while a strong wire in the upper end, passing through a
small hole in a thin brass plate, keeps the shaft truly vertical.
About thirty *radii*, of equal length, made of sash glass, cut
in narrow slips, issue horizontally from the circumference of
the board; the ends most distant from the center being about
four inches apart.   On the end of every one a brass thimble
is fixed.

If now the wire of a bottle, electrified in the common way,
be brought near the circumference of this wheel, it will attract the nearest thimble, and so put the wheel in motion.
That thimble, in passing by, receives a spark, and thereby
being electrified, is repelled, and so driven forwards, while
a second being attracted approaches the wire, receives a spark,
and is driven after the first; and so on, till the wheel has
gone once round; when the thimbles before electrified approaching the wire, instead of being attracted, as they were
at first, are repelled, and the motion presently ceases.

But if another bottle, which had been charged through
the coating, be placed near the same wheel, its wire will attract the thimble repelled by the first, and thereby double
the force that carries the wheel round; and not only taking
out the fire that had been communicated by the thimbles to
the

the firſt bottle, but even robbing them of their natural quantity, inſtead of being repelled when they come again towards
the firſt bottle, they are more ſtrongly attracted; ſo that the
wheel mends its pace, till it goes with great rapidity, twelve
or fifteen rounds in a minute, and with ſuch ſtrength, that
the weight of 100 Spaniſh dollars, with which we once loaded it, did not in the leaſt ſeem to retard its motion. This is
called an ELECTRICAL JACK, and if a large fowl was ſpitted
on the upper ſhaft, it would be carried round before a fire,
with a motion fit for roaſting.

BUT this wheel, continues the Doctor, like thoſe driven
by wind, moves by a foreign force, to wit that of the bottles.

THE SELF MOVING WHEEL, though conſtructed on the
ſame principles, appears more ſurpriſing. It is made of a
thin round plate of window glaſs, ſeventeen inches diameter,
well gilt on both ſides, all but two inches next the edge.
Two ſmall hemiſpheres of wood are then fixed with cement
to the middle of the upper and under ſides, centrally oppoſite;
and in each of them a thick ſtrong wire, eight or ten inches
long, which together makes the axis of the wheel. It turns
horizontally, on a point at the lower end of its axis, which
reſts on a bit of braſs, cemented within a glaſs ſalt cellar.
The upper end of its axis paſſes through a hole in a thin braſs
plate, cemented to a long and ſtrong piece of glaſs; which
keeps it ſix or eight inches diſtant from any non electric, and
has a ſmall ball of wax or metal on its top, to keep in the
fire.

IN a circle on the table which ſupports the wheel, are fixed
twelve ſmall pillars of glaſs, at about eleven inches diſtance,
with

with a thimble on the top of each. On the edge of the wheel is a fmall leaden bullet, communicating by a wire with the gilding of the upper furface of the wheel; and about fix inches from it, is another bullet, communicating, in like manner, with the under furface. When the wheel is to be charged by the upper furface, a communication muft be made from the under furface to the table.

When it is well charged, it begins to move. The bullet neareft to a pillar moves towards the thimble on that pillar, and, paffing by, electrifies it, and then pufhes itfelf from it. The fucceeding bullet, which communicates with the other furface of the glafs, more ftrongly attracts that thimble, on account of its being electrified before by the other bullet, and thus the wheel increafes its motion, till the refiftance of the air regulates it. It will go half an hour, and make, one minute with another, twenty turns in a minute, which is 600 turns in the whole, the bullet of the upper furface giving in each turn twelve fparks to the thimbles, which makes 7200 fparks, and the bullet of the under furface receiving as many from the thimble, thofe bullets moving in the time near 2500 feet. The thimbles are well fixed, and in fo exact a circle, that the bullets may pafs within a very fmall diftance of each of them.

If inftead of two bullets, you put eight, four communicating with the upper furface, and four with the under furface, placed alternately (which eight, at about fix inches diftance, complete the circumference) the force and fwiftnefs will be greatly increafed, the wheel making fifty turns in a minute, but then it will not continue moving fo long.

These

THESE wheels, the Doctor adds, may be applied perhaps to the ringing of chimes, and moving light made orreries. *

A PHIAL makes the moft beautiful appearance when it is charged without any coating on the outfide, by putting the hand, or any conductor, to it, for then, at whatever part of the jar the difcharge is made, the fire will be feen to branch from it in moft beautiful ramifications all over the jar, and the light will be fo intenfe, that the minuteft of the branches may be feen in open day light.

THE difcharge of a large electrical battery is rather an aw-ful than a pleafing experiment, and the effects of it, in rend-ing various bodies, in firing gun powder, in melting wires, and in imitating all the effects of lightning, never fail to be viewed with aftonifhment. In order to fire gun powder, it muft be made up into a fmall catridge, with blunt wires in-ferted at each end, and brought within half an inch of each other, through which the fhock muft pafs: or a very fmall wire may be drawn through the center of it, and the explo-fion will be made by its melting. A common jar will eafily ftrike a hole through a thick cover of a book, or many leaves of paper, and it is curious to obferve the bur raifed on both fides, as if the fire had darted both ways from the center.

A CONSIDERABLE number of experiments with an electrical battery, fome of which exhibit fine appearances, will be par-ticularly defcribed in the laft part of the work.

* Franklin's Letters, p. 28, &c.

SECTION

SECTION III.

ENTERTAINING EXPERIMENTS MADE BY A COMBINA-
TION OF PHILOSOPHICAL INSTRUMENTS.

IN order to exhibit some of the finest electrical experiments, the operator must call to his aid other philosophical instruments, particularly the condensing machine, and the air pump.

IF the fountain made by condensed air be insulated, and be made to emit one stream, that stream will be broken into a thousand, and equally dispersed over a great space of ground, when the fountain is electrified; and by only laying a finger upon the conductor, and taking it off again, the operator may command either the single stream, or the divided stream at pleasure.   In the dark, the electrified stream appears quite luminous.

The greatest quantity of electric light is seen in *vacuo*. Take a tall receiver very dry, and in the top of it insert with cement a wire not very acutely pointed.   Then exhaust the receiver, and present the knob of the wire to the conductor, and every spark will pass through the the vacuum in a broad stream of light, visible through the whole length of the re-

4 D   ceiver,

ceiver, be it ever fo tall.   This ftream often divides itfelf into
a variety of beautiful rivulets, which are continually changing their courfe, uniting and dividing again, in a moft pleafing manner.   If a jar be difcharged through this vacuum, it
gives the appearance of a very denfe body of fire, darting directly through the center of the vacuum, without ever touching the fides; whereas, when a fingle fpark paffes through,
it generally goes more or lefs to the fide, and a finger put on
the outfide of the glafs will draw it wherever a perfon pleafes.
If the veffel be grafped by both hands, every fpark is felt like
the pulfation of a great artery, and all the fire makes towards
the hands.   This pulfation is felt at fome diftance from the
receiver; and in the dark, a light is feen betwixt the hands
and the glafs.

ALL this while the pointed wire is fuppofed to be electrified pofitively; if it be electrified negatively, the appearance
is remarkably different.   Inftead of ftreams of fire, nothing is
feen but one uniform luminous appearance, like a white
cloud, or the milky way in a clear ftar-light night.   It feldom reaches the whole length of the veffel, but is generally
only like a lucid ball at the end of the wire.

A VERY beautiful appearance of electric light in a darkened
room may alfo be produced by inferting a fmall phial into the
neck of a tall receiver, fo that the external furface of the
glafs may be expofed to the vacuum.   The phial muft be
coated on the infide, and while it is charging, at every fpark
taken from the conductor into the infide, a flafh of light is
feen to dart, at the fame time, from every part of the external
furface of the jar, fo as quite to fill the receiver. Upon making
                                                            the

the difcharge, the light is feen to return in a much clofer body, the whole coming at once.

But the moft beautiful of all the experiments that can be exhibited by the electric light is Mr. Canton's AURORA BO-REALIS, of which the following is but an imperfect defcription. Make a torricellian vacuum in a glafs tube, about three feet long, and feal it hermetically, whereby it will be always ready for ufe. Let one end of this tube be held in the hand, and the other applied to the conductor, and immediately the whole tube will be illuminated, from end to end; and when taken from the conductor, will continue luminous without interruption for a confiderable time, very often above a quarter of an hour. If, after this, it be drawn through the hand either way, the light will be uncommonly intenfe, and without the leaft interruption from one hand to the other, even to its whole length. After this operation, which difcharges it in a great meafure, it will ftill flafh at intervals, though it be held only at one extremity, and quite ftill; but if it be grafped by the other hand, at the fame time, in a different place, ftrong flafhes of light will hardly ever fail to dart from one end to the other; and this will continue twenty four hours, and perhaps much longer, without frefh excitation. Small and long glafs tubes exhaufted of air, and bent in many irregular crooks and angles, will, when properly electrified in the dark, beautifully reprefent flafhes of lightning.

I fhall conclude this defcription of entertaining experiments with an account of the manner in which Dr. Franklin and his friends clofed the year 1748. The hot weather coming on,

when

when electrical experiments were not fo agreeable, they put an end to them for that feafon, as the Doctor fays, fomewhat hu-mouroufly in a party of pleafure on the banks of the Skuylkil. Firft, fpirits were fired by a fpark fent from fide to fide through the river, without any other conductor then the water. A turkey was killed for their dinner by the electrical fhock, and roafted by the electrical jack, before a fire kindled by the elec-trified bottle, when the healths of all the famous electricians in England, Holland, France, and Germany, were drunk in electrified bumpers, under a difcharge of guns from the electrical battery. *

HAPPY would the author of this treatife be to fee all the great electricians of Europe, or even thofe in England, upon fuch an occafion, and efpecially after having made difcoveries in electricity of equal importance with thofe made in Phila-delphia in the year referred to. With pleafure would he obey a fummons to fuch a rendezvous, though it were to ferve the illuftrious company in the capacity of operator, or even in the more humble office of waiter. Chearfulnefs and focial intercourfe do, both of them, admirably fuit, and promote the true fpirit of philofophy.

* Franklin's Letters, p. 35.

PART

PART VIII.

NEW EXPERIMENTS IN ELECTRICITY, MADE IN
THE YEAR 1766.

I SHALL, in the laſt part of this work, preſent my reader
with an account of ſuch new experiments in electricity as
this undertaking has led me to make. I hope the peruſal of
this work may ſuggeſt many more, and more conſiderable
ones to my readers, and then I ſhall not think that I have
written in vain.

To make this account the more uſeful to ſuch perſons as
may be willing to enter into philoſophical inveſtigations, I
ſhall not fail to report the real views with which every experiment was made, falſe and imperfect as they often were.
I was always greatly pleaſed with the extreme exactneſs and
ſimplicity of Mr. Grey, and ſhall, therefore, imitate his art-
leſs manner. And though an account of experiments drawn
up on this plan be leſs calculated to do an author honour as a
philoſopher; it will, probably, contribute more to make
other

other perfons philofophers, which is a thing of much more confequence to the public.

MANY modeft and ingenious perfons may be engaged to attempt philofophical inveftigations, when they fee, that it requires no more fagacity to find new truths than they themfelves are mafters of; and when they fee that many difcoveries have been made by mere accident, which may prove as favourable to them as to others. Whereas it is a great difcouragement to young and enterprifing geniufes, to fee philofophers propofing that firft, which they themfelves attained to laft; firft laying down the propofitions which were the refult of all their experiments, and then relating the facts, as if every thing had been done to verify a true preconceived theory.

THIS fynthetic method is, certainly, the moft expeditious way of making a perfon underftand a branch of fcience; but the analytic method, in which difcoveries were actually made, is moft favourable to the progrefs of knowledge.

I HAVE, indeed, endeavoured to make the whole preceding hiftory of electricity ufeful in this view, by not contenting myfelf with informing the reader what difcoveries have been made; but, wherever it could be done, acquainting him *how* they were made, and what the authors of them had in view when they made them. In general, this has not been difficult to do, the facts being recent, and moft of the perfons concerned now living. And perhaps, in no branch of fcience has there been lefs owning to genius, and more to accident; fo that no perfon, who will give a little attention to the fubject, need be without hopes of adding fomething to the common ftock of electrical difcoveries. Nay it would be extraordinary

ordinary, if, in a great number of experiments, in which things were put into a variety of new situations, no new fact, worth communicating to the public, should arise.

The method I propose will, likewise, give the most pleasure to those persons, who delight in tracing the real progress of the human mind, in the investigation of truth, and the acquisition of knowledge; as I hope it will carry with it sufficient evidence of its own authenticity. For this progress, we may assure ourselves; has, in all cases, been by easy steps, even when it has been the most rapid. Were it possible to trace the succession of ideas in the mind of Sir Isaac Newton, during the time that he made his greatest discoveries, I make no doubt but our amazement at the extent of his genius would a little subside. But if, when a man publishes discoveries, he, either through design, or through habit, omit the intermediate steps by which he himself arrived at them; it is no wonder that his speculations confound others, and that the generality of mankind stand amazed at his reach of thought. If a man ascend to the top of a building by the help of a common ladder, but cut away most of the steps after he has done with them, leaving only every ninth or tenth step; the view of the ladder, in the condition in which he has been pleased to exhibit it, gives us a prodigious, but an unjust idea of the man who could have made use of it. But if he had intended that any body should follow him, he should have left the ladder as he constructed it, or perhaps as he found it, for it might have been a mere accident that threw it in his way. It is possible he had even better have destroyed it entirely

tirely; as, in some cases, a person would more easily make a new ladder of his own, than repair an old and damaged one.

THAT Sir Isaac Newton himself owed something to a casual turn of thought, the history of his astronomical discoveries informs us; and where we see him most in the character of an experimental philosopher, as in his optical inquiries (though the method of his treatise on that subject is by no means purely analytical) we may easily conceive that many persons, of equal patience and industry (which are not called qualities of the understanding) might have done what he did. And were it possible to see in what manner he was first led to those speculations, the very steps by which he pursued them, the time that he spent in making experiments, and all the unsuccessful and insignificant ones that he made in the course of them; as our pleasure of one kind would be increased, our admiration would probably decrease. Indeed he himself used candidly to acknowledge, that if he had done more than other men, it was owing rather to a habit of *patient thinking*, than to any thing else.

I DO not say these things to detract from the merit of the great Sir Isaac Newton, but I think that the interests of science have suffered by the excessive admiration and wonder, with which several first rate philosophers are considered; and that an opinion of the greater equality of mankind, in point of genius, and powers of understanding, would be of real service in the present age. It would bring more labourers into the common field; and something more, at least, would certainly be done in consequence of it. For though I by no means think that philosophical discoveries are at a stand, I

think

think the progress might be quickened, if studious and modest persons, instead of confining themselves to the discoveries of others, could be brought to entertain the idea, that it was possible to make discoveries themselves. And, perhaps, nothing would tend more effectually to introduce that idea, which is at present very remote from the minds of many, in which it ought to have a place, than a faithful history of the manner in which philosophical discoveries have actually been made by others.

THAT this fidelity has been preserved in the following narrative, I make no doubt of its being its own voucher. Its imperfections will be a sufficient evidence. The same fidelity will also oblige me to relate several facts as appearing new to myself, which the course of the preceding history will show to have been discovered by others, though I was not then aware of it. Of such after discoveries, however, I have mentioned only those which, it will be seen, I have pursued something farther than the original authors, having attended to circumstances overlooked by them; or, at least, having made the experiments with much more exactness; so that the reader may expect something really new under every article. And the experiments which prove the same thing will be found considerably different from those of others, and to furnish additional arguments of the same general propositions. This repetition of old discoveries, and this variety in the experiments by which they were made, were both occasioned by a situation which is more or less common to every electrician in England; whereby we are ignorant of a great deal of what has been done by others.

4 E

In

In the following narrative will alſo be found an account not only of experiments which are complete, which exhibit ſome new fact, and from which ſomething relating to the general theory of electricity may be concluded; but alſo ſome that are incomplete, which produced no new appearance, and from which nothing poſitive could be concluded. If electricians in general had done this, they would have ſaved one another a great deal of uſeleſs labour, and would have had more time for making experiments really new, and which might have terminated in conſiderable diſcoveries. Beſides, if things be really put into new ſituations, though nothing *poſitive* can be inferred from the experiment, at leaſt ſomething *negative* may; and this cannot be ſaid to be of no importance in ſcience; nor ſtrictly ſpeaking to be no new truth. A ſufficient number of theſe experiments may, in many caſes, lay a foundation for probable and poſitive concluſions.

I make no apology for leaving ſo many of theſe experiments imperfect, and for publiſhing this account of them before they have been purſued ſo far as it may, perhaps, be thought they deſerve. I rather think the generality of philoſophers ought to make an apology to the public, for delaying the communication of their experiments and diſcoveries ſo long as they have done. It is poſſible I may never have any more leiſure or opportunity to purſue them, and others may better command both; whereby the diſcoveries will be ſooner brought to their maturity, and the progreſs of this branch of philoſophy accelerated. The genuine ſpirit of philoſophy is, ſurely, not that of mechanicks, who make the moſt of every little improvement in their arts, and never divulge them till

they

they can make no more advantage of them themselves. If I could this day communicate to any fellow-labourer a hint, which it was more probable he could immediately pursue to advantage than myself, I would not defer it till to-morrow. Nor do I think it is any great boast of a philosophical indifference to fame to make this declaration. The great Sir Isaac Newton seems to have had no idea of the pursuit of fame. He deferred the communication of his important discoveries through real modesty, thinking it impertinent to trouble the public with any thing imperfect. I make no pretensions to that kind of modesty; whether it be of a true or a false kind, I think it manifestly injurious to the progress of knowledge. Like those who contended in one of the games of antient Greece, I shall immediately deliver my torch to any person who can carry it with more dexterity. If others do the same, it may come into my hands again several times before we reach the goal.

It may be said, that I ought, at least, to have waited till I had seen the connection of my new experiments with those that were made before, and have shown that they were agreeable to some general theory of electricity. But when the facts are before the public, others are as capable of showing that connection, and of deducing a general theory from them as myself. If but the most inconsiderable part, of the temple of science be well laid out, or a single stone proper for, and belonging to it be collected; though at present it be ever so much detached from the rest of the building, its connection and relative importance will appear in due time, when the intermediate parts shall be completed. Every *fact* has a

real,

real, though unseen connection with every other fact : and when all the facts belonging to any branch of science are collected, the system will form itself. In the mean time, our guessing at the system may be some guide to us in the discovery of the facts; but, at present, let us pay no attention to the system in any other view; and let us mutually communicate every new fact we discover, without troubling ourselves about the system to which it may be reduced.

I think I shall give the most distinct view of the few things that I have observed, in the short course of my electrical experiments, if I relate them pretty nearly in the order in which they occurred, only taking care not to intermix things of a very different nature. The earliest date in my experiments is the beginning of the year 1766; when, in consequence of forming an acquaintance with some gentlemen who have distinguished themselves for their discoveries in electricity, and of undertaking to write the preceding history, my attention was first turned towards making some original experiments in this part of Natural Philosophy, which had served for my occasional amusement some time before.

SECTION

SECTION I.

EXPERIMENTS ON EXCITATION, PARTICULARLY OF
TUBES IN WHICH AIR IS CONDENSED, AND OF LARGE
GLASS GLOBES.

FINDING by my own experiments, and thofe of others, that a glafs tube, out of which the air was exhaufted, difcovered no fign of electricity outwards, but that all its effects were obferved on the infide; I imagined that, if the air was condenfed in the tube, it would operate more ftrongly on the outfide; fo that an additional atmofphere would give it a double virtue. But the refult was the very reverfe of my expectations.

SOMETIME in the month of January, when the weather was dry and frofty, I took a glafs tube, fuch as is generally ufed for electrifying, about two feet and a half in length, and an inch in diameter, It was clofed at one end, and by means of a brafs cap at the other, I fitted a condenfing engine to it; and when the tube was very dry, and in excellent order for making experiments, I began to throw in more air. At every ftroke of the pifton I endeavoured to excite the tube,

but

but found its virtue diminished. It was obliged to be brought nearer than before to attract light bodies, and gave less light when rubbed in the dark; till, as near as I could judge, I had got one additional atmosphere into the tube; when its power was scarce discernible. Letting out the air by degrees, I observed it gradually recovered its power. It attracted light bodies at a greater distance; it gave louder snappings and more light in the dark; and when the additional air was wholly let out, its power was immediately as great as it had been before any air was thrown in. This I tried several times with the same success.

COMMUNICATING these experiments to Dr. Franklin and Dr. Watson, thy suggested to me, that the non-excitation of the tube above-mentioned might be owing to moisture introduced along with the air, and adhering to the inside of the tube. This conjecture was rendered more probable by ano her experiment I had made in the mean time.

REPEATING my attempts to excite the tube above-mentioned, I found that, after very hard rubbing, it began to act a little; and that its virtue encreased with the labour. Thinking it might be the warmth which produced this effect, I held the tube to the fire, and found that when it was pretty hot, it would act almost as well as when it contained no more than its usual quantity of air. I conjectured that the warmth might expel the moisture from the sides of the glass, or make the enclosed air capable of holding a greater quantity of water in a state of perfect solution.

WILLING to determine whether the additional quantity of

air

air, with the moisture it occasioned, acted in all respects like a non-electric coating, I tried the experiment with condensed air, that Dr. Desaguliers did with sand.   After condensing the air, and finding the excitation of it impossible, as usual, I let the air suddenly out, to see whether the tube would then show any effect of the preceding friction; but it had not acquired the smallest degree of electricity; though the first stroke of the rubber, immediately afterwards, made it give sparks to the finger at the distance of two and three inches. Perhaps the degree of moisture it had contracted was very slight, and expelled by the act of excitation.

UPON being desired to repeat this experiment with a particular view to the moisture; I observed that not the least cloudiness could be perceived to adhere to the glass, at the time that it was absolutely incapable of excitation.   When one part of the tube was made warm, and the other left cold, the same stroke of the rubber would excite the warm part, without in the least affecting the rest.   But still the cold part of the tube appeared not in the least more cloudy than that which was warm; and the moment the air was let out, the first stroke of the rubber made the whole strongly electrical.

WILLING to ascertain whether the condensing of air necessarily introduced more moisture in a glass vessel than the air could hold in perfect solution, I constructed a glass condenser in such a manner, that I could charge and discharge small phials in the inside of it; concluding, that if the additional air brought more additional moisture, it would be impossible to charge a phial at all in those circumstances; whereas, if

the

the air was free from moisture, it would make the phial hold a greater charge, double in two atmospheres, treble in three, &c. Accordingly, I charged a tube about three quarters of an inch in diameter, and coated about eight inches, in the glass vessel, containing about two atmospheres; and it received a much greater charge than it could be made to take in the open air, and as near as could be judged, by the report and flash, twice as great.   At last the tube burst by a spontaneous discharge, after being charged and discharged three or four times, in the condensed air.   It is not at all probable, that it could have been broke by any charge it could have held in the open air.   This experiment seemed to determine, that there was no very great degree of moisture introduced into the glass vessel by the condensation of air.

I afterwards found that experiments on condensed air had been made by Mr. Du Faye and others, but not with all the circumstances above-mentioned.

Some of my electrical friends are of opinion, that the reason why a tube with condensed air in it cannot be excited is, that the dense air within prevents the electric fluid from being forced out of the inside of the tube, without which none can be forced into the outside; and that heating the tube makes the air within less electrical, and the tube also; in consequence of which, it more easily parts with the fluid on one side, and admits it on the other.   But upon this principle how can a solid stick of glass be excited?

Imagining that a greater quantity of electric fire would be produced from the friction of larger globes than those of the usual size, I provided myself, on the 24th of April 1766,

with

with a globe feventeen inches and a half in diameter. It had only one neck, and was made exceeding well; only being rather too large for the mouth of the furnace, a fmall coal had ftuck to its equatorial diameter, which, when it was ftruck off, made a fmall hole in it. This, in fome meafure, dif-figured the globe, but I never imagined it could prevent its excitation in any great degree; fo that I ftill indulged hopes of acquiring, by its means, a prodigious power of electricity. But what was my furprize when, after I had got it mounted in the beft manner poffible, and after trying, for hours to-gether, every method of friction, in the moft favourable cir-cumftances for excitation, I could fcarce get the appearance of fire from it; the fparks from the prime conductor being barely vifible.

ACQUAINTING Dr. Franklin with my difappointment, he advifed me to get the firft coat of the globe taken off with emery; as it had often been obferved, that many globes would not work well, till after a confiderable time, when the glafs-houfe coat, as it may be called, is worn off. This operation I accordingly performed upon it, and incredibly la-borious it proved; which greatly increafed my difappoint-ment, when I found that it had all been labour in vain, for the globe had no more electrical power than before.

DISPAIRING of making any thing of this globe, I laid it afide, and, on the 22d of May, got another, about fourteen inches in diameter. In blowing this globe, every circum-ftance that I could imagine had, in the leaft, contributed to my ill fuccefs with the former, was carefully avoided. The former was made late in the week, when the metal had been long in fufion; becaufe I had been told, that globes made in

4 F

that

that ſtate of the metal were always the beſt for electrical purpoſes.   This was blown early in the week, when the workmen ſay the metal is moſt tranſparent, and freeſt from all
kinds of imperfection.   The former was warmed, in the
courſe of making it, in a place in which wood and coals were
frequently thrown, to keep up the heat.   This was kept free
from the fumes of any fewel whatever.   Nothing could be
finer than the metal of this globe, nothing more perfect in
its form.   It was alſo very well mounted, and I did not
doubt of ſucceſs.   But, after all, this globe, if poſſible, gave
leſs fire than the former.   I had recourſe to every method of
excitation that I had ever heard of, or could myſelf imagine,
but all in vain.   The thing looked like enchantment.

WHILST I was thinking over every thing that I could
imagine might poſſibly be the cauſe of my ill ſucceſs with
theſe globes, I recollected, that another globe, which I had
got made for a friend, in the ſame ſtate of the metal with my
laſt, and only an inch and half leſs in diameter, acted exceeding well; and that there was no other apparent difference
between them, but that his had two necks, and an axis
quite through it; whereas mine had only one neck, and no
axis at all.   Willing to try every thing, I reſolved to get the
braſs cap of my globe perforated, and a ſmall wire introduced,
to ſerve inſtead of an axis.   This was done; but, in making
the perforation, it happened, unfortunately, as I then thought,
but the moſt fortunately in the world as it proved, that a
lump of hard cement, about the bigneſs of a ſmall wallnut
was puſhed into the inſide of the globe.   Vexatious as this
circumſtance was, I was impatient to try my new experiment, and immediately began to whirl the globe, with this

ſuccedanium

succedanium of an axis, though the cement was all the while
rattling, and fouling the infide.

I HAD not whirled the globe long, in thefe circumftances,
before I plainly perceived that its power increafed.   After
fome time, it was pretty confiderable, and I did not doubt
but it was owing to the axis ;   nay I had formed a pretty
plaufible theory, to account for an axis being neceffary to a
globe of fuch a fize.   Willing, however, to verify the fact,
and afcertain my new hypothefis, I took out the wire ;   but,
to my furprife, found the virtue of the globe not at all diminifhed.   On the contrary, it continued increafing, and by
the time that the cement was well broken, and difperfed, fo
as to have given a kind of lining to the globe, its power was
exceeding ftrong, and it acted as well as any globe I had ever
feen.   In this ftate, I obferved, that after exciting any part
of the furface, the fmall pieces of cement in the infide, to the
diftance of about two inches, would jump from the finger, or
any conductor, prefented on the outfide.

HAVING, in this unexpected manner, made a perfect cure
of this fmaller globe, I remounted the larger, and confidering,
that the cement could probably act only as any other electric
lining, I introduced into it fome pounded fulphur, mixed
with fome flower of brimftone ;   and found that, as foon as
there was enough to render it femi-opaque, it acted very
well.

IN this ftate, the appearance of the globe was, in feveral
refpects, very remarkable.   The part that was rubbed had
none of the fulphur upon it, except thofe places where the
polifh had been, in fome meafure, taken off by the emery,
in the firft operation.   Thefe being circular, the brimftone

lay

lay upon them, like the belts of Jupiter. The hemisphere opposite to the neck had twice as much sulphur upon it as the other; and, in both hemispheres, the sulphur lay thicker, as it receded from the equatorial diameter.

I AFTERWARDS put as much more sulphur into it, which doubled the lining equably every where; but left two or three great heaps, in particular parts of the equatorial diameter, where it was rubbed, and where I could perceive no defect of polish. Whirling the globe, upon this, I found the virtue almost quite gone, and even the amalgam could not revive it. Endeavouring to take the sulphur out of the globe, I broke a great hole into it; and also the new globe was broken the same day, by a lump of hard cement, in the inside, falling from the top to the bottom. These accidents rendered my experiments incomplete.

I THEN proposed to get another large globe, with one neck, and a large hole in the opposite side; by means of which I could easily put different substances into it, and take them out again, in order to find the cause of the appearances above-mentioned. But apprehending this course of experiments might prove a little too expensive, and after all, terminate in nothing, I unwillingly desisted.

I HAD afterwards, however, an opportunity of trying a few experiments upon a glass spheroid, about twelve inches in diameter, in one end of which I had got a pretty large hole made, through which I could introduce my hand into the inside, and which, when it was whirled, I could leave open or closed at pleasure. The other end was furnished with a brass cap and pulley, to adapt it to the machine. The experiments I

made

made with it could not afcertain what I wanted to difcover, namely the reafon of the non-excitation of the large globes in the circumftances above-mentioned ; becaufe this fpheroid was excited without any electric lining. But, however, it is poffible, that the reader may think them, on other accounts, worth communicating.

I first put my handkerchief, and various other non-electric fubftances, into the infide ; and obferved, that while they kept in one piece, and did not difperfe to different fides of the globe at the fame time, the excitation was very little impeded ; but when my handkerchief &c. opened, and fpread over the furface of the fpheroid, the excitation was nearly ftopped.

I put a quantity of quickfilver into it ; and while the motion of the fpheroid was moderate, and allowed the quickfilver to remain at the bottom, or to afcend but a little height on one fide, the excitation was much impeded ; and ftill more if it fpread further, over a confiderable part of the infide furface, though without partaking of its motion. But when the motion of the fpheroid was rapid, and entirely communicated to the quickfilver, fo that they had no motion with refpect to one another, the virtue always revived a little ; but ftill the excitation was inconfiderable.

Pounded glafs, and other electrics had no fenfible effect ; neither had any number of quills, or a great quantity of downy feathers ; though they inftantly difperfed themfelves, and covered all the equatorial parts of the fpheroid.

I shall add to this fection, on the fubject of excitation, that I once whirled a very thin globe, about fix or feven inches in diameter, which was made to weigh air, and not
                                                            one

one fourth part so thick as a common Florence flask. It was excited very powerfully by a piece of leather which had been soaked in a mixture of tallow and bees wax, and into which a quantity of amalgam had been worked. With this globe I could make my common jar discharge itself over more than five inches of the external surface, which I reckon to be a considerable proof of its power. It seems to follow from this experiment, that the thinness of glass globes, or tubes, is by no means any obstruction to their electric power.

In the course of these experiments I had read Mr. Bergman's account of his curing a globe by a lining of melted sulphur, and had proposed to try that in the last place, on account of the disagreeable operation; but found it superseded in the manner described above.

SECTION

SECTION   II.

EXPERIMENTS WHICH PROVE A CURRENT OF AIR
FROM THE POINTS OF BODIES ELECTRIFIED EITHER
POSITIVELY OR NEGATIVELY.

DURING a courfe of electrical experiments, made to
divert fome of my friends, one of the company happened to prefent a pointed wire to my hand, as I was ftanding upon the ftool; when I was furprifed to perceive a cool
blaft proceeding from it; though, according to Dr. Franklin's theory, the current of the fluid went from my hand to
the point. I then prefented my noftrils to the point, and
perceived the fame ftrong phofphoreal fmell, as if the point
had been electrified pofitively. Thefe facts made me entertain fome doubts about the direction of the current, and the
principles of Dr. Franklin's theory, and led me to the following courfe of experiments; which prove nothing againft
that theory, but eftablifh a real current of air from the points
of all electrified bodies.

To afcertain the courfe of the electric fluid, I began with
difcharging large fhocks through a quantity of water, in
which

which duſt was floating; and I electrified pointed wires, inſerted into cloſe receivers full of ſmoke; &c. but theſe determining nothing, I, at laſt, recollected, that, of all things, flame is the leaſt ſenſibly affected with electrical attraction or repulſion, but moſt eaſily with the leaſt breath of air; and did not then doubt, but that the current of air would be in the direction of the fluid, being, as it were, impelled by it.

WITH theſe views, on the 25th of February, I preſented the flame of a candle to a pointed wire, electrified negatively, as well as poſitively. The blaſt was ſo ſtrong (in both caſes alike) as to lay bare the greateſt part of the wick, the flame being driven from the point; and ſometimes a pretty large candle would be actually blown out by the blaſt. But, in all caſes, the effect was the ſame whether the electric fluid iſſued out of the point, or entered it.

PLACING the flame between two points, one of which communicated with the prime conductor electrified poſitively, and the other with the floor, the flame was blown from that which communicated with the conductor upon the other, but not to ſo great a diſtance as if the other had been away. Changing the points, the effect was ſtill the ſame, whether that which communicated with the conductor was the more ſharp, or the more blunt of the two, the flame always receding from it.

REVERSING this experiment, and making one of the points communicate with the rubber, and the other with the floor, the former always blew the flame towards the latter. It was evident, however, that the point which communicated with the floor blew likewiſe; for it counteracted the other, and

would,

would, when brought near the flame, raife it almoft perpendicular, when it had been blown quite afide by the other.

PLACING the flame between two points, one of which
communicated with the rubber, and the other with the conductor, it was equally affected by both, being always blown
from the point which was neareft to it.

IT was very obfervable, that, notwithftanding the current
of air from the points affected the flame fo remarkably;
yet a fmall portion of it, when it was brought very near the
point, would be ftrongly attracted by it, at the fame time
that the greateft part of the flame was, by the current of air,
blown the contrary way. This effect was always the fame,
whether the point was electrified pofitively or negatively;
though, I fancied that the negative point attracted the flame
more fenfibly than the other.

To make the points blow more ftrongly, in a direct line,
I fometimes inclofed them in fmall tubes of glafs.

AFTERWARDS I diverfified this experiment in the following manner. I charged the infide of a fmall jar pofitively;
then fetting it upon a glafs ftand, in contact with a pointed
wire, I placed the flame of a candle within an inch of the
point, and touched the wire of the jar, with a brafs rod
which I held in my hand. At every touch the flame was
blown ftrongly from the point. Sometimes it would be
blown out; but another point, being held oppofite to it,
would fupport the flame; and more ftrongly, if that point was
joined with the rod with which I touched the wire of the jar.
Charging the jar negatively in the infide, all the effects were
the very fame. Difcharging the jar through the points, with

the

the flame in a right line between them, it was difturbed, but not blown to one fide more than the other.

To take off all the effect of electrical attraction and repulfion, and leave the current of air to act fingly, I interpofed pieces of brafs wire communicating with the earth, between the points of the wire and the flame; and found the blaft to be rather increafed than diminifhed thereby.

HAVING communicated thefe experiments to Dr. Franklin, he advifed me to try the force of this current upon *paper vanes*, fuch as he has defcribed in his letters : for, with him, they feemed to turn one way or the other indifferently, juft as they happened to fet out. Accordingly I took a cork, and ftuck into the fides of it thirteen vanes, each being half a card, well dried, and each proceeding from the center of the cork. Into the cork I ftuck a needle, by which I fufpended the whole on a magnet.

THESE vanes I held two or three inches from the point of a wire, communicating with the outfide coating of the jar, placed upon an electric ftand, in the manner defcribed above; and obferved, that whenever I took a fpark from the wire communicating with the infide, the vanes were ftrongly blown upon, and made to turn, as if the current of air had flowed from the point; at the fame time that, according to Dr. Franklin's theory, the electric fluid was entering it. If they were made to turn the contrary way, the current foon ftopped them, and never failed to bring them back, and make them move as before.

WHEN wires communicating with the floor were placed between the vanes and the point, to take of all the electrical at-

traction

traction and repulsion, the vanes still moved as briskly as ever.

When the jar was charged pretty high, the motion might be made so swift, that the separate vanes could hardly be distinguished, as the whole set turned round.

I moreover observed, that the vanes were turned very briskly, not only when held near the point, but also when held any where within the distance of six or seven inches from the sides of the wire, which I made sometimes of a considerable length. The stream would turn the vanes one way on one side of the wire, and the contrary way on the other; and being removed quickly to the different sides, the direction of their course might be changed several times, in the discharge of one small jar.

I made points to project two ways at the same time, and observed, that the stream was the same from both, and also when the points were made to project at right angles from one another. In this position of the wires, it was amusing to observe, that the vanes would move one way, when held near one of the wires; and immediately turn about and move the contrary way, if removed near the other.

Hitherto I had made my vanes of exceeding dry paper, in order to make them less affected by electrical attraction and repulsion, that so the current of air might be the more indisputable; but Mr. Canton desiring me to try vanes that were conductors, I first dipped my paper vanes in water, and afterwards made a set of tinsel, of the same form with the other. These vanes, being conductors of electricity, promoted a freer current of the electric matter, and consequently, occasioning a greater motion to be given to the air, they

whirled

whirled about with more rapidity than the former. When they were infulated, they were affected juft as the dry paper vanes had been.

WITH thefe vanes, I diverfified the experiment in a manner which fhowed the famenefs of the current, notwithftanding the change of electricity, in a clearer manner than before. I infulated a jar, with a wire projecting from the coating, and held the tinfel vanes near the extremity of it. All the time the jar was charging, the vanes turned with great rapidity, as if by a blaft from the point. Keeping the jar, the pointed wire, and the vanes in the fame fituation, the gradual difcharge of the jar, made by now and then touching the wire which communicated with the infide, made the vanes ftill turn the fame way, and, as far as could be perceived, with the fame force.

To diverfify this experiment, I placed a charged jar upon a ftool which had glafs feet, a pointed wire projecting from the coating, a quantity of brafs duft before the point, and a brafs chain communicating with the ground on the other fide of the duft. In this fituation, every attempt to difcharge the jar threw a confiderable quantity of the duft from the point, being raifed about feven or eight inches, and blown to a confiderable diftance. Removing the pointed wire from the coating of the jar, and connecting it with the chain, the fame attempt to difcharge it blew the duft upon the jar. Ufing two points, one at the jar, and the other at the chain, the duft was difturbed, and raifed up, but not blown one way more than the other. Fine flour anfwered nearly as well.

LASTLY, I made the experiment of the current with vanes in the form of a fmoke jack, which anfwered as well as the

others.

others.   They were moved when held more than a foot a-
bove the point, and likewise at a confiderable diftance below
it, when it was turned downwards.

After thefe experiments, I read in Mr. Wilfon's treatife
on electricity, that the vanes would not turn in vacuo.   This
I tried, and found it to be true, and at the fame time I found,
they would not turn in a clofe receiver, not exhaufted, where
the air was confined, and had not a free circulation.

The current of air from the points of bodies, electrified
plus or *minus*, is not more difficult to be accounted for on
Dr. Franklin's hypothefis of pofitive and negative electricity,
than any other cafe of electrical repulfion.   The particles of
the atmofphere, near the points of electrified bodies, having,
by their means, become poffeffed of more or lefs than their
natural fhare of the electric fluid, muft, according to the rule
above-mentioned, retire to places where they can difcharge or
replenifh themfelves, as occafion may require.   If it be afked
why the particles of the atmofphere do not, in the fame manner, recede from all the parts of the electrified body, as well
as from the points; it is anfwered, that, as the preffure of
the atmofphere will prevent a vacuum, and as electrical attraction and repulfion are moft powerful at the points of
bodies, on account of the eafier entrance or exit of the fluid
at the points (upon whatever principle that effect depends)
the electrified atmofphere (whether negative or pofitive makes
no difference) muft fly off at the points preferably to any other
places, and the weight of the atmofphere will force the air of
the neighbouring places upon the flatter parts of the electrified conductor, notwithftanding the real endeavour it may
have to recede from it.

SECTION

SECTION III.

EXPERIMENTS ON MEPHITIC AIR, AND CHARCOAL.

I HAVE related several instances of self-deception in other persons : to show that I do not mean to spare myself, I shall now relate one of my own. However it is a mistake that I should not have troubled the reader with, had it not terminated in a real discovery.

HAVING read, and finding by my own experiments, that a candle would not burn in air that had passed through a charcoal fire, or through the lungs of animals, or in any of that air which the chymists call *mephitic*; I was considering what kind of change it underwent, by passing through the fire, or through the lungs, &c. and whether it was not possible to restore it to its original state, by some operation or mixture. For this purpose, I gave a great degree of intestine motion to it; I threw a quantity of electric matter from the point of a conductor into it, and performed various other operations upon it, but without any effect.

AMONG other random experiments, I dipped a charged phial into it; but, though I could not perceive that it pro-

duced

duced any effect upon the air, I was furprifed to find, upon taking out the phial, that it was wholly difcharged. I imagined, however, at that time, that the difcharge was owing to fome imperfection in the manner of making the experiment, which I could not eafily remedy.

AFTERWARDS, concluding, from fome experiments of my own, and others of Dr. Macbride that this mephitic air was not any thing that had ever been common air, but a fluid *fui generis*, that had feveral properties very different from thofe of common air; I thought of refuming my experiments, to afcertain whether it was not different from common air with refpect to electricity; mephitic air being perhaps a conductor, as common air was a non-conductor.

ACCORDINGLY, fome time in the month of January 1766, I filled a receiver, which was open at the top, and which held about three pints, with air from my lungs; and, with every precaution I could think of, dipped a fmall charged phial into it; and, upon taking it out after continuing there about two feconds, found it quite difcharged, juft as might have been expected from dipping it into water. I repeated the experiment feveral times, and always with the fame fuccefs. Left I fhould have difcharged the phial, fome way or other, in paffing the neck of the receiver; I changed the air, and then introduced the charged phial as before, feveral times; but never found it in the leaft difcharged. Hence I concluded, that, in the former cafe, it muft have been difcharged by the quality of the air in the receiver.

As there ftill remained fome fufpicion, that the phial might have been difcharged by the *moifture* which was mixed with the air as expired from my lungs (though I did not
doubt

doubt but that what moisture there was in it was quickly attracted by the sides of the glass vessel, so as to leave the center free from it, and imagined that no quantity of moisture which the air could be supposed to contain could be able to discharge the phial so quickly) I immediately repeated the experiment, with air which I thought could not be supposed to contain any moisture.

I FILLED the same receiver with air from the center of a charcoal fire, and dipping the charged phial into it, as I had done before, found it as effectually discharged as in the former case. I repeated this experiment also a great number of times, and always with the same success. Upon changing the air also, and then introducing the phial, I found it not discharged. There seemed, therefore, to be no doubt, but that air in which flame would not subsist was a conductor of electricity.

I WAS confirmed in this opinion by placing a coated phial in a receiver filled with mephitic air, and finding that it was absolutely impossible to charge it in the smallest degree, in that situation; though it was charged very well, the moment I took it out of that air, without any wiping; which seemed to show, that the incapacity of charging had not been owing to any moist vapours that adhered to the glass.

I WAS also confirmed in my opinion of the conducting power of mephitic air, by considering, that all metallic bodies, which are the most perfect conductors we know, consist of a vitrifiable earth, and what the chymists call *phlogiston*, or probably nothing more than this same mephitic air in a fixed state.

IN

In this confidence of having made a complete difcovery, and one of fome importance, after having repeated the above-mentioned experiments, and many others to the fame purpofe, a great number of times; I once happened to change the mephitic for common air in my receiver, without wiping the infide of the glafs, which had been my ufual method, though I had done it with no other view than more effectually to change the air; but now, though the air was fufficiently changed, by moving the receiver feveral times up and down, I had no fooner dipped the charged phial into it, than I found it difcharged, in the fame manner as it had been done in the mephitic air. The difcharge had, after all, been made by the moiftture on the infide. To afcertain this, I moiftened the infide of the receiver with a fpunge, and found that no phial could remain charged in it for the leaft fpace of time. This fhowed, that the experiments above recited were am-biguous.

Dr. Franklin, to whom I had communicated thefe experiments, recollected them when he was this laft fummer at Pyrmont, where a large body of mephitic air always lies upon the furface of the medicinal fpring (for this air is evidently fpecifically heavier than common air, and does not eafily mix with it) but not having a proper apparatus, and the company there making experiments inconvenient, he did nothing that was decifive; though, from the little that he had an opportunity of doing, he imagined it was not a conductor.

These experiments on mephitic air, deceitful or, at leaft, ambiguous as they were, led, however, to a difcovery, which may poffibly throw fome new light upon fome of the moft fundamental principles of electricity; and which certain-

ly

ly ſtrengthen the ſuſpicion, that mephitic air may ſtill be a conductor, at leaſt in its fixed ſtate.

FINDING that I could make nothing of mephitic air itſelf (for I had endeavoured to procure it in a variety of ways, but without ſucceſs) I conſidered that it was from *charcoal* that I procured it in the greateſt quantity, and thought I would try charcoal itſelf in ſubſtance. Accordingly, on May the 4th. 1766, I tried the charcoal, in a variety of ways and ſtates; and found it to be, what I had ſuſpected, an excellent conductor of electricity.

PRESENTING a piece of charcoal to the prime conductor, together with my finger, or a piece of braſs wire, I conſtantly obſerved, that the electric ſpark ſtruck the charcoal before either of the other conductors, if it happened to be advanced ever ſo little before them. Having a very rough ſurface, the charcoal did not take a denſe ſpark from the conductor, till it was made a little ſmooth, and brought within about half an inch; when, to all appearance, it did quite as well as any piece of metal, there being a conſtant ſtream of denſe, and white electric fire between the conductor and it. I tried the charcoal in every ſtate of heat or cold, and found no alteration of its conducting power.

I PLACED a great number of pieces of charcoal, not leſs than twelve or twenty, of various ſizes, in a circuit, and diſcharged a common jar through them; when, to all appearance, the diſcharge was as perfect, as if ſo many pieces of metal had been placed in the ſame manner. Two of the pieces, about the middle of the circuit, I placed about an inch and a half from one another; but, upon the diſcharge, the ſpark paſſed the interval very full and ſtrong. A piece

of

of charcoal alfo made the difcharge at the wire with one fpark, but the report was not fo loud as when the difcharge was made with a piece of metal. It was obfervable, that a black grofs fmoke rofe from between each of the pieces of charcoal, at the moment of the difcharge; but the ignition was momentary, and the fire could not be perceived on the charcoal.

To make the experiment of the conducting power of charcoal in the moft indifputable manner, I took a piece of baked wood, which I had often ufed for the purpofe of in-fulation, being an excellent non-conductor, and putting it into a long glafs tube, I thruft it into the fire, and converted it into charcoal. In this operation, a very great quantity of grofs fmoke rofe from it, fo that, feeming to part with more of its moifture, one would have expected it would have come out a better non-conductor; but, upon trial, its electric pro-perty was quite gone, and it was become a very good con-ductor.

The experiments above-mentioned were firft made with *wood charcoal*, of which I found pieces of very different de-grees of conducting power; but the moft perfect conductors I have found of this kind are fome pieces of *pit charcoal*. Thefe feem to be, in all refpects, as perfect conductors as me-tals. They receive a ftrong bright fpark from the prime conductor, though feldom at above an inch diftance, on ac-count of the roughnefs on their furface, which cannot be taken off; and in difcharging a jar through them, or with them, no perfon can imagine any difference between them and metal, either in the colour of the electric fpark, or the found made by the explofion. When they are broken, they

4 H 2   exhibit

exhibit an appearance which very much refembles that of broken fteel. There is however a great variety in the electrical properties of different. pieces of this kind of charcoal; and for want of proper opportunity I have not yet fucceeded in afcertaining, with fufficient certainty, the circumftances, in the preparation, &c. on which this variety depends..

I would have preferred the examination of wood charcoal on many accounts; particularly, as the fame fubftance is, in this cafe, converted from a perfect electric to a perfect conductor; and all the degrees of conducting power may be found in different fpecimens of it; whereas pit coal is itfelf a conductor, though an imperfect one: but not having any opportunity, I procured fpecimens of all the varieties I could imagine in the fame heap of pit charcoal, with refpect to their nearnefs or diftance from the furface, &c; but though I examined them with all the care and attention that I could apply, and in every method that I could think of, the differences were fo exceedingly fmall, if any, that I could not fix upon any circumftance that I could depend upon for the caufe of them.

Even common cinders from an open fire, of the kind of coals which we generally burn, I find to be very little inferior to charcoal; which is fuffered to flame, but covered very clofe as foon as it is well burnt, and before any afhes are formed. Coals and cinders from a common fire, being a very commodious fubject for experiments, I did not fail to make as many upon them as I could imagine would be of any ufe; except that I had no opportunity of trying a fufficient variety of coals. I took feveral out of the fire after they had done blazing, fome of which I covered with afhes, fome I quench-

ed

ed in water, and some I left to cool in the open air. I also reduced some of the coals to cinders in a glass vessel, without suffering them to flame; and I treated in the very same manner various pieces of oak, cut from the same plank; but when I examined them, I found their differences, with respect to their power of conducting electricity, very inconsiderable, if any. I thought the cinder of a coal which we call kennel, and which is remarkable for flaming much while it burns, to be a better conductor than a cinder from a common coal: but the difference might be owing to its more uniform texture, and smoother surface. Charcoal made of coals which yield a strong sulphureous smell when they are burnt, and of which the charcoal itself is not quite divested, was, to all appearance, as good a conductor as that of the other kind, which is more esteemed.

In this course of experiments I have found myself much at a loss for a sufficiently accurate method of ascertaining the difference of conducting substances, and I wish that electricians would endeavour to find such a measure. One of the best that I am acquainted with, and which I applied among others on this occasion, is by the residuum of discharges, measured by Mr. Lane's electrometer. It is well known, that the worse the conductors are that form the circuit, the greater residuum will be left in a jar after a discharge; and Mr. Lane's electrometer, which measures an explosion, will likewise measure the residuum. To apply this method with accuracy, I put pieces of charcoal, &c. of the same length into the circuit, I used the very same chain in every experiment, and the same disposition of every part of the apparatus; I also made the explosions exactly equal, and after every discharge compleated the circuit

circuit by the chain before I took the residuum; and lastly, I was careful to take up the same time in each operation, which I repeated very often. This method of measuring the conducting power of substances I learned of Mr. Lane, to whom, if what I have said be not quite intelligible, I must refer my reader. On the other hand, if I have said too much of it, that ingenious electrician must blame his own backwardness to publish an account of his electrometer and its various uses.

In the prosecution of these experiments on charcoal, I burned a piece which I had found to be a most excellent conductor, first between two crucibles, and then in the open fire, and tried it at different times till it was almost burned away; but, contrary, to my expectations, I found its property very little diminished. I was, likewise, surprized to find that *soot* whether of wood coal, or pit coal hardly conducted at all. I made five or six inches of the soot of pit coal part of the electric circuit, which compleated the communication between the inside and outside of a charged jar for several seconds; and yet found the charge not much diminished. A piece of wood soot, which is a firm shining substance, which does not soil the fingers, and which seems to break in a polish in several places, would hardly conduct any part of a charge in the least sensible degree. When rubbed against my hand, or my waistcoat in frosty weather (though it was difficult to find any part of it that was large and smooth enough for the purpose) I more than once thought it attracted the thread of trial. The snuff of a candle would not conduct a shock, though it was placed in the middle of the circuit, and it was easily set on fire by the explosion of a small jar.

But notwithstanding my want of success, I make no doubt,

but

but that any perſon of tolerable ſagacity, who has an opportunity of making experiments in a laboratory, where he
could reduce to a coal all kinds of ſubſtances, in every variety
of method, might very ſoon aſcertain what it is that makes
charcoal a conductor of electricity. In all the methods in
which I could make charcoal, the fume of the bodies was ſuffered to eſcape; but let trials be made of ſubſtances reduced
to a char without any communication with the open air, or
where the vapours emitted from them ſhall meet with different degrees of reſiſtance to their eſcape, aſcertained by actual preſſure.

CHARCOAL, beſides its property of conducting electricity,
is, on many other accounts, a very remarkable ſubſtance;
being indeſtructible by any method beſides burning in the
open air; and yet it ſeems not to have been ſufficiently ſtudied
by any chymiſt. A proper examination of it promiſes very
fair, not only to aſcertain the cauſe of its conducting, and,
perhaps, of all conducting powers; but to be an opening to
various other important diſcoveries in chymiſtry and Natural
Philoſophy; and the ſubject ſeems to be fairly within our
reach.

PIT COAL, and probably all other ſubſtances, at the ſame
time that they loſe much of their weight, increaſe conſiderably
in their bulk in the operation of charring. Does it not ſeem to
follow from hence, that its conducting power may poſſibly
be owing to the largeneſs of its pores, agreeable to the hypotheſis of Dr. Franklin, that electric ſubſtances have exceeding ſmall pores, which diſpoſes them to break with a poliſh.

OR, ſince the calces of metals, which are electric bodies,
become metals, and conductors, by being fuſed in contact
with

with charcoal; are not metals themfelves conductors of electricity, in confequence of fomething they get from the charcoal? Is not this the mephitic air; as the modern chymifts
fuppofe, that this is all that the metallic calces require to
their revivification?

THIS courfe of experiments, however, evidently overturns
one of the earlieft, and, hitherto, univerfally received maxims in electricity, viz. that *water* and *metals* are conductors,
and all other bodies non-conductors: for we have here a
fubftance, which is clearly neither water, nor a metal, and
yet a good conductor.

SECTION

SECTION IV.

Experiments on the CONDUCTING POWER of
VARIOUS SUBSTANCES.

FINDING some contrariety of opinion among electricians about the nature of ICE, some saying it was a conductor of electricity, and others a non-conductor, so as even to be capable of being charged like glass, I took the opportunity of a pretty severe frost, in the month of February, to assure myself of the fact.

IN order to this, I took a large piece of ice, washed it very clean, and scraped off all the sharp points about it. After this, when it was again perfectly frozen, I insulated it, at night, in the open air, whither I had carried my machine on purpose, at the same time that it was freezing intensely.

WHEN, by drawing a feather over its surface, I found it to be perfectly dry, I electrified it, and fetched large sparks, not less than an inch in length, from all parts of it. I charged a jar at it, almost as well as at the prime conductor; I also

4 I   discharged

discharged the jar through it, and along the surface of it, in several places; so that I had no doubt, but that ice was, nearly, as good a conductor of electricity as water. To try the same to more advantage, I took a charged jar into the open fields; and, by means of a great length of chain, discharged it along a large surface of ice on a pond, whilst the surface was very dry, and the frost continued very intense. But the ice being not so good a conductor as metal, if the chain communicating with the outside of the jar happened to lie five or six inches from the knob of the wire communicating with the inside, the fire would strike to the chain, along the surface of the ice, without entering it.

Snow is evidently not so good a conductor as ice; probably because its parts do not lie in contact with one another, as those of ice.

Finding also that electricians were not perfectly agreed about the conducting power of *hot glass*, and that the methods which had been used to prove it were liable to objection; since, when the electricity was communicated along the outside of the glass, it might be said that the hot air, and not the hot glass was the conductor; it occurred to me, that the following experiment would determine this affair, in a more satisfactory manner than it had hitherto been done.

I procured a glass tube, about four feet long; and, by means of mercury in the inside, and tinfoil on the outside, I charged about nine inches of the lower part of it. Then carefully slipping off the tinfoil, and pouring out the mercury, I heated the charged part of the glass red hot; and found, upon replacing the coating, that it was discharged.

I MADE

I MADE the experiment a fecond time, with the fame fuc-cefs; fo that I had no doubt, but that glafs, when red hot, was pervious to the electric fluid. It could not have gone round from the infide to the outfide, without going over a furface of fix feet of glafs, the greateft part of which was kept very cold, and all exceeding dry.

THAT the charge had not been loft by changing the quick-filver was evident: for when I repeated that part of the expe-riment, without heating the glafs, the charge was found to be very little diminifhed.

SOME time after, when I was preparing fome baked wood for the purpofes of infulation, I found, that if I ufed them foon after they were taken out of the oven, they would not anfwer my purpofe at all. The electricity went off by them to the floor. But when they had ftood, in the very fame fituation, till they were cold, they infulated very well.

UPON this, I made a piece of baked wood, which I had formerly ufed for infulation, pretty hot; and when it was fo hot, that I could hardly hold it in my hand, it took a flender fpark from the conductor, about an inch long; but it would not difcharge a jar at once. It did it however filently, pretty much like moift wood.

THE confideration of the conducting power of charcoal, and the manner in which it is made, namely, by burning inflammable fubftances in a clofe place, and generally without flaming, led me to make a few experiments on the conduc-ting power of the effluvia of flaming bodies, at the very time of their emiffion: for whatever thofe effluvia be, they feemed in fome meafure to contain the conducting principle. That fubftance, or principle, or whatever it be called which can

only

only be exhaled in flame, and not in fmoke, does in this cafe make a body a conductor; whereas if it be fuffered to efcape, the conducting power is diminifhed.

THE conducting power of the flame of a candle was obferved very early; but it was not compared with that of other things, and it had by fome been fuppofed to be nothing more than the heat communicated to the neighbouring air. The experiments I am going to recite feem to overturn this hypothefis, and to eftablifh that mentioned above.

MARCH the 14th. a fmall charged phial held not longer than a fecond within two or three inches of the flame of a candle, either above or below it, where the heat was altogether inconfiderable, and the rarefaction of the air in a manner nothing, was totally difcharged. The event was the fame when I ufed the flame of a wax candle, or the flame of fpirits of wine. When it was held much nearer to a red hot poker, it was not difcharged near fo foon; and when it was held exceeding near to a piece of red hot glafs, it was not difcharged at all, except by one explofion, feemingly conducted by the hot glafs. Similar experiments were made by placing the candle, the poker, and the hot glafs near the prime conductor. It was alfo found, that the fmall phial above-mentioned could not be difcharged in the focus of a concave mirrour. About the fame time I obferved, that this phial was much fooner difcharged by being drawn through fteam, than through fmoke.

BUT the fmall jar above-mentioned was difcharged in thefe experiments filently; and though they feemed to be clearly in favour of the conducting power of the effluvia, which pafs off in flame, there was nothing very ftriking in them; but

afterwards,

afterwards, when I had constructed an electrical battery, I repeated the experiments in a much more striking and convincing manner.

DECEMBER the 15th. I brought the flame of a candle between two brass knobs, one communicating with the inside, and the other with the outside of the battery; and observed, that as the flame advanced towards them, it began to be put into a quivering motion, exceeding quick, and was strongly drawn both ways towards each knob, leaving the wick bare at the top; and as soon as the flame was quite between the rods, the battery discharged at once, at the distance of three inches and an half. This is a very fine experiment. The interposition of the flame between the two brass rods is like putting fire to a train of gunpowder, which explodes immediately.

To compare the conducting power of these effluvia with other substances, I made the dense vapour from the spout of a boiling kettle pass between the two brass rods abovementioned, when they were very near the distance at which the battery explodes through the air, but it did not promote the discharge; neither would the smoke of rosin, nor that of a candle just blown out: but when I advanced the ignited wick towards the rods, it was ventilated exceeding briskly; and when it was put between them, when separated to about the distance of an inch, the discharge was made, and the candle blown in again.

To compare the conducting power of flame with that of other bodies, which had more heat but less effluvia, I put a red hot poker between the two rods, but it did not promote the discharge of the battery till they were brought within a-
                                                                    bout

bout an inch and an half of one another; so that the explosion was made at about twice the usual distance, allowing for
the space occupied by the poker itself; and yet the air in the
neighbourhood of the poker was more than ten times hotter
than in the neighbourhood of the candle, considering the diftance at which they were held from the rods. Both sides of
the hot poker were marked with an imperfect circle, like
those that were impressed on each of the knobs; an account
of which will be given hereafter.

I THEN interposed a piece of red hot glass, which has as
great a heat as the iron, but emits less effluvia; but it
did not promote the discharge till the brass rods were brought
within an inch of one another, which was so near, that the
glass almost touched them both.

As I was diversifying the experiments concerning the passage of the electric explosion over the surfaces of various bodies, as will be mentioned hereafter, I accidentally discovered
how exceedingly poor a conductor is *oil* of every kind; infomuch that I think it ought rather to be classed among electric substances; and what I said page 435 of all bodies in a
state of fluidity, except air, being conductors, ought to be
retracted; though at that time I had no doubt of its being
true; and imagined that oil did not differ very much from
water with respect to its conducting power. Also the remark
upon Dr. Watson's experiment with oil of turpentine, p. 77,
must be retracted, as unjust. I had been led into the mistake
by some experiments of Mr. Wilson, who has some where
advanced the proposition above-mentioned; and argues that
the tourmalin is possessed of a fixed kind of electricity, incapable of being conducted away, because it retains the separate

rate power of each of its sides, though surrounded with melted grease; whereas I find, that nothing of an oily nature will
conduct electricity.

LAYING a chain which communicated with the outside of
my battery in a dish of melted tallow, I brought a brass rod
communicating with the inside towards it, in order to make
the discharge, by transmitting the explosion over the surface
without entering it; when I was surprised to find, not only
that the electric matter would not take the surface, but that,
though it attracted a column of tallow at the distance of about
three quarters of an inch (which was thicker in proportion as
the rod was brought near the surface) and though I continued
amusing myself with this column of tallow a considerable
time; in which state it formed a complete communication
between both sides of the battery, yet the charge was very
little dissipated. I repeated this experiment, with the same
event, with oil of olives, the thinnest oil of turpentine, and
even ether. A plate of common oil of olives connected
the inside and outside of the battery for near ten minutes,
without my being able to perceive that the charge was more
dissipated, than it would have been without that communication. Ether is the lightest fluid in nature next to air; yet,
being properly an oil, it proved no better a conductor than
the most tenacious. I was most surprised that the ether did
not take fire by this treatment, as nothing is more inflammable; and if the electric matter can pass through it, nothing
fires so soon.

FROM these experiments, and those above-mentioned, on
ice, I concluded, that fluidity, as such, contributes nothing
to the conducting power of substances, separate from the heat
which

which makes them fluid.   To complete my experiments on oils, I filled phials with all kinds of oils, according to their chymical diftinctions, including the fineft *effential oils*, the ftrongly *empyreumatic*, and thofe that are termed mineral, as *oil of amber*; and found them all incapable of giving a fhock.   But I found that this method of trying the conducting power of. fubftances, viz. by inclofing them in phials, and endeavouring to give fhocks by them is very inaccurate, fhowing them to be better conductors than they really are.   Pounded glafs, flour of brimftone, and other electric fubftances gave a confiderable fhock; but a bottle containing nothing but air gave a greater fhock than any of them; though the wire inferted into it was very blunt, and was kept in the center of the bottle.   Finding, by thefe experiments, that oil plainly conducted much lefs than air, I endeavoured to charge a plate of oil like a plate of glafs; and for this purpofe I perforated a glafs falver, and thereby gave a coating of tinfoil to both fides of a quantity of oil poured into it; but the brim of the falver would not contain enough to give it a fufficient thicknefs; otherwife, I make no doubt, but that a fhock might be given by it better than by air.

I shall juft mention upon this fubject, what I lately obferved, and do not know whether it has been noticed by any writer, that *ice of oil*, contrary to ice of water, is fpecifically heavier than the fluid fubftance, and finks in it.

Finding fo great an agreement, with refpect to electric properties, in this whole chymical clafs of bodies, I began a kind of courfe of chymical electricity; but had not leifure, or opportunity to purfue it as it deferved.   The few hints that I collected may poffibly be of fervice to future inquirers; and

for

for this reason I shall note them just as they occurred, though they contain little that is remarkable.

ALL *saline substances* that I examined proved, in general, pretty good conductors. I tried most of them by making the discharge of the battery through them when insulated; which appears to me to be a very good method, indeed the only one that can well be depended upon. In discharging the battery with a piece of *alum*, the explosion was attended with a peculiar hissing noise, like that of a squib. *Rock salt* conducted pretty well, but not quite so well as the alum. The electric spark upon it was peculiarly red. *Sal ammoniac* exceeded them both in its conducting powers, but it would not take the least sensible spark; so that it seemed made up of an infinite number of the finest points. *Volatile sal ammoniac* I only tried in a phial, when it gave a small shock. *Salt petre* did not conduct so well as sal ammoniac. Endeavouring to make the electric explosion pass over its surface, it was dispersed into a great number of fragments in all directions with considerable violence, some flying against my face. *Selenitic salt* conducted a shock but poorly. *Vitriolated tartar* gave a small shock. *White Sugar* seems to be an exception to this rule: for it may be fairly said to be no conductor; as the charge of the battery would hardly pass through it in the least degree.

THE *metallic salts* in general conducted better than other neutrals: *blue and green vitriol* conducted very well, though they would not transmit a shock.

THAT ores in which the metal is really in a metalline state should be very good conductors might naturally be expected. Thus a piece of gold ore from Mexico was hardly to be distinguished, in this respect, from the metal itself; and a piece

4 K   of

of silver ore from Potosi, though mixed with pyrites, conducted very well. But even ores in which the metal is
mineralized with sulphur or arsenic, as the ores of lead and
tin, and *cinnabar* the ore of quicksilver were little, if at all inferior to them. The cinnabar that I tried was factitious; but
there can be no doubt of its being the same as the native.
When I made the explosion of the battery pass through it, it
was rent into many pieces, and the fragments dispersed in all
directions. Ores, however, that contain nothing but the
earth of the metal conduct electricity but little better than o-
ther stones; though I thought that all the specimens of iron
ore that I tried conducted better than marble.

I EXAMINED some *black sand* that came from the coast of
Africa, which is a good iron ore, and part of which is affected by the magnet as much as steel filings; and found it to
conduct electricity, but not a shock. Separating with a magnet all that would be easily attracted by it (about one sixth of
the whole) it conducted a shock very well. The rest would
hardly conduct at all.

THOUGH I think I may venture to say, that the true and
proper ores of the more valuable metals might be known by
their property of conducting electricity, I cannot say that e-
lectricity will furnish any rule to ascertain the value of the
different ores of the same metal. I tried two pieces of copper ore, one the most valuable that is known, and another of
only about half the value; but they were hardly to be distinguished from one another in their conducting power.

BLACK *lead* in a pencil conducted a shock seemingly like
metal or charcoal. A small lump of it took as full and strong
a spark from the prime conductor as a brass knob.

ALL

ALL the *stony substances* that I tried conducted very well, though dry and warm. Even a piece of polished *agate*, though semi-pellucid, received the electric spark into its substance ; though it would pass over about three quarters of an inch of its surface to reach the finger that held it, and it discharged the battery but slowly. *Limestone*, and *lime* just burnt were equally imperfect conductors, hardly to be distinguished from one another, *Lapis hæmatites*, and *touchstone* both conducted pretty well ; as did a piece of *gypsum*, and *plaister of Paris*, only the latter, having a smoother surface, took a stronger spark. A piece of *slate*, such as is commonly used to write on, was a much better conductor than a piece of free stone, which conducted very poorly. Marbles also conducted considerably better than free stone. I found very little difference among any of the specimens of marble that I tried. in which was a piece of Egyptian granite. A piece of Spanish chalk, which is a *talk*, conducted pretty much like marble.

A LARGE piece of white *spar*, with a tinge of blue, and semi-transparent, would hardly conduct in the least degree. I took pretty strong sparks from the prime conductor while it was in contact with it.

A PIECE of *pyrites* of a black colour took sparks at a considerable distance from the prime conductor, like some of the inferior pieces of charcoal. Another piece of pyrites, which had been part of a regular sphere, consisting of a shining metallic matter, did not conduct near so well, though much better than any other stony substance. It was a kind of medium betwixt a stone and an ore.

A PIECE of *asbestos* from Scotland, just as it is taken from

its

its bed, would not conduct. It was in contact with the conductor, while I took sparks at the distance of half an inch with a moderate electrification.

OF *liquid substances*, *oil of vitriol* conducted pretty well, and the most highly rectified *spirit of wine* gave a shock much like water, but perhaps not quite so well.

THIS course of experiments on the conducting power of substances, according to their chymical classes, would, probably, be very useful, if pursued with care. Those mentioned above were generally single experiments, which are not so much to be depended upon.

THERE are some other *mixed substances* whose conducting power I have tried, and because I think it would not be easy to say, *a priori*, to which of the two classes they belong, I shall just mention the result of my experiments upon them, nearly in the order in which they were made.

DRY *glue*, which is an animal substance, is a conductor of electricity, but does not conduct a shock.

POUNDED *glass* mixed with the white of an egg, and which had stood till it was perfectly dry, was a conductor. I had put it upon some broken jars, thinking that the composition would be an electric substance, and that it would make the jars hold a charge again.

PAINT, made of white lead and oil, very old, and dry, proved a conductor. I tried it in a china vessel which had been firmly pieced with it. A part of the vessel, through which there was no crack, would receive a charge very well; but a piece in which there was a crack, and which had been filled with this cement, could not be charged at all.

SECTION

SECTION V.

EXPERIMENTS ON THE DIFFUSION OF ELECTRICITY OVER THE SURFACES OF GLASS TUBES, CONTAINING A NEW METHOD OF GIVING THE ELECTRIC SHOCK.

IT had been obferved by many electricians, that new globes are often difficult to excite; but I have made fome experiments, which prove this fact, and other differences between new and old glafs, in a more diftinct manner than any thing elfe I have yet met with; but they leave the caufe ftill unexplained.

THE moft remarkable property of new flint glafs is the eafy diffufion of electricity over its furface. I have feveral times got tubes made two or three yards long, terminating in folid rods. Thefe I have taken almoft warm from the furnace, in the fineft weather poffible, have immediately infulated them, and hanging pith balls at one extremity, have always found, that they would feparate the moment that the wire of a charged phial was applied to the other end. This I had reafon to think would be the cafe at almoft any diftance at which the experiment could be made. I have even charged a phial very fenfibly, when it was held clofe to the glafs, at the diftance of a yard from the wire of a charged phial, held clofe

to

to another part of it, the coatings of both phials being held in my hands. When the same tubes were a few months older, I found that the electric virtue could not be diffused along their surfaces farther than about half a yard.

SOME tubes, which I have tried the day they were made, I have found impossible to be excited in the least degree, even with the use of oiled silk and amalgam, for an hour together; when a single stroke of the same rubber has rendered other tubes highly electrical, and two or three have made them to emit spontaneous pencils. The same new tubes, upon being much rubbed, have begun to be excited, and in a few days have acted pretty well.

BUT that the first coat of new glass is, in some measure, a conductor of electricity, was most evident from some experiments which I made with long and very thin tubes, which were blown some time in the month of March. These, to amuse myself, I coated in different places, and the diffusion of electricity, from the coated part to that which was not coated, appeared to me very extraordinary. I think my reader will not be displeased if I relate a few of the particulars.

I procured a tube, open at both ends, about a yard in length, but of very unequal width. About three inches of the middle part of it I coated on both sides; and charging it, by means of a wire introduced at one of the ends; I perceived, not only that the part through which I had introduced the wire was strongly electrical on the outside, but that at the opposite end, where there was neither coating nor wire, the fire crackled under my fingers, as I drew the tube through them, and a flame seemed to issue continually out at both the ends, while it was at

rest

rest and charged.   N. B. One end of this tube was broken, and rough, the other was smooth.

I PROCURED another tube, about an inch in diameter, and very thin.   It was about three feet and a half in length, and closed at one end.   About nine inches below the mouth, I coated three inches of it, both on the inside and outside. This part I charged, and then observed the whole tube, to the very extremity of it, to be strongly electrical, crackling very loud when I drew my hand along it, and giving sparks, as from an excited tube, at about the distance of an inch, all the way.

To give the reader a better idea of these experiments, I have given a drawing [Pl. i. fig. 7.] of one of the tubes with which they were made.   It is open at one end, and the part [a] is coated.

AFTER drawing the whole tube through my hand, all the electricity on the outside was discharged; but, upon putting my finger within the mouth of the tube, an effort to discharge itself seemed to be produced, which showed itself by a light streaming visibly from the coating, both towards the finger and likewise as vigorously towards the opposite end of the tube.   After this I found all the outside of the tube loaded with electricity as before, which might be taken off, and revived again many times, with the same original charge; only it was weaker every time.

HOLDING this tube by the coated part, and presenting the uncoated outside, near the close end of the tube, to the prime conductor, the inside became charged as well as the outside; and, upon introducing a wire, a considerable explosion was made.

THE

THE difcharge made the outfide ftrongly electrical, and by taking this electricity off, the tube was changed again very fenfibly.

HOLDING it by the uncoated part, and prefenting the coated part to the conductor, the infide became charged as before.

HAVING firft perfectly difcharged this tube, I clofed the open end with cement, made of bees wax and turpentine, an inch or more in thicknefs; but ftill, by applying the outfide of the tube (either the coated or the uncoated part) to the conductor, I found it manifeftly charged, but not quite fo high as when the end was left open, though the difference was not great.

I PROVIDED myfelf with another tube, about an inch and a quarter wide, and three feet long; but it was drawn out one foot more very fmall; and another foot at the extremity was folid, fo that it was in all five feet long. I coated about four inches of this tube, two feet below the mouth of it. The balls being hung at the extremity of this tube, or rather of the folid rod in which it terminated, they feparated the moment I began to charge the coated part. The difcharge brought them together, though not immediately, but a fecond difcharge would generally do it.

THE refiduums of any of thefe tubes, of which fo fmall a part was coated, were very confiderable. I thought that all of them might be equal to the firft difcharge. In the laft mentioned tube, there was a refiduum after a great number of difcharges, I believe twenty or thirty.

IMAGINING that the diffufion on the furfaces of the tubes above-mentioned depended upon the newnefs of the glafs, I preferved them fix or feven months; having obferved, by

examining

examining them at proper intervals in the mean time, that this property, and others depending upon it gradually leſſened; and before this time it was quite gone. There was no diffusion of electricity over their ſurfaces, and they were as eaſily excited as other tubes, at the ſame time that they received a very good charge.

At length, by ſome accident or other, all the tubes on which I had made theſe experiments were broken, except one, which was cloſed at one end, and which, indeed, was the moſt remarkable of them all. Upon this tube, in the month of November, I began to renew my experiments, comparing it with others which I got made at that time, in order to aſcertain on what circumſtances this diffuſion of electricity depended. Theſe I ſhall diſtinctly relate, noting the time when each experiment was made, and every other circumſtance which I can imagine could poſſibly have any influence in the caſe.

November the 13th. I once more endeavoured to repeat the experiments above-mentioned with the old thin tube, with as much care and precaution as poſſible, but without the leaſt ſucceſs. At the ſame time I charged two other thin tubes, one cloſed, and the other open, after they had been made about ſix weeks, but without being uſed in the mean time, and they anſwered exactly as the former tube had done, when it was new. The charge from a ſmall coated part diffuſed itſelf all over the tube; ſo that at the diſtance of a yard from the coating, it gave ſparks to the finger of an inch in length, and in all reſpects exhibited the appearance of a tube freſh excited. On this occaſion I firſt obſerved, what afterwards drew my attention in a more particular manner, that

4 L   when

when my finger was brought to the tube about two inches above the coating [as at *b*. Pl. i. fig. 7.] it difcharged a great quantity of that diffufed electricity; and my whole arm was violently fhocked.

NOVEMBER the 19th. After heating the old tube, and endeavouring to repeat the former experiments, both while very warm, and after it was cold again, but to as little pur-pofe as before; I took it to the glafs houfe, and got it made red hot all over, fo that it would eafily bend any way; and as foon as it was cold, I tried the old experiments, and found that it had completely recovered its former property. Charging a fmall coated part, the electricity was diffufed to the end of the tube, over three feet of dry glafs, and it gave fparks at the diftance of an inch in any part of it, exactly as if it had been excited with the beft rubber. When it was drawn through my hand, whereby that diffufion was taken off, it prefently returned again; and the extremity of the tube would get loaded while its communication with the coating had been cut off, by my hand being conftantly held on the middle of it.

I ALSO obferved, that the middle part of this tube, which had been ofteneft heated, in melting the whole over again, one half at a time, had a much ftronger diffufion than the o-ther parts. It was no fooner taken off, than it appeared again, fo that it gave a continual ftream of fire.

THE quantity of *refiduum* after a difcharge of this tube was prodigious, fo that the outfide coating would, immediately after, give almoft a conftant ftream of fire to any conductor prefented to it, for a confiderable time.

THIS

This tube was now, as it had been at the firft, abfolutely incapable of being excited with the beft rubber.

January the 6th. 1767. Examining all the tubes with which I had made the experiments of the diffufion, I found that property either quite, or very nearly gone. One of them I reftored by heating it red hot. Another I heated only at the end moft remote from the coating; but there was no diffufion upon it, when the coated part was charged; the part which had not been made red hot intercepting it.

November the 24th. In order to determine whether this property of diffufion depended in any meafure upon the fmoothnefs of the furface, I made a circular part of one of the thin tubes, about half a yard beyond the coated part, quite rough with emery, about three inches in length; but this did not prevent the diffufion in the leaft; both that rough part, and the fmooth glafs beyond it were as much loaded with electricity as the reft.

I then took the polifh off a line the whole length of the tube, from the coating to the extremity of it; but ftill the effect was the fame: and I make no doubt would have been fo if I had made all the furface rough.

In order to afcertain whether this property depended upon the thinnefs of the tubes, I got one made of a twelfth of an inch thick, and ufed it immediately; the diffufion was very fenfible, and it was incapable of being excited. This, however, was not always the cafe with tubes of fo great a thicknefs.

November the 25th. Willing to carry this experiment a little further, I got another tube four feet long, and of the eighth of an inch thick. I coated a fmall part of it in the fame manner as I had done the others, namely about three inches, at the

distance

diſtance of nine inches below the orifice, and obſerved the diffuſion to be very remarkable in proportion to the charge it received, which was very moderate.   It could not be ſenſibly excited in the leaſt degree; except that, in the dark, an exceeding ſmall light was viſible near the finger, when it touched any part of it, immediately after excitation; but not the leaſt ſnapping could be perceived, nor any thing felt with the finger.

To find whether this property depended upon the kind, as well as the newneſs of the glaſs, I, afterwards, coated a part of a very thin glaſs of the common bottle metal, but I found no diffuſion upon it at all.   It was what is commonly called a ſinging glaſs. I would have purſued this experiment by trying the ſame glaſs in other forms, and by trying other-kinds of glaſs, but I had no opportunity.

I OBSERVED, in all the tubes which had the diffuſion, that in drawing my hand from the extremity of them towards the coating, after they were charged, ſo as to take off the diffuſion, there was a conſiderable noiſe at the orifice, as if the tube had been gradually diſcharging itſelf, and this operation did apparently leſſen the charge.

IN the dark the electric fire ſeemed to pour perpetually from the open end, or both the ends, if they were both open; and whenever I drew my hand over it, the fire ſtreamed from the coating towards my hand in a very beautiful manner.

IT was very remarkable, that, the firſt time I charged any of theſe tubes after they had ſtood a while, the diffuſion was the moſt conſiderable, and that it leſſened every ſucceſſive charge; till, at laſt, it was exceeding ſmall; but after the

tube

tube had ftood a few hours uncharged, it was as vigorous as ever.

December the 1ft. I, for the firft time, took particular notice, that, in charging a thin tube, and afterwards holding the coating in one hand, and drawing my other hand, fo as to grafp the tube; beginning at that end which was moft remote from the coating; that fometimes, when my hand came near the coating [as at *b*. Plate i. fig. 7.] I received a very confiderable fhock through both my arms and in my breaft, exactly like what is felt from the Leyden phial.

The fame day, I felt a fimilar fhock from another thin tube; and what was more remarkable, I did not receive it till the third time of drawing my hand over the tube, having miffed the ftroke the two firft times; though I moved my hand, as near as I could judge, in the fame manner. This fhock was not very great, but fenfible in both arms.

December the 3d. I received another fhock, the third time of drawing my hand over the tube, and much more violent than the laft; it affecting both my arms and breaft. At this time I obferved very exactly, that my hand was near two inches and an half from the coating, and that a ftrong light was vifible under my hand, and extended to the coating. The diffufion at this time had not been very great, and the tube feemed to be about half difcharged after the fhock.

At that time I could not think of any plaufible theory to account for this fhock; but prefently after I accidentally received another fhock, in fome refpects fimilar to this, the theory of which I have been fo happy as to inveftigate, and which may throw fome light upon this.

December

DECEMBER the 21ft. I made a torricellian vacuum in a tube about a yard in length [Plate i. fig. 8.] and holding one end of it in my hand, I prefented a part near the other end to the prime conductor; and obferved, that, while the electric fire was pouring along the whole length of it, I felt fome peculiarly fmart twitches every now and then in my hand, juft fuch as are felt when a thin uncoated phial is held in the hand, while it is charged at the prime conductor, but more pungent. On removing the tube from the prime conductor, it threw out fpontaneous fparks from the place where it had touched the conductor, exactly like thofe which iffue from the wire of an over charged phial; but they were longer, and much more beautiful. Then, bringing my other hand near the place where the tube had touched the conductor, I received a very confiderable fhock in both my arms and breaft, exactly like that which I had received before from the thin tubes; and, as with them, the fhock was rather ftronger in the hand which was brought to the tube than in that which held it. If, without bringing my other hand to the tube, I only prefented it to the table, or any other conductor, it would throw out from the fame place feveral ftrong fparks, attended with a flafh of light, which filled the whole length of the tube. Thefe fparks refembled thofe which iffue from the wire of a charged phial, when it is prefented to the like imperfect conductors, and at the fame time held in the hand.

I AFTERWARDS obferved, that the ftrongeft fhock which this tube could give was felt when one hand continued in the place where it held the tube, in order to charge it, and the other was made to touch the tube, an inch or two above it [as at c. Pl. i. fig. 8.] and at the inftant of the ftroke, a very

denfe

denfe fpark of electric fire was feen darting the whole length of the tube. When three perfons befides myfelf joined hands, it fhook all our arms greatly.

The tube could not be difcharged by putting one hand fo near the other, unlefs that part of the tube had been brought to the conductor in charging it; and if any particular part of the tube only, had been brought to the conductor, the difcharge could not be made without touching that part.

When the tube had given a fhock from any one place, it would give one or two more fmaller fhocks, from other places.

The experiments I made with this tube being certain and invariable, and the fhock I received from the other tubes precarious, I gave more particular attention to this, in order to afcertain the nature of this fhock; thinking that, if I could accomplifh this, it would affift me in the inveftigation of the other. Accordingly, I coated about fix inches, near each end of the tube, [a and b Pl. i. fig. 8.] leaving the fpace of about half a yard of uncoated glafs between the coatings; and obferved, that when I held one of thefe coatings in my hand, and prefented the other to the prime conductor, it always received a confiderable charge, and was difcharged in one bright fpark at the diftance of above an inch, and fometimes two inches, if, befides the coated part, I had likewife prefented the uncoated part to the prime conductor; and fometimes the uncoated part would difcharge itfelf by a bright flafh to the lower coating, leaving the coated part charged as before. If I held the tube by the middle, where there was no coating, and prefented one of the coatings to the conductor, it received a pretty good charge.

I THEN

I THEN ſtood upon an inſulated ſtool, and preſenting one of the coatings to be charged, while I held the other; I obſerved, that it received not more than one fourth part of the charge it had before; upon which I immediately concluded, that the lower coating muſt have been charged negatively, whilſt the upper was charged poſitively. This was quite confirmed by obſerving, that ſparks could be drawn from my body, while I ſtood upon the ſtool preſenting the tube to be charged, but no longer than till the tube had received its full charge; and that then the exploſion was as great as it had been when I ſtood upon the floor.

WHEN I inſulated the tube, by placing it in a glaſs veſſel, it was ſtill leſs capable of taking a charge than when I ſtood upon a ſtool and held it, this method making a more perfect inſulation. If any conductor was preſented to the lower coating while the other was held to the prime conductor, ſparks iſſued from it very plentifully, till it had got a conſiderable charge; when thoſe ſparks entirely ceaſed, and the tube, upon trial, gave a very great exploſion.

THESE experiments make the theory of this new method of giving the electric ſhock pretty obvious. The electric matter thrown upon the upper coating repels an equal quantity from the inſide of the tube oppoſite to it; which, paſſing freely through the vacuum, (as is viſible in the dark) is accumulated on the inſide of the other extremity of the tube, and thereby repels a quantity from the lower coating: ſo that the two coatings being in oppoſite ſtates, though on the ſame ſide of the tube, the ſame kind of ſhock is given by them, as if they had been on oppoſite ſides.

BEING

BEING fully satisfied with the experiments made by these two coatings, and the theory of them, I amused myself with coating the middle part of the tube in various ways.

WHEN the three coatings were about the same size, and placed at equal intervals; which ever of them was held in the hand, the other two were charged and discharged separately. If the coating of one of the ends was held in the hand, and the other two were charged, the greatest explosion was from that which was discharged first. If those two coatings were placed near one another, they were both discharged by the attempt to discharge either of them, and a flash of light was seen betwixt them both. In this case the explosion was sometimes made at the distance of two inches and a half.

WHEN the middle coating was made very large, and placed contiguous to the upper, the explosion was less; a spontaneous discharge being soon made to the lower coating.

WHEN the middle coating was taken away, it often happened that, in drawing the whole tube over the prime conductor, beginning at the upper coating; when it came to the lower, by which I held it, a spark would dart to it from all the uncoated part of the tube, which discharged the electricity of that part, while the upper coating still retained its proper charge.

WHEN this spontaneous discharge was not made, the explosion might be made at twice, once at the naked glass, near the lower coating, and again at the upper coating. If the discharge was first made at the upper coating, there remained very little for the lower part of the tube. And if the explosion was made about the middle of the tube, the whole was discharged at once, and in a very beautiful manner.

4 M

I MUST

I MUST leave my reader to compare the theory of this shock with that given by the long and open tubes, as I am not able to do it to my entire satisfaction without more experiments; which, as I observed, are precarious, and which I had not leisure to attend to.

As this course of experiments was begun by an accidental observation of the different electric properties of new and old glass, I shall (after this long excursion, which I little foresaw) conclude with an experiment or two relating more immediately to the original subject of them.

IMAGINING that the difference between new and old glass might be owing to the larger superficial pores of the former, which made it approach to the nature of a conductor, and which contracted with time; I thought it might possibly be determined by the experiment of the metallic tinge, the wider pores receiving it better than the smaller; and I was not disappointed in my expectations. November the 19th. I several times laid two glass tubes, one a very old one, and the other quite new, close together, with a piece of leaf gold or copper between them; and though I varied the disposition of them in every way that I could think of, and changed the tubes for others; I always found the new glass to receive a much fairer, more beautiful, and indelible impression than the old glass. Twice the quantity of the metal was in all the cases struck into it.

SECTION

SECTION VI.

EXPERIMENTS TO VERIFY SEVERAL PARTICULARS OF SIGNIOR BECCARIA's THEORY OF ELECTRICITY; PARTICULARLY CONCERNING THE ELECTRIC MATTER CARRYING INTO ITS PATH LIGHT SUBSTANCES TO ASSIST ITS PASSAGE.

BEING greatly ftruck with Signior Beccaria's theory concerning the paffage of the electric matter from the earth to the clouds, previous to a thunder ftorm, and thinking his experiments to prove the power of electricity to conduct into its path light fubftances that could affift its paffage not quite fatisfactory, I endeavoured to afcertain the fact in a better manner, and fhall lay before my readers the refult of my experiments.

NOVEMBER the 9th. I difcharged frequent fhocks, both of a common jar, and another of three fquare feet, through trains of brafs duft, laid on a ftool of baked wood, making interruptions in various parts of the train; and always found the brafs duft fcattered in the intervals, fo as to connect

the

the two disjoined ends of the train: but then it was likewise scattered nearly as much from almost all other parts of the train, and in all directions. The scattering from the train itself was, probably, occasioned by small electric sparks between the particles of the dust; which, causing a vacuum in the air, drove all that light matter to a considerable distance. But the particles of the dust which were strowed in the intervals of the train, some of which were, at least, three inches, could hardly be conveyed in that manner.

WHEN small trains were laid, the dispersion was the most considerable, and a light was very visible in the dark, illuminating the whole circuit. It made no difference, in any of these experiments, which way the shock was discharged.

WHEN I laid a considerable quantity of the dust at the ends of two pieces of chain, through which the shock passed, at the distance of about three inches from one another, the dust was always dispersed over the whole interval, but chiefly laterally; so that the greatest quantity of it lay in arches, extending both ways, and leaving very little of it in the middle of the path. It is probable, that the electric power would have spread it equably, but that the vacuum made in the air, by the passage of the fluid from one heap of dust to the other, dispersed it from the middle part.

I THEN insulated a jar of three square feet, and upon an adjoining glass stand laid a heap of brass dust; and at the distance of seven or eight inches a brass rod communicating with the outside of the jar. Upon bringing another rod, communicating with the inside, upon the heap of dust, it was dispersed in a beautiful manner, but not one way more than another.

However,

However, it presently reached the rod communicating with the outside.

MAKING two heaps, about eight inches asunder, I brought one rod communicating with the inside upon one of them, and another rod communicating with the outside upon the other. Both the heaps were dispersed in all directions, and soon met ; presently after which the jar was discharged, by means of this dispersed dust, in one full explosion. When the two heaps were too far asunder to promote a full discharge at once, a gradual discharge was made through the scattered particles of the dust.

WHEN one heap of dust was laid in the center of the stand, and the two rods were made to approach on each side of it, they each attracted the dust from the side of the heap next to them, and repelled it again in all directions. When they came very near the heap, the discharge was made through it, without giving it any particular motion.

ALL these experiments show that light bodies, possessed of a considerable share of electricity, disperse in all directions, carrying the electric matter to places not abounding with it ; and that they sometimes promote a sudden discharge of great quantities of that matter from places where it was lodged, to places where there was a defect of it. But an accident led me to a much more beautiful, and perhaps a more satisfactory manner of demonstrating the last part of this proposition, than any that I hit upon while I was pursuing my experiments with that design.

DECEMBER the 11th. Hanging a drop of water upon the knob of a brass rod communicating with the inside of my battery, in order to observe what variety it might occasion in

the

the circular spots, which will be mentioned hereafter; I was greatly surprised to find the explosion made all at once, at the distance of two inches.

I, AFTERWARDS, put some brass dust upon a plate of metal communicating with the inside of the battery, and making the discharge through the dust, it exploded at the distance of an inch and a half. The dust rose towards the discharging rod, and from thence was dispersed in all directions.

THESE experiments are the more remarkable, as they demonstrate so great a difference between the distance at which the battery may be made to discharge at once, by the help of these light bodies, and without them. The discharge of the battery by the knobs of brass rods, in the open air, is at the distance of about half an inch; but, by this means, we see it made at three or four times that distance.

UNLESS a person try the following experiment, he will hardly conceive the extreme probability of the clouds and the rain being possessed of an electric virtue, in order to their uniform dispersion, according to Signior Beccaria's theory. Put a quantity of brass dust into a coated jar, and when it is charged, invert it, and throw some of the dust out. It is very pleasing to see with what exact uniformity it will be spread over any flat surface, and fall just like rain or snow. In no other method can it be spread so equably.

IT is taken for granted by Signior Beccaria and others, that persons are sometimes killed by lightning without being really touched by it, a vacuum of air only being suddenly made near them, and the air rushing out of their lungs to supply it; and with so much violence, that they could never recover their breath. As a proof of this, he says, that the

lungs

lungs of such persons are found flaccid; whereas, when they are properly killed by the electric shock, the lungs are found inflated. This account always appeared to me highly improbable. It determined me, however, to make a few experiments, in order, if possible, to ascertain the fact with some degree of exactness. The result was as follows.

DECEMBER the 18th. I placed that part of an egg shell in which is a bladder of air within an inch of the place where I made the explosion of the battery, on the surface of some quicksilver; when the bladder was instantly burst, and the greatest part of it torn quite away. The shell was quite dry; so that the bladder could not stretch in the least.

IT is evident from this experiment, that there is a sensible expansion of the neighbouring air, to fill a vacuum made by the electric explosion; but that this is so considerable as to occasion the suffocation and death of any animal, is, I think, very improbable from the following facts.

I PUT a cork as slightly as possible in the mouth of a small phial; but, though I held it exceeding near the place of the explosion, it was not drawn out.

I MADE the explosion pass over the surface of a moist bladder, stretched on the mouth of a galley pot; but it produced no sensible effect upon it. I also held at one time the bill of a robin red breast, and at another time the nose of a mouse near the electric explosion, but they did not seem to be at all affected by it. In order to examine the state of the lungs, I killed small animals by shocks discharged both through the brain, and through the lungs; but when they were dissected there appeared no difference. The lungs were in the very same state as when they were killed in another manner.

To

To thefe mifcellaneous experiments, intended to verify feveral particulars of Signior Beccaria's theory of electricity, I fhall add a fmall fet, which, though they were begun before I had feen that author, are in fome refpects fimilar to his curious experiment of difcharging a plate of glafs hanging by a filken ftring without giving motion to it; his being defigned to afcertain the effect of the difcharge upon the glafs, and mine refpecting the conducting fubftances that formed the circuit.

OCTOBER the 7th. To determine, if poffible, the direction of the electric fluid in an electric explofion, I hung feveral brafs balls by filken ftrings, and difcharged fhocks through them, when they were as much at reft as I could make them; but I could not perceive that any motion was given to them by the ftroke. Afterwards, I difcharged a jar a great number of times through fmall globules of quickfilver, laid on a fmooth piece of glafs; but could not perceive that they were driven one way more than another, though they were often thrown into diforder; probably by the repulfion of the air, occafioned by the vacuum of the explofion.

I THEN placed four cork or pith balls, at equal diftances, upon a ftool of baked wood, with a piece of chain at the fame diftance from the outermoft balls; and obferved, that, upon every attempt to make a difcharge, the two middle balls were driven clofe together, while the two outermoft were each of them attracted to the piece of chain that was next to it. Then, laying a great number of bits of threads in the fame manner upon the ftool, feveral of the pieces that lay near the chain ftuck to them, and a great number of thofe that lay in the middle were driven together in a heap.

THE

THE attraction to the chains I attribute to the electricity given to them by their connection with the jar, which would be greatly encreafed by the attempt to make an explofion ; and the crowding together of the pieces in the middle of the circuit, I attribute to the current of air blowing them together from both the extremities of the chain. Thus part of the flame of a candle next to an electrified point will be attracted by the power of electricity, while the reft of the flame will be repelled from it by the current of air.

THESE experiments led me to make a difcharge through an infulated bell, in order to obferve in what manner it would be affected by the electric fhock only, when it was not touched by any thing elfe. Accordingly, I made the difcharge of the battery through it feveral times ; and by each explofion it was made to ring, as loud as it could be made to do with a pretty fmart ftroke of ones finger nail.

I ALSO made a difcharge of the battery through the external coating of a glafs jar, but without touching it with the difcharging rods ; and it plainly produced the fame tone, as when it was rung by percuffion.

SECTION VII.

VARIOUS EXPERIMENTS RELATING TO CHARGING AND
DISCHARGING GLASS JARS AND BATTERIES.

AS several things have occurred, in the course of my experiments, relating to charging and discharging both
common jars and large electrical batteries, which I have not
feen in the writings of any electricians; and as fome of the
facts are not eafily accounted for, I fhall mention a few of the
more remarkable of them, juft as they happened.

APRIL the 28th. As I was amufing myfelf with charging
three jars of the ordinary fize, while they ftood upon a metal
plate on the table, with their wires at different diftances from
the fame prime conductor, which was fixed on pillars of baked wood; I obferved, that whenever one of the jars, which
ftood next to the conductor, difcharged itfelf, the others
would difcharge themfelves too; though they were far from
having received their full charge, being placed at a greater diftance from the common conductor, and confequently having
taken

taken but few sparks, in comparison of that which stood the nearest.

A variety of experiments seem to show, that, while a jar continues charged, the electric matter is continually insinuating itself farther and farther into the substance of the glass, so that the hazard of its bursting is the greatest some time after the charging is over.

May the 26th. After having charged forty one jars together, each containing about a square foot of coated glass, I let them stand about a minute and a half, while I was adjusting some part of the apparatus, in order to make the discharge; when they exploded by the bursting of one of the number. I observed also, that the jar which was burst was at a considerable distance from the place where I saw the flash at the wires.   It was also broke through in two different places.

For the same reason, there is no being sure that jars, which have stood one discharge, will bear another equally high.   I am confident that several of mine have burst with a much less charge than they had actually held before.

June the 29th.   A jar of an ordinary size, which had been in constant use for several months, and which had discharged itself more than a hundred times without any injury, at length burst, as I was discharging it at the prime conductor.   The hole was at a different place from that at which the discharge was made, but this does not always happen.   The tip of my little finger happened to lie very lightly on the place, and I felt it was burst by a small pricking, as of a pin, though the explosion at the conductor was nearly equal to that of any other discharge.   The coating of a jar contiguous to one that is burst is always melted by the explosion.

June the 25th. A small thin phial, which had been charged singly as high as it could bear, so as to have discharged itself, and also in conjunction with four others of its own size, burst by a spontaneous explosion, when it was charged in conjunction with a battery.

I had never heard of a jar bursting in more than one place, or more than one jar in a battery bursting at the same time; but I have often found, to my cost, that this event is very possible. In this case, there must be a discharge at more places than one at the same time: and, besides, it seems to follow from it, that whenever there is a solicitation, as we may say, to discharge at one place, the effort to discharge at every other place is encreased at the same time.

It has frequently happened with me, that jars have been burst at the instant that I was making the discharge in the common way; and when I have come to charge them again, they have appeared to be burst, in some place of the battery where I never expected it. Two instances of this kind happened in the explosion mentioned above; but the most remarkable fact of this kind happened the 31st. of May, when a battery of about forty jars, each containing a square foot of coated glass discharged itself.

Upon examination, I found that six of the jars were burst, one had the tinfoil coating on the outside quite melted, in a circular spot about half an inch in diameter; and in the inside it was burned quite black, near an inch and an half. A second was melted, on the outside, about three quarters of an inch in diameter, and the black spot in the inside was two inches. A third had one hole made in the form of a star, more small cracks like *radii* proceeding from a center than

could

could be counted.   And there was hardly one of the jars that was burſt with a ſingle hole.   Some were burſt in ſeven or eight different places, of which ſome were very remote from others; but generally there was one principal hole, and ſeveral ſmaller, but independent ones, in the neighbourhood of it, as within half an inch, an inch, or two inches from it.

JUNE the 14th.   The above-mentioned battery diſcharged itſelf once more; when three jars were burſt, and one of them, beſides its principal hole, had a circular row of fractures, quite round the hole, at the diſtance of about half an inch.   This appearance ſtruck me as ſomething very remarkable, but ſome light may perhaps be thrown upon it by a ſubſequent courſe of experiments.   Each of the ſmaller fractures was about a tenth of an inch in length.

NOVEMBER the 17th. Having charged both my batteries, one of them, at that time, of thirty one ſquare feet, and the other of thirty two, I made the exploſion; when one jar in each battery was afterwards found to be broken.   They broke at the inſtant of the diſcharge, ſo that I did not ſuſpect what had happened.   Both the batteries had frequently borne a much higher charge.   In one of the ſmaller jars, the coating, beſides being burſt oppoſite to the hole, was rent about an inch and an half along a crack that was made from it.

WHEN jars diſpoſed in batteries have been burſt in this manner, I have never failed to obſerve one circumſtance which appears to me truly remarkable.   It is, that though, in this caſe, ſeveral paſſages be opened for reſtoring the equilibrium of the electric fluid, yet the whole ſeems to paſs in the circuit that is formed for it externally.   At leaſt, the effect of the exploſion is not ſenſibly diminiſhed, upon any

ſubſtances

substances that are exposed to it. This I had a fair opportunity of observing when I was transmitting the explosion of the battery through wires of different metals. I found that the utmost force of the battery would do little more than melt a piece of silver wire on which I was trying it, and yet it was, at one time, totally dispersed by an explosion, in making which three jars were broke in different parts of the battery.

The most remarkable fracture I remember, was of a jar an eighth of an inch thick, and which was therefore, for a long time, thought to be too thick for use. This jar, however, which had never held but a moderate charge, burst spontaneously; when there was found in it one hole like what is commonly observed, from which extended two cracks that met on the opposite side of the jar, so that it came in two parts: but, besides this, there were two other holes, barely visible to the naked eye, at some inches distance from the principal hole, and considerably distant from one another. Yet these holes, when examined with a microscope, were plainly fractures, like others made in the same manner, having a white speck in the middle. One of them was above the external coating, but not above the internal.

I had frequently been much surprised at the great distance at which several of my jars would discharge themselves, one of five inches being very common. This induced me to try at what distance I could make that spontaneous discharge.

February 21st. I got a jar, eight inches and a half in depth, and three in diameter. Finding it discharged itself very easily, when coated in the usual manner, that is, about four inches from the top; I cut the coating away, till I had brought it within two inches and a quarter from the bottom;

when

when it still retained the same property; and, at length, it burst by a discharge through a white speck of unvitrified matter, an inch and three quarters above the top of the coating.

I then procured a jar made blue with zaffre, seven inches and a half high, and two inches and an half in diameter. I coated of it only one inch and a quarter from the bottom, and yet it discharged itself very readily. I afterwards, by degrees, cut the coating down to little more than half an inch from the bottom, and still the discharge continued to be made as before. This property it retained till the month of October following, when it was broke by an accident.

I have another blue jar, of nearly the same size with the former, wich is almost full of brass dust, but has no coating at all on the outside. Yet, if I set this jar upon the table, in contact with a single piece of brass chain, going quite round it, and lying upon the table, it will discharge itself the whole length of the glass. N. B. The manner in which the uncoated part of these jars becomes charged exhibits an exceeding beautiful appearance, especially in the dark; the fire flashing from the top of the coating, in the form of branches of trees, first on one side of the glass, and then on the other, and growing larger and larger till they go over the top.

I have made some experiments, to try how thick a plate of glass may be charged, but I have not been able to ascertain this circumstance with any degree of exactness: I only found, that I was not able to give the least charge to a plate of glass half an inch thick, when it was not warmed. It was the bottom of a large glass tumbler; but meeting with it only upon a journey at the house of an ingenious electrician, I had no opportunity of making many experiments upon it. I imagine

gine that warming it would have made it capable of being charged. Glafs of a quarter of an inch thick will hold a pretty good charge.

Mr. Kinnersley's experiments, p. 426. leaving me no reafon to doubt, but that Florence flafks were capable of receiving a charge like any other thin glafs, which might be made a conductor by heat, I imagined I could foon construct a very strong and very cheap electrical battery out of them. Accordingly I procured a few, for a fpecimen, but was greatly furprized to find that the electricity went through them, when quite cold, like water through a fieve, without making any fracture in them: for they continued to hold the fame fmall charge, which was different in different flafks. Mentioning this difappointment to Mr. Canton, he informed me, that he had met with the fame, and that the permeability of this kind of glafs to the electric fluid was owing to fmall unvitrified parts which may be feen in them. I thought it might be of ufe to publifh this fact, as it may prevent other perfons from being difappointed in the fame expectations.

As glafs had generally been charged when it was fmooth, and electrics which had the property of rough glafs, when excited, were exceeding difficult to charge; I had the curiofity to try what might be done with *rough glafs* itfelf. Accordingly, I firft made part of a jar rough, connecting the infide with the outfide coating; thinking that the roughnefs might poffibly promote a fpontaneous difcharge; but I found it was not made in that place preferably to any other. I afterwards took the polifh from all that part of the outfide of a jar that was above the coating, but it was charged and difcharged exactly as before. Laftly I made a plate of glafs

rough

rough on both sides, taking off all its polish, and found that it received a charge as well as a smooth plate.

The manner in which tubes and plates of glass have broken, when I have failed to strike a metallic tinge into them, by the discharge of an electrical battery, have sometimes been attended with circumstances which I cannot easily account for.   The following are the facts.

December the 3d. Endeavouring to fix a metallic tinge upon a flat piece of glass, it was broken by the explosion, parallel to the line along which the metal lay, at about an inch distance, but not where the tinge itself was.

In attempting to give a metallic tinge to a part of a long glass tube, it broke, though not in the place where the tinge was made, but on the opposite side, which was shattered all to pieces.   The leaf gold had been bound tight to the glass, under a piece of pasteboard, which covered the gold, but not all the tube.   Another tube also was broken in large fragments where the metal had been put on, but into small splinters on the opposite side; and for the space of six or seven inches further, it was not broken at all on the side of the metal, but very much on the other side.

At another time, in attempting the same thing with another glass tube; the end of it, which was near a foot distant from the place where the metal was laid, and which was a little cracked in an oblique direction, broke off in a round piece.

As few experiments have been published about *melting wires*, and procuring globules of metal by electrical discharges, and as several things have occurred in my attempts that way, which

4 O   perhaps

perhaps have not occurred to other perfons, I fhall mention a few of the moft material circumftances. They will, at leaft, ferve as a direction to thofe who may be difpofed to attempt the fame thing.

I HAD frequently attempted to procure thofe beautiful *globules of metal*, mentioned, p. 293, fome of which I had feen with Mr. Canton; and for that purpofe had made the difcharge through fmall wires laid in the bottom of china bowls, &c. but always without fuccefs. At length, I thought of inclofing the wires in fmall tubes; and this expedient fully anfwered my purpofe: for, November the 12th. difcharging a battery of thirty two fquare feet through an iron wire inclofed in a fmall glafs tube, I found innumerable globules of the metal, of very different fizes. The whole piece melted was about two inches. Breaking the glafs tube, I found the infide furface uniformly covered with thofe globules, and a black duft, both fixed into the glafs; fo that they could not be feparated, without tearing away part of the glafs.

THINKING to avoid this inconvenience, I fixed the fmall wire in the center of a glafs tube, of, at leaft, a quarter of an inch in diameter; but, upon the difcharge, this tube, though much wider than the former, was uniformly covered with the globules, and the black duft, which ftuck very faft to it, though the metal did not feem to have penetrated into the fubftance of the glafs. When the tube was broken, I fcraped off part of the black lining, and the part next the glafs looked like a thin plate of the metal.

IMAGINING that the melted metal would not adhere fo clofely to a conducting fubftance, as it had to the glafs, I

inclofed

inclofed the wire in a paper tube a quarter of an inch wide. Upon opening it after the difcharge, it was found uniformly covered with that black duft, and the ftain was every where indelible. Sparks of fire had been feen three feet from the place of difcharge, but no part of the metal could be found.

I THEN confined the wire clofer, wrapping it tight in paper. Upon the difcharge, a great number of fparks were feen, for about a fecond of time, a quarter of a yard from the paper, which was burned through in feveral places. Very few pieces could be found; but thofe were pretty large and irregular. I now found, that, in order to produce thefe globules, the charge muft be moderate, that when the charge was very high, the whole fubftance of the wire was difperfed in particles too fmall to be found; and, on the other hand, when the charge was not fufficient, the metal was melted into fragments too large to form themfelves into regular globules.

WITH the fame battery I once melted a piece of iron wire one feventieth of an inch in diameter, when a piece of it was thrown quite acrofs the table, to the diftance of about fix feet; where it fell upon a bureau, then tumbled down to the ground, and continued glowing hot all the time. At other times, fpa ks from melted iron have been thrown three yards, in oppofite directions, from the place of the fufion, and continued a fenfible fpace of time red hot upon the floor.

AT another time I had a very fine opportunity of obferving what part of the conductors which form an electric circuit are moft affected by the explofion: for, upon difcharging a battery of fifty one fquare feet through an iron wire nine inches long, the whole of it was glowing hot and continued

4 O 2

fo

ſo for ſome ſeconds; the middle part growing cool firſt, while both the extremities were ſenſibly red. Upon examining it afterwards, both the extremities were found quite melted; an inch or two of the part next to them were exceeding brittle, and crumbled into ſmall pieces upon being handled; while the middle part remained pretty firm, but had quite loſt its poliſh, ſo that it looked darker than before.

SECTION

SECTION VIII.

Experiments on ANIMALS.

A S I have conftructed an electrical battery of confiderably greater force than any other that I have yet heard
of, and as I have fometimes expofed animals to the fhock of
it, and have particularly attended to feveral circumftances,
which have been overlooked, or mifapprehended by others;
it may not be improper to relate a few of the cafes, in which
the facts were, in any refpect, new, or worth notice.

June the 4th. I killed a rat with the difcharge of two jars,
each containing three fquare feet of coated glafs. The animal died immediately, after being univerfally convulfed, at
the inftant of the ftroke. After fome time, it was carefully
diffected; but there was no internal injury perceived, particularly no extravafation, either in the abdomen, thorax, or
brain.

June the 19th. I killed a pretty large kitten with the difcharge of a battery of thirty three fquare feet; but no other
effect was obferved, except that a red fpot was found on the
pericranium, where the fire entered. I endeavoured to bring
it

it to life, by diftending the lungs, blowing with a quill into the trachea, but to no purpofe. The heart beat a fhort time after the ftroke, but refpiration ceafed immediately.

June the 21ft. I killed a fmall fhrew with the difcharge of a battery of thirty fix fquare feet, but no other effect was perceived, except that the hair of the forehead was finged, and in part torn off. There was no extravafation any where, though the animal was fo fmall, and the force with which it was killed fo great. This fact, and many others of a fimiliar nature, make me fufpect fome miftake, in cafes where larger animals are faid to have had all their blood veffels burft by a much inferior force.

In all the accounts that I have met with of animals killed by the electric fhock, the victims were either fmall quadrupeds, or fowls; and they are all reprefented as killed fo fuddenly, that it could not be feen how they were affected previous to their expiration. In fome of my experiments, the great force of my battery has afforded me a pretty fair opportunity of obferving in what manner the animal fyftem is affected by the electric fhock, the animals which I have expofed to it being pretty large; fo that a better judgement may be formed of their fenfations, and confequently of the immediate caufe of their death, by external figns. I do not pretend to draw any conclufion myfelf from the following facts. I have only noted them as carefully as I could, for the ufe of phyficians and anatomifts.

June 26th. I difcharged a battery of thirty eight fquare feet of coated glafs, through the head, and out at the tail of *a full grown cat*, three or four years old. At that inftant, fhe was violently convulfed all over. After a fhort refpite, there

came

came on fmaller convulfions, in various mufcles, particularly on the fides; which terminated in a violent convulfive refpiration, attended with a rattling in the throat. This continued five minutes, without any motion that could be called breathing, but was fucceeded by an exceeding quick refpiration, which continued near half an hour. Towards the end of this time, fhe was able to move her head, and fore feet, fo as to pufh herfelf backwards on the floor: but fhe was not able to move her hind feet in the leaft, notwithftanding the fhock had not paffed through them. While fhe continued in this condition, I gave her a fecond ftroke, which was attended, as before, with the violent convulfion, the fhort refpite, and the convulfive refpiration; in which, after continuing about a minute, fhe died.

BEING willing to try, for once, the effect of a much greater fhock than that which killed the cat upon a larger animal, I gave an explofion of fixty two fquare feet of coated glafs to a dog of the fize of a common cur. The moment he was ftruck, which was on the head (but, not having a very good light, I could not tell exactly where) all his limbs were extended, he fell backwards, and lay without any motion, or fign of life, for about a minute. Then followed convulfions, but not very violent, in all his limbs; and after that a convulfive refpiration, attended with a fmall rattling in the throat. In about four minutes from the time that he was ftruck, he was able to move, though he did not offer to walk till about half an hour after; in all which time, he kept difcharging a great quantity of faliva; and there was alfo a great flux of rheum from his eyes, on which he kept putting his feet; though in other refpects he fay perfectly liftlefs. He never opened
his

his eyes all the evening in which he was ſtruck, and the next morning he appeared to be quite blind, though ſeemingly well in every other reſpect.

HAVING diſpatched the dog, by ſhooting him through the hinder part of his head, I examined one of his eyes (both of which had an uniform bluiſh caſt, like a film over the pupil) and found all the three humours perfectly tranſparent, and, as far as could be judged, in their right ſtate; but the *cornea* was throughout white and opaque, like a bit of griſtle, and remarkably thick.

BEFORE this experiment, I had imagined, that animals ſtruck blind by lightning had probably a *gutta ſerena*, on account of the concuſſion which is ſeemingly given to the nervous ſyſtem by the electric ſhock; whereas this caſe was evidently an inflammation, occaſioned by the exploſion being made ſo near the eyes, terminating in a ſpecies of the *albugo*; but which I ſuppoſe would have been incurable. One of the eyes of this dog was affected a little more than the other; owing, probably, to the ſtroke being made a little nearer to one eye than the other. I intended to give the ſtroke about an inch above the eyes.

IN order to aſcertain the effects of electricity on an animal body, I, after this, began a courſe of experiments on the conducting power of its conſtituent parts; and for ſome time imagined that a piece of ſpinal marrow of an ox conducted ſenſibly worſe than the muſcular fleſh; but after a great number of trials with pieces of ſpinal marrow from various animals, and pieces of muſcular fleſh, of the ſame ſize and form, and in various ſtates of moiſture and dryneſs, I gave up that opinion as fallacious; but I cannot help wiſhing the ex-

peri ments

periments were refumed with fome more accurate meafure of conducting power than hath yet been contrived.

BEING willing to obferve, if poffible, the immediate effect of the electric fhock on the heart and lungs of animals, I gave, June the 5th. a fhock from fix fquare feet to a frog, in which the thorax had been previoufly laid open, fo that the pulfation of the heart might be feen. Upon receiving the ftroke, the lungs were inftantly inflated; and, together with the other contents of the thorax, thrown quite out of the body. The heart, however, continued to beat, through very languidly, and there was no other fign of life for about ten minutes. After that, a motion was firft perceived under its jaws; which was propagated, by degrees, to the mufcles of the fides; and at laft the creature feemed as if it would have come to life, if it had not been fo much mangled. The ftroke entered the head, and went out at the hind feet.

JUNE the 6th. I difcharged a battery of thirty three fquare feet through the head and whole extended body of another frog. Immediately upon receiving the ftroke, there was, as it were, a momentary diftention of all the mufcles of the body, and it remained fhrivelled up in a moft furprifing manner. For about five minutes there appeared no fign of life, and the pulfation of the heart could not be felt with the finger. But afterwards, there firft appeared a motion under the jaws, then all along the fides, attended with convulfive motions of the other parts, and in about an hour, it became, to all appearance, as well as ever.

THE fame day, I gave the fame ftroke to two other frogs. They were affected in the fame manner, and perfectly recovered in lefs than three hours.

4 P   THESE

THESE facts furprized me very much. I attribute the reco-
very of the frogs partly to the moifture, which always feems
to cover their body, and which might tranfmit a good part
of the fhock; and partly to that provifion in their conftitution,
whereby they can fubfift a long time without breathing. To
afcertain this, I would have given the fhock to toads, ferpents,
fifhes, &c. and various other exanguious animals, but I had
not an opportunity. Befides, it is paying dear for philfophi-
cal difcoveries, to purchafe them at the expence of humanity.

SECTION

SECTION IX.

EXPERIMENTS ON THE CIRCULAR SPOTS MADE ON
PIECES OF METAL BY LARGE ELECTRICAL EXPLOSIONS.

IN the courses of experiments with which I shall present
the reader in this and the two following sections, I can
pretend to no sort of merit. I was unavoidably led to them
in the use of a very great force of electricity. The first new
appearance was, in all the cases, perfectly accidental, and engaged me to pursue the train; and the results are so far from
favouring any particular theory or hypothesis of my own, that
I cannot perfectly reconcile the various phenomena to any
hypothesis.

FROM the first of my giving any particular attention to
electrical experiments, I entertained a confused notion, that
a person would stand the best chance for hitting upon some
new discovery by a greater force than had hitherto been used.
Considering the prodigious number of electrical machines that
were in the hands of so many ingenious men, in different

4 P 2

parts

parts of the world, I imagined that all that could be done *in little* had been tried; and that the ufual experiments had been diverfified and combined in almoft every method poffible; whereas, fince electrical machines, I obferved, had, of late years, been gradually reduced into lefs and lefs compafs, a great power of electricity would be almoft a new thing, and might therefore fupply the means of new experiments. Even Dr. Franklin's force, I confidered, was fmall, in comparifon of what might eafily be raifed, and without a very great expence.

WITH thefe general and random expectations, I kept gradually increafing my quantity of coated glafs, till I got a battery of thirty, forty, fixty, and at length near eighty fquare feet; and the reader will, in fome meafure, have feen already, that I was not wholly difappointed in it. The following courfes of experiments are more remarkable inftances of the advantage I derived from this power of electricity.

THE firft remarkable fact that I was by this means led to difcover, is that of the circles with which pieces of metal that receive electrical explofions are marked. I pretend not to account for this phenomenon, by any theory of electricity. But, to enable the reader to account for it, I fhall faithfully relate all the circumftances, and varieties in which it has been exhibited, and the obfervations I have made upon it; and this I cannot do better than by writing the narrative, in the order in which the appearances occurred.

JUNE the 13th. 1766. After difcharging a battery, of about forty fquare feet, with a fmooth brafs knob, I accidentally obferved upon it *a pretty large circular fpot*, the center of which feemed to be fuperficially melted, in a great number

of

of dots, larger near the center, and smaller at a distance from it. Beyond this central spot was a circle of black dust, which was easily wiped off; but, what I was most struck with was, that, after an interruption of melted places, there was an entire and exact circle of shining dots, consisting of places superficially melted, like those at the center. The appearance of the whole, exclusive of the black dust, is represented, plate i. fig. 5. No. 1.

June the 14th. I took the spot upon smooth pieces of lead, and silver. It was, in both cases, like that on the brass knob, only the central spot on the silver consisted of dots disposed with the utmost exactness, like *radii* from the center of a circle, each of which terminated a little short of the external circle.

Examining the spots with a microscope, both the shining dots that formed the central spot, and those which formed the external circle, appeared evidently to consist of *cavities*, resembling those of the moon, as they appear through a telescope, the edges projecting shadows into them, when they were held in the sun.

The most beautiful appearance of this kind was exhibited by a spot, which I took on a gold watch case. Besides the cavities, there were, in several places of the spot, hollow *bubbles* of the metal, which must have been raised when it was in a state of fusion. These looked very beautiful when examined with a microscope in the sun, and were easily distinguished from the cavities, by having their radiant points (which were very remarkable, and dazzling to the eye) on opposite sides to those of the cavities, with respect to the sun. The whole progress seems to have been first a fusion,

then

then an attraction of the liquid metal, which helped to form the bubbles; and laſtly the burſting of the bubbles, which left the cavities. N. B. By this exploſion half an inch of a ſteel wire, one ſeventieth of an inch in diameter, was melted, and entirely diſperſed. In the diſperſion, ſparks of it were ſeen red hot, above half a yard from the place where the wire had lain. This circumſtance I have frequently obſerved ſince. I have alſo ſeen ſuch ſparks fly from large braſs knobs, at the inſtant of ſeveral exploſions.

I took the ſpot upon poliſhed pieces of ſeveral metals, with the charge of the ſame battery, and obſerved that the cavities in them were ſome of them deeper than others, as I thought, in the following order, beginning with the deepeſt, *tin, lead, braſs, gold, ſteel, iron, copper, ſilver.*

I will not be very poſitive as to the order of ſome of the metals, but ſilver was evidently not affected a fourth part ſo much as gold, and much leſs than any of the others. The circles were marked as plain, but the impreſſion was more ſuperficial. Qu. Is this owing to the heat being ſooner diffuſed equably through a piece of ſilver, than through the ſubſtance of any other metal?

I thought there might poſſibly be ſome difference in the circles on metals which had been a long time in a ſolid form, and thoſe which had been lately fluid; and to aſcertain it, I made the exploſion between a piece of lead juſt ſolid after melting, and another ſmooth piece, that I had kept a conſiderable time. The piece of freſh lead was melted more than the other, but there was no other difference between them.

THE

The semi-metals, as *bismuth* and *zink*, received the same impreffion as the proper metals; being melted about as much as iron.

I made three difcharges between a piece of highly polifhed fteel, and a piece of very fmooth iron; and in all the cafes thought the fteel was more deeply melted than the iron. I mention this experiment more particularly, on account of the fingular, and beautiful appearance of the circular fpot upon the fteel in two of the difcharges. A circular fpot, of about an eighth of an inch in diameter, was uniformly melted, and pretty well defined; and there was a fpace round this central fpot, of the fame breath, uniformly filled with fmall melted places; but in one of them twice as large as in the other. They exhibited the exact appearance of a planet furrounded by a denfe atmofphere; fuch as, I think, I remember feeing the figures of, in the plates of Burnet's theory of the earth. The other circle upon the fteel was a common one.

When the kitten above-mentioned was killed, there was no circular fpot, or any fufion on the brafs knob. I have always found it the moft perfect, when the circuit has been compofed of the beft conductors, and had the feweft intervals.

June the 19th. Hanging the cafe of one watch upon the brafs knobs communicating with the infide of the battery, and receiving the explofion from it upon the cafe of another watch, which was fometimes of the fame, and fometimes of a different metal, and meafuring the circles afterwards; I found them to be very nearly of the fame diameter. The fmall varieties feemed to be accidental; or at leaft did not depend, either upon the metal, or the direction of the electric fluid. But I thought it pretty evident, from a great number of experiments,

periments,

periments, that the metal which communicated with the outfide of the battery, and which I held in my hand to take the explofion, was marked the more diftinctly of the two.

IT feemed that, when the battery was charged very high, the central fpot was the moft irregular, many of the dots which compofed it fpreading into the outer circle, and fome dots appearing beyond the outer circle, and very much effacing it; fo that the beft way to procure a diftinct circle, is to take a moderate charge of a very large battery. This may be the reafon why the outer circle cannot be perceived, when only fmall jars are ufed; the circumference of the circle being very fmall, and the charge generally too high. In a very weak charge, it is too faint to be perceived. I have fometimes, however, feen a very diftinct circle made by only two jars, each containing half a fquare foot of coated glafs.

THE *diameter* of the fpot feems to depend upon the quantity of coated glafs; but in what proportion, I have not yet accurately afcertained.

I HAVE obferved a good deal of variety in the external circles. Sometimes they have confifted of pretty large dots, difpofed at nearly equal diftances, in an exact circle; which, in the fpaces betwixt each large dot, was completed by fmaller dots, vifible only by a microfcope. But, generally, the external circle confifts of a fpace full of dots, placed irregularly, but fo that a line drawn through the midft of them makes a pretty exact circle round the central fpot.

PRESENTLY after I had obferved the fingle circle, I imagined that, whatever was the caufe of the appearance, it was not improbable, but that two or more concentric circles might be procured, if a greater quantity of coated glafs was

ufed,

used, or perhaps if the explosion was received upon metals that were more easily fused than brass. Accordingly, June the 27th. taking the moderate charge of a battery, confisting of about thirty eight square feet, upon a piece of tin, I firft observed a second outer circle, at the same distance from the firft, as the firft was from the central spot. It confisted of very fine points hardly visible, except when held in an advantageous light; but the appearance of the whole was very beautiful, such as is represented, plate i. fig. 5. No. 2.

June the 28th. I got another double circle, on a flat pewter standish, much plainer than the former, the outer being about the same distance from the inner, as the inner was from the outside of the central spot.

Having hitherto found the circles the most distinct on metals that melt with the least degree of heat, I soon after procured a piece of that composition which melts in boiling water; and having charged sixty square feet of coated glass, I received the explosion with it, and found, what I was endeavouring to get, *three concentric circles*; the outermost of which was not quite so far from the next to it, as that was from the innermost. All the space within the firft circle was melted; but the space was very well defined, and by no means like a central spot, which in this case was quite obliterated. The appearance of these three concentric circles is represented, plate i. fig. 5. No. 3.

I have several times since found parts of three concentric circles upon brass knobs, when I have used no more than thirty square feet of coated glass. They seem to be more easily perceived, when the knobs are a little tarnished: for then the small dots, in which the the metal is melted, are

more

more easily distinguished, especially when they are held in a proper light with respect to the sun.

I MADE many attempts to make these circles larger than I had usually found them upon pieces of metal, chiefly by means of worse conductors; thinking that the electric matter not being so well conducted, and passing with less rapidity, would spread wider. This was probably the case, but then it is likewise probable, that I wanted force to make such an in pression visible. For this purpose, however, I received the explosion between two pieces of raw flesh, two potatoes, two moist bladders, and things of a similar nature, but without any effect whatever; no mark at all, or, at least, nothing regular remaining upon them. When I took the explosion upon a piece of wood charcoal, it seemed to be melted, and run in small heaps, within the space of about the usual diameter of a circular spot; and when I took it upon a piece of pit charcoal, a piece seemed to be struck out of it, and a hole was left in it; but there was no regular circle upon either of them; nor was there any sensible ignition in either case.

AT one time I laid a piece of *lead ore* scraped very smooth upon the wires of the battery, and took the explosion with a piece of tin ore scraped in the same manner; but though I examined the places with a microscope, I could not be sure that there was any part melted, much less any regular circular spot; but there lay on both of them a yellow matter, like sulphur, round the place where the explosion was taken, and a very disagreeable smell was excited. This probably arose from a mixture of the sulphur of the lead ore, and the arsenic of the tin ore.

I RECEIVED the explosion in *vacuo*, at the distance of about

three

three inches; but found no regular circular fpot, owing, probably, to the two interruptions I was obliged, in this cafe, to make in the circuit, one in the air, and the other within the receiver; by means of which the effect of both would be weakened, the whole force being, as it were, divided between them : for in all fuch cafes, though both the explofions were made in the open air, I found the circles lefs perfect.

AFTERWARDS, I contrived to make the explofion in one additional atmofphere of condenfed air, but the circles were fmaller, and lefs diftinct than the other two circles, which I was obliged, at the fame time, to make at the other interruption of the circuit, in the open air. The denfer air would probably confine the electric matter within a narrower compafs; in the fame manner as the common air prevents that diffufion of it which is remarkable in *vacuo*.

THE diftance at which the difcharge was made occafioned no difference in the diameter of thefe circular fpots. When, by putting a drop of water upon the brafs rod communicating with the infide of the battery, I made the difcharge at the diftance of two inches, the fpot was juft the fame as if it had been received, as ufual, at the diftance of half an inch, i. e. about a quarter of an inch in diameter.

I ALWAYS found that if the explofion was obliged to pafs through any bad conductor before it reached the metal, the impreffion it made upon it was contracted, and deeper than if it had been received immediately by the metal. This was evident when paper, a piece of bladder, or varnifh were put upon the brafs rods with which the difcharges were made; though a very thin coating of varnifh or moifture did not entirely prevent the appearance of the circles.

IN

IN making a courfe of experiments with bad conductors, and in ufing various methods to promote the difcharge of the battery at greater diftances than ufual, I was peculiarly ftruck with fome phenomena which occurred in the ufe of water.

I PUT a drop of water, about a quarter of an inch in diameter, upon the brafs rod communicating with the infide of the battery, and took the explofion directly over it. The difcharge was made at the diftance of about an inch, and the extremity of the drop was marked with a moft beautiful circle, exceedingly well defined on the infide, and vanifhing gradually outwards, like a fine fhade in drawing. But what ftruck me moft in the appearance was, that, in this circle, there was no central fpot.

NOT knowing what this new circumftance was owing to, I wetted a piece of fmooth copper, which lay upon the wires of the battery, and taking the explofion upon it, I only found a long ftreak at the edge of the wetted place, well defined on the fide of the water, but vanifhing gradually on the oppofite fide, as in the former cafe. In this, and other fimilar experiments, I obferved that the electric matter avoided the water, and would go a greater way in the air, in order to come at the metal.

I THEN laid more water upon the copper, but fo as only to moiften it; for the furface, being convex, would not allow it to lie in any great quantity; and upon taking the explofion, I found no circle, but feveral beautiful circular fpots melted very deep, one of which was much larger than the reft. Thefe experiments feem to fhow, that the electric matter meets with a confiderable refiftance in paffing through water, which confines its excurfion more than the air; and that, by

fuch

such a condensation, its force is greatly increased, so as to leave deeper impressions upon the metal than when it had passed only through the air, in like manner, if two pieces of metal be placed, nearly in contact, or if they be light, and one of them lie upon the other; the impression made upon both of them, by the discharge of the battery passing through them, will be considerably deeper, than it would have been if the electric matter had not been confined to so small a compass as the points in contact.

As I observed before, I do not pretend to account for the formation of these concentric circles. All that can be concluded from the appearance is, that the electric matter issues in the form of hollow cylinders, and that these cylinders are formed of other smaller and solid ones; since all the circles are made of round dots. Or these might have been hollow, but being so small, the metal could not show that circumstance; being liquified, so as to fill up that small cavity. But what it is that disposes the electric matter to issue in this manner, or on what property of the fluid this phenomenon ultimately depends, I cannot form a conjecture worth communicating to the public. I do not despair, however, but that the communication of these imperfect experiments, and the repetition of them by other persons, may be a means of investigating the cause, in complete and satisfactory manner; and I hope in a short time. A few facts of a similar nature may possibly throw some light upon this subject.

THE manner in which several of the jars, mentioned in a former section, were broken seems to be analogous to the formation of these circles. I mean those that were pierced with a number of small holes in the neighbourhood of the principal
one;

one; but more especially that which was broke with an intire circle of small and independent fractures round the principal hole.

THE remarkable story of the five peasants of whom the first, third, and fifth were killed by lightning, as they were walking in a right line; and which was mentioned before, as analogous to a fact observed by Mr. Monnier, will perhaps be thought more analogous to this. For supposing the diameter of the concentric circles formed by lightning to be sufficiently great, and the central spot to fall upon the third person, the two on each side of him would escape, by being in the first interval round the central spot; while the two who walked first and last would fall into the circumference of the first circle.

COMMUNICATING this experiment to Mr. Price, he suggested to me, that the circles called *fairy rings*, which consist of grass of deeper green in pasture fields, and which have by some been imagined to be occasioned by lightning might be analogous to the circles above-mentioned, but that they want a central spot. I have since examined one of these rings. It was about a yard in diameter, the ring itself about a quarter of a yard broad, and equally so in the whole circumference; but there was no appearance of any thing to correspond to the central spot; which, except in that with water above-mentioned, I have never failed to observe in my experiments, on whatever form of a surface I have taken the circle.

I HAVE since met with a curious article in the Philosophical Transactions, relating to those fairy circles, communicated by Mr. Jessop, which confirms the supposition of their being occasioned by lightning, and with which I shall therefore conclude this section.

" I

" I have often been puzzled to give an account of thofe
" phenomena, which are commonly called fairy-circles. I
" have feen many of them, and thofe of two forts; one fort
" bare, of feven or eight yards diameter, making a round
" path fomething more than a foot broad, with green grafs
" in the middle; the others like them, but of feveral big-
" neffes, and encompaffed with a circumference of grafs,
" about the fame breadth, much frefher and greener than
" that in the middle. But my worthy friend Mr. Walker,
" gave me full fatisfaction from his own experience; it was
" his chance one day, to walk out among fome mowing
" grafs (in which he had been but a little while before) after
" a great ftorm of thunder and lightning; which feemed by
" the noife and flafhes to have been very near him: he pre-
" fently obferved a round circle, of about four or five yards
" diameter, the rim whereof was about a foot broad, newly
" burnt bare, as the colour and brittlenefs of the grafs roots
" did plainly teftify. He knew not what to afcribe it unto
" but the lightning, which, befides the odd capricios re-
" markable in that fire in particular, might without any
" wonder, like all other fires, move round, and burn more
" in the extremities than the middle. After the grafs was
" mowed, the next year it came up more frefh and green in
" the place burnt, than in the middle, and at mowing-time
" was much taller and ranker." *

* Phil. Tranf. abridged, Vol. 2. p. 182.

SECTION

SECTION X.

EXPERIMENTS ON THE EFFECTS OF THE ELECTRICAL EXPLOSION DISCHARGED THROUGH A BRASS CHAIN, AND OTHER METALLIC SUBSTANCES.

FROM the very firſt uſe of my battery, I had obſerved a very *black ſmoke*, or *duſt* to ariſe upon every diſcharge, even when no wire was melted, and the braſs chain I made uſe of was of a conſiderable thickneſs. Of this circumſtance, however, I only made a ſlight memorandum, as what I could not then account for, and paid no particular attention to it; till on the 13th. of June 1766, I was ſtruck with another caſual appearance, as I was intent upon the experiments relating to the circles above-mentioned.

I OBSERVED, that a piece of white paper, on which lay the chain I was uſing to make the diſcharge, was marked with a *black ſtain*, as if it had been burnt, wherever the links had touched it. Yet I could not then think that it could be burnt by ſo thick a chain. I imagined the chain muſt have been dirty, and the dirt have been ſhaken off by the ſtroke. Still however I neglected the experiment till, ob-

serving

ſerving a very ſtriking appearance of the ſame kind, on the 1ſt. of September following, I was determined to attend to the circumſtances of it a little more particularly than I had done.

I MADE my chain very clean, and wrapping it in white paper, I made a diſcharge of about forty ſquare feet through it, and found the ſtain wherever it had touched the paper.

SOME time after, I wrapped the paper, in the ſame manner, round a piece of braſs wire; but, making a diſcharge through it, ſaw no ſtain. To aſcertain whether this appearance depended upon the diſcontinuity of the metallic circuit; on the 13th. of the ſame month, I ſtretched the chain with a conſiderable weight, and found the paper, on which it lay as the ſhock paſſed through it, hardly marked at all.

FINDING that it depended upon the diſcontinuity, I laid the chain upon white paper, making each extremity faſt with pins ſtuck through the links; and when I had made the diſcharge, obſerved that the black ſtains were oppoſite to the *body of the wire* that formed the chain, and not to the *intervals*, as I had ſometimes ſuſpected.

SEPTEMBER the 18th. Obſerving that a pretty conſiderable quantity of black matter was left upon the paper, on every diſcharge with the ſame chain; I imagined it muſt have loſt weight by the operation, and to aſcertain this circumſtance, I took another chain not ſo thick as that I had uſed before. It was five feet four inches long, and weighed exactly one ounce, ſeventeen penny weights, four grains. After the diſcharge, I found it had loſt exactly half a grain of its weight. The ſhock had only paſſed through a part of it, the reſt lying on a heap. I then diſcharged the ſame ſhock through its whole length, and weighing it, found it had loſt juſt another

4 R

half

half grain. By repeated experiments I found, afterwards, that the fureſt way to ſtrike off part of its weight, was to make the ſhock paſs through a ſmall part of its length, and that when a conſiderable length was uſed the event was uncertain.

N. B. THESE and all the following experiments, except where the contrary is expreſſed, were made with a battery of *thirty two ſquare feet*, that force appearing to be ſufficient, and the charging of it not taking up much time. At the time of both the above-mentioned diſcharges, an iron wire of one ſeventieth of an inch in diameter was made red hot, but, but was not melted.

OBSERVING how deep a ſtain was made by the links of a thick braſs chain, I had the curioſity to try what would be the conſequence of ſending a ſhock through a piece cf charcoal. accordingly I took a ſmall piece, about half an inch in length, and found that, in the diſcharge, it was all blown to duſt. The paſteboard on which it lay was torn, the charcoal being forced into it, ſo that the impreſſion appeared on the other ſide. The blackneſs was ſpread to a great diſtance, and the tinge every where indelible.

SEPTEMBER the 21ſt. In making the mark above-mentioned, on part of the ſheet of paper, on which I had written an account of the experiment to Dr. Franklin, I happened to lay the chain ſo as to make it return at a ſharp angle, in order to impreſs the form of a letter on the paper; and obſerved that, upon the diſcharge, the part of the chain that had been doubled was diſplaced, and pulled about two inches towards the reſt of the chain. At this I was ſurpriſed, as I thought it lay ſo, as that it could not ſlide by its own weight. Upon this I repeated the experiment with more accuracy. I ſtretched

ed the whole chain along a table, laying it double all the way, and making it return by a very sharp angle. The confequence always was, that the chain was shortened about two inches, and sometimes more; as if a sudden pull had been given to it by both the ends.

CONSIDERING that this pull must have been given to it by the several links suddenly repelling one another, at the instant of the explosion, I compared the links with the black marks that were made by them upon the table, and found that each link had been pulled from the place on which it had lain, and most of all, at the greatest distance from the place of the explosion.

CONVINCED that the chain had been shortened by the mutual repulsion of the links, I endeavoured to measure with exactness how much the shortening was, in a given length of chain. To do this, I measured two feet four inches of the chain, as it lay upon the table, in one straight line, without any return, one end being fixed and the other moveable; and found that, upon discharging sixty four square feet through it, it was shortened a quarter of an inch in its whole length. I had contrived that the suddenness of the motion should not throw one part of the chain upon the other.

SUSPECTING that the black smoke, which rose at every discharge, might come, not from the chain, but from the paper, or the table on which it lay, and which was probably burnt by the contact of it, I let the chain hang freely in the air; but, upon making the discharge, I observed the same black gross smoke that had before risen from the paper or the table. It was therefore part of the metal itself, which had been converted into that black dust.

4 R 2   To

To give my reader a better idea of the mark made upon white paper by a chain, through which the electric ſhock is tranſmitted, I laid a chain upon the original drawing of plate I. for the engraver to copy as exactly as he could; and he has ſucceeded pretty well. The breadth of the ſpots are about the mean thickneſs of the wire of the chain, and [a, b] marks the place to which that part of the chain which was returned was thrown back, by the ſudden repulſion of the links.

I HAD before obſerved the electric ſparks betwixt each link to be moſt intenſely bright, ſo as, ſometimes, to make the whole chain appear like one flame in the dark; but the appearance of the chain at the inſtant of the ſhock, as it hung freely in the air, was exceedingly beautiful; the ſparks being the largeſt and brighteſt at the bottom, and ſmaller, by degrees, towards the top, where they were ſcarcely viſible; the weight of the lower links having brought them ſo much nearer together.

SEPTEMBER the 26th. Being ſtill in ſome doubt whether the blackneſs that was left on the paper came from the burning of the paper, or ſome thing that was thrown from the chain; I once more hung the chain freely in air, and put under it, but ſo as not to touch it, a piece of white paper, on which I alſo laid a few pieces of down, to obſerve whether they would be affected by any electric attraction or repulſion. On making the diſcharge, the down was all diſperſed, and the paper was marked with a black ſtain, near the length of an inch; which was the diſtance at which the two parts of the chain hung from one another, a little above the paper. Some parts of the ſtain were deeper than others, the whole mark conſiſting of four different ſpots of a deeper black, joined by fainter

ſtreaks,

ſtreaks, anſwering to four links of the chain, which hung
nearly parallel to the paper. The ſtain could not be wiped
off with a handkerchief, though it was not ſo deep as when
the chain had touched the paper. Thus I was ſatisfied, that
a conſiderable part, at leaſt, of the blackneſs had come from
the chain.

SEPTEMBER the 27th. Willing to aſcertain more exactly
what part of the chain, the ſolid links, or the intervals, was
moſt affected by the ſhock; I dipped it in water, and laying
it quite wet upon a piece of white paper, diſcharged a ſhock
through it. Part of the water was thrown into my face,
being ſcattered in all directions, and all the chain left inſtantly and perfectly dry. The paper was very much ſtained
for the ſpace of an inch broad, wherever the chain lay; not
equally, but as if it had been handled with dirty fingers.
The ſtain was indelible, and where the chain was returned,
a hole was ſtruck quite through the paper.

To determine whether the paper, in the above-mentioned
inſtances, had really been *burnt,* as well as *ſtained.* I laid a
part of the chain, at the time of the laſt diſcharge, upon three
half crowns; and found they were all melted, in the places
where the chain had touched them. The marks made by the
fuſion were about the breadth of the chain, and ſo deep that
nothing but a tool could efface them.

To determine, if poſſible, more ſenſibly what it was that
made the black tinge, I laid the chain upon my hand, when
I had a moderate charge; and it was marked juſt like the
paper. I felt a kind of pricking or burning at the inſtant of
the exploſion, and the painful ſenſation continued a ſmall
ſpace of time.

I MADE

I MADE no doubt but that with a heat that melted metals, I could eafily contrive to fire gunpowder; but, though I laid the chain upon the grains, and rammed the powder about the chain put through a quill, I could not fucceed. In the firft cafe, the powder was difperfed; and in the fecond, the quill was burft, and there was a fmell, as after an explofion of gun powder, but no actual firing of it.

HITHERTO I had always put the chain in contact with bodies that were conductors. I was now willing to try what would be the confequence of laying it in contact with electrics. Accordingly, I dipped the chain in *melted rofin*, till it had got a coating of a confiderable thicknefs. When it was quite ftiff, I laid it carefully, without bending, upon white paper, and made the difcharge through it. The rofin was inftantly difperfed from all the outfide of the chain, it being left as clean as if none had ever been put on. That with which the holes in the chain had been filled, having been impelled in almoft all directions, was beaten to powder; which, however, hung together, but was perfectly opaque; whereas it had been quite tranfparent, before this ftroke. I felt fome of the rofin fly in my face. The ftain upon the paper was very deep, containing a good deal of rofin, and feveral holes were ftruck through the paper on which it was laid. A half crown, on which part of the chain had lain, was melted, and fo deeply ftained with the rofin, that it could not eafily be cleaned.

I NEXT laid the chain upon a *piece of glafs*; and confidering how both the half crown and the rofin had been affected, expected it would have been broken to pieces; but, inftead of that, the glafs was marked in the moft beautiful manner,

                                                    wherever

wherever the chain had touched it; every spot the width and colour of the link. The metal might be scraped off the glass at the outside of the marks; but in the middle part it was forced within the pores of the glass; at least nothing I could do would force it off. On the outside of this metallic tinge was the black dust, which was easily wiped off.

I HAVE since given the same tinge to glass with a silver chain, and small pieces of other metals; but could not do it with large pieces. They were melted where they touched one another, but the glass was not tinged.

OCTOBER the 7th. I had the curiosity to try, whether I could not give a tinge to glass with quicksilver. In order to this, I laid some globules in a right line, and laid a thin piece of glass upon them, to flatten them, and bring them nearer into contact with one another. Both the slips of glass were shattered in a thousand pieces, and dispersed all over the room, several of them flying in my face; though no part of the quicksilver could be found, except what adhered to some fragments of the glass, to which it had given a kind of uniform whiteness; but no distinct globules could be seen, and it was easily wiped off, so that no part of it was fixed in the glass. My head ached all the remainder of the day, which I attributed to the fumes of the mercury.

SEPTEMBER the 28th. Having dipped the chain in water, and found it instantly dispersed, I wished to see what would be the effect of discharging a shock through a chain quite covered with water. Accordingly, little imagining the consequence, I laid the chain upon a piece of white paper, in the bottom of a china dish, and poured in water just sufficient to cover it. Also, under one part of the chain, and in the

water,

water, I put a half crown. Upon the explosion, the water was blown about the room, to a great distance, the half crown was melted in two places, the dish broken into many pieces, and the part that lay immediately under the chain into very small fragments. The paper was a little stained, and the water, I could perceive, had been a little fouled by the black dust.

BEING certain that the dish must have been broken by the the concussion given to the water by the electric spark under it, in the manner in which Signior Beccaria's tubes were broken (though I had not seen his work at that time, but had seen the experiment at Mr. Lane's) I had the curiosity to try what would be the effect of making a discharge through the chain hanging freely in water. I therefore got a tin vessel, holding a quart, and letting the chain hang three inches and a half below the surface of the water, made the discharge. The electric sparks appeared intensely bright in the water, all along the chain; some of the water was thrown out, and the vessel appeared to have been pressed with some force upon a book, which I had put under it, a visible impression being made upon it. The vessel must have received a great concussion: for the dust had been shaken from the bottom upon the book, though I had carried the vessel up and down the room, without perceiving that any dust adhered to it.

I WAS willing to repeat this experiment with some variation of circumstances, and fastened a piece of small silver wire to two pieces of strong brass wire, and plunged the whole an inch or two under the surface of the water. Upon the discharge, the silver wire was melted, at least snapped asunder, the vessel had been pressed downwards more violently than

before,

before, a confiderable quantity of the water was thrown a-
bout the table, and fome was dafhed perpendicularly upwards,
againft the top of the room; where there were five wet places,
each about the bignefs of a half crown. I have fince frequently melted wires under water.

September the 29th. I made a difcharge through three
pieces of the fame chain, each being a different circuit. They
all left their impreffion upon the paper, and nearly equal. They
Alfo three out of four pieces made pretty equal marks, but
the fourth failed intirely.

At another time, a chain, which communicated with the
outfide of the battery, but which made no part of the circuit,
made the black ftain on a piece of white paper on which it
accidentally lay, almoft as deep as the chain that formed the
circuit. I was then melting a piece of wire, which had the
fame effect as ufing a bad conductor. The fame thing has
frequently happened fince.

November the 12th. I put a chain through a glafs tube,
fo wide as that it could only touch one fide; and upon the
difcharge, obferved four fets of marks, made by the metal
being driven into the glafs; as if four chains had been in
the tube, and all had received the fhock. Two of the rows,
on one of which I imagine the chain had lain, were better
marked than the other two, but all were very plain.

The laft thing that engaged my attention with refpect to
this courfe of experiments, was that *black duft* which I have
obferved to be difcharged from the brafs chain, and other
pieces of metal. As it was fo extremely light as to rife like
a cloud in the air, fo as fometimes to be vifible near the top
of the room; I concluded that it could not be the metal itfelf,

4 S   but

but probably the *calx*, or the calx and *phlogiston*, in another kind of union than that which conftitutes a metal; and that the electric explofion reduced metals to their conftituent principles as effectually as any operation by fire could do it, and in much lefs time. I was confirmed in this opinion by finding, in the firft place, that this black duft collected from a brafs chain would not conduct electricity, which is known to be a property of the calces of metals, and alfo by the refult of fome of the following experiments.

CONSIDERING this black duft as a proof of calcination, and obferving it to be produced when I made the explofions for the circular fpots between gold and filver watch cafes, as was related above; I began to think I had made a calcination of thofe metals, which all the chymifts fay is impoffible: but the following experiments convinced me, that it could only be the alloy that was in them which had yielded the black duft or calx.

SENSIBLE that my experiments with thefe metals would conclude nothing, unlefs I got the fpecimens quite pure, I firft procured a fmall quantity of *grain gold*, which I was informed was the pureft that the goldfmiths know, and difcharged an explofion of the the battery through a train of the pieces, an inch and a half in length, laid on a piece of white paper. Only two of the larger grains could be found after the explofion. Two leaves of paper were burnt, or torn through in feveral places, and more would probably have been torn in the fame manner, if I had ufed more. But what I principally attended to was the tinge that was given to the paper, with a view to which I had made the experiment. The paper was ftained near an inch on each fide of

the

the train, with black intermixed with red, making an odd motley appearance.

WITH the fame view, I laid a fimilar train of bits cut with a knife from a piece of as pure filver as I could procure. They were difperfed, and the paper burnt through, in the fame manner as with the gold; and the fpace of about an inch on each fide of the train was ftained with black intermixed with a deep yellow, which was confiderably different from the tinge made by the fufion of the gold.

THE blacknefs in thefe tinges convinced me, that there had been a calcination of fome part of the metal; but I was convinced it muft have been fome alloy, by an experiment I prefently after made with a piece of leaf gold; which, I believe, is generally the pureft that can be got. A fmall flip of this I put through a quill, letting a part hang out at each end; and when I had made the difcharge through it, I found the quill tinged with a beautiful *vermillion* red, without the leaft intermixture of black. When I difperfed a flip of leaf brafs in the fame manner, the greater part of the tinge was black, with a little brown mixed with it in a few places.

I MAKE no doubt but that if I could avoid the black duft, in thefe experiments I could make each of the metals give a tinge of its genuine colour, as they are excellently defcribed by Mr. Delaval, in a late paper of the Philofophical Tranfactions.

IN order to afcertain whether the black duft was a pure calx, or contained a portion of the metal, I procured a fmall quantity of it, by fending an explofion through fome pieces of iron wire, fometimes put into a quill, fometimes laid upon white paper, and fometimes upon glafs, or inclofed in glafs

tubes;

tubes; but could never be quite sure that there was any part of it that was not affected with the magnet, which the mere calx would not have been. But the black dust could not be the pure metal, for then it would either have been bright immediately after fusion, or have given a blue colour, which is the proper production of iron.

SOME of the experiments with the brass chain, related in this section, are similar to one of Mr. Wilson's, mentioned p. 94, concerning bodies placed without the electric circuit being affected with the explosion. As to the cause of this, and the other appearances above-mentioned, I have no conjecture worth communicating to the public. I have only pursued the analogy of facts, and that not very far. Others may compare them, pursue them farther, and ascertain their causes.

SECTION

SECTION XI.

EXPERIMENTS ON THE PASSAGE OF THE ELECTRICAL EXPLOSION OVER THE SURFACE OF SOME CONDUCTING SUBSTANCES, WITHOUT ENTERING THEM.

I OBSERVED, in relating the experiments on ice, that, in my attempts to afcertain its conducting power, I sometimes faw the flafh of the electrical explofion ftrike directly to the chain, along the furface of the ice. But as this paffage on the furface was produced only by a common jar, it was not much greater than the diftance at which the difcharge was ufually made, and the appearance did not ftrike me. But afterwards the fame phenomenon occurred in the ufe of my battery, where the paffage over the furface fo far exceeded the ufual diftance of a common difcharge, that it engaged my attention in a very particular manner, and produced fome pleafing experiments ; which I fhall recite in the manner, and nearly in the order in which they happened.

DECEMBER the 11th. Thinking to make a circular fpot on a piece of raw flefh, I took a leg of mutton, and laying
the

the chain that communicated with the outfide of the battery over the fhank of it, took the explofion on the outward membrane, about feven inches from the chain; but was greatly furprifed to obferve the electric fire not to enter the flefh, but to pafs, in a body, along the furface of it, to come to the chain.

THINKING that this effect might be occafioned by the fatty membrane on which the explofion was taken, I again laid the chain, in the fame manner, over the fhank, and took the explofion upon the fibres of the mufcles, where they had been cut from the reft of the body; but ftill the fire avoided entering the flefh, made a circuit of near an inch round the edge of the joint, and paffed along the furface, to come to the chain as before, though the diftance was near eleven inches.

IMAGINING this effect was promoted by the chain lying lightly on the furface of the flefh, and therefore not really in contact with it; I took another explofion, when the hook of the chain was thruft into the flefh; on which the fire entered the mutton, and, as I held it in my hands, both my arms were violently fhocked up to my fhoulders; whereas, in the cafes of the electric fire paffing over the furface of the flefh, my fingers, happening to touch the chain, were only affected with a flight pricking, or fuperficial burning, which has been explained before.

THIS phenomenon being fo remarkable, and the battery by this means difcharging at a diftance about twenty times greater than it could ufually be made to do, I thought to try other fubftances, of a conducting power fimilar to that of raw flefh; and of thefe, *water* was the moft obvious. Accordingly,

ingly, the next day, I laid a brafs rod communicating with
the outfide of the battery very near the furface of a quantity
of water (to refemble the chain lying upon the furface of the
flefh, without being in contact with it) and, by means of a-
nother rod furnifhed with knobs, made a difcharge on the
furface of the water, at the diftance of feveral inches from
any part of the rod ; when the electric fire ftruck down to the
water, and, without entering it, paffed vifibly over its furface, till it arrived at that part of the rod which was neareft
to the water, and the explofion was exceeding loud. If the
diftance at which I made the difcharge exceeded feven or
eight inches, the electric fire entered the water, making a
beautiful ftar upon its furface, and yielding a very dull found.

THE refemblance between this paffage of the electric matter over the furface of the water, and that which Dr. Stukeley
fuppofed to fweep the furface of the earth, when a confiderable quantity of it is difcharged to the clouds during an earthquake, immediately fuggefted to me, that the water over
which it paffed, and which was vifibly thrown into a tremulous motion, muft receive a concuffion, refembling that which
is given to the waters of the fea on fuch an occafion.

To try this, myfelf, and other perfons who were prefent,
put our hands into the water, at the time that the electrical flafh
above-mentioned paffed over its furface ; and we felt a fudden
concuffion given to them, exactly like that which is fuppofed
to affect fhips at fea during an earthquake. This percuffion
was felt in various parts of the water, but was ftrongeft near
the place where the explofion was made.

AFTERWARDS, I made the explofion of a jar, containing
three fquare feet of coated glafs, at fome diftance below the
furface

furface of the water, fo as to be vifible in the water, and we felt the fame concuffion that we had done before, when the fire of the battery paffed over the furface, only much weaker. The flafh of electric fire in the water does certainly difplace fome of it, and thereby give a fudden concuffion to the reft; and the fimilarity of the effect is a confiderable evidence of a fimilarity in the caufe.

I AFTERWARDS made the fire of a jar pafs through the water, making a fpace of about a foot part of the circuit; when, putting our hands in its paffage, they were affected, but in a very different manner from what they were before: for this evidently affected the nerves and mufcles of the hand internally, and occafioned a fmall degree of the fame kind of convulfion which is felt by the electric fhock itfelf; whereas the other was a mere percuffion, affecting the furface of the hand. Both fenfations were, indeed, felt moft fenfibly at the furface of the water, though our hands were, in fome meafure, affected by both as low as we could put them.

BEING willing to experience what kind of a fenfation this paffage over the furface would occafion, I laid a chain in contact with the outfide of a jar lightly on my finger, and fometimes kept it at a fmall diftance, by means of a thin piece of glafs; and, if I made the difcharge at the diftance of about three inches, the electric fire was vifible on the furface of the finger, giving it a fudden concuffion, which feemed to make it vibrate to the very bone; and when it happened to pafs on that fide of the finger which was oppofite to the eye, the whole feemed perfectly tranfparent in the dark. If I took the diftance much larger, the fire entered the finger, occafioning a very different fenfation from the former. The one

was

was like a blow, but of a very peculiar kind, whereas the o-
ther is well known to be a convulfion.

I THEN ventured to put my fingers upon a piece of the fpinal
marrow of an ox, while the explofion of the battery was paffing over it, when I felt only a flight pricking, or percuffion
on each fide of my finger; and the fenfation continued for
fome time. This fenfation did not extend at all beyond the
place of percuffion; but afterwards, putting two of my fingers on the fame piece of fpinal marrow, when the charge of
the battery was confiderably ftronger, I received a concuffion
which affected my whole hand, but it was with a kind of a
vibratory motion.

PLEASED with this refemblance of the earthquake, I endeavoured to imitate that great natural phenomenon in other
refpects; and it being frofty weather, I took a plate of ice,
and placed two fticks, about three inches high, on their ends,
fo that they would juft ftand with eafe; and upon another
part of the ice I placed a bottle, from the cork of which was
fufpended a brafs ball by a fine thread. Then, making the
electric flafh pafs over the furface of the ice, which it did
with a very loud report, the nearer pillar fell down, while
the more remote ftood; and the ball, which had hung nearly
ftill, immediately began to make vibrations about an inch in
length, and nearly in a right line from the place of the flafh.

I AFTERWARDS diverfified this apparatus, erecting more pillars, and fufpending more pendulums, &c. fometimes upon
bladders ftretched on the mouth of open veffels; and at other
times, on wet boards fwimming in a veffel of water.
This laft method feemed to anfwer the beft of any; for the
board reprefenting the earth, and the water the fea, the

4 T   phenomena

phenomena of them both during an earthquake may be imitated at the same time; pillars, &c. being erected upon the board, and the electric flash being made to pass either over the board, over the water, or over them both. This makes an exceeding fine experiment.

WHEN I first made this experiment of the electric flash passing over the surface of the water, I thought it necessary, that neither the piece of metal communicating with the out side, nor that communicating with the inside of the jars should touch the water immediately before the discharge. But I afterwards found, that the experiment would answer, though either, or even both of them were dipped in the water : for in this case the explosion would still prefer the surface to the water itself, if the distance was not very great; and would even pass at a greater distance along the surface, when there was a nearer passage from one rod to the other in the water.

JUST before the discharge, both the rods were observed to attract the water very strongly. It was thrown upon the rod communicating with the outside when it was laid near half an inch above the surface. When I put a drop of water on the rod communicating with the inside, the discharge was made at the distance of about two inches from the surface of the water, the fire first descending perpendicularly, and then passing along the surface; and if the rod communicating with the outside had a drop of water upon it, it might be placed higher over the water than if it had not. At the time of the explosion, this drop was elongated, and promoted the discharge very considerably.

MY attention was next drawn to the kind of impression which was made upon the water by the passage of the electric

fluid

fluid in this manner. To afcertain this, I firft placed a fhilling level with the water, to receive the explofion before it
paffed along the furface; and obferved that it was melted,
but only about half as much as I imagined it would have
been in the common way. There was no regular circular
fpot. And I could never perceive that the brafs rod which
communicated with the outfide of the battery was at all melted by the explofion.

JUDGING from the concuffion given to the whole body of
the water over which the fhock paffed, I thought that the
trace of it might poffibly be preferved on the furface of foft
pafte; and accordingly I made the explofion pafs over the
furface of fome, and plainly obferved, that the part under
the paffage was depreffed; the electric matter having repelled
it. The impreffion was not deeper where the explofion firft
fell than in any other part of the track.

To diftinguifh more accurately between the effect of the
electric matter when it properly enters the water, and when
it only paffes over the furface, I fpread a little water, exceeding thin, upon the furface of a fmooth piece of flate; but,
though the explofion paffed over the furface, with its ufual
violence, I could not perceive that it had occafioned the leaft
degree of evaporation; which Signior Beccaria found to be the
confequence of making the electrical explofion through water
in fuch circumftances.

WHEN the explofion paffed over the furface of the plate of
ice, in the experiment of the earthquake above-mentioned,
the ice feemed to be melted, both where the chain had been
laid, and alfo along the tract over which the explofion had
paffed. But this melting, if it was fuch, was not uniform;

      but

but looked as if a chain with fmall links had been laid hot upon it; and the impreffion was not at all deeper where the explofion was firft received.

WHEN the explofion paffed over the furface of a green leaf, the leaf was rent in two directions; the longer in the track of the explofion, and the other at right angles to it.

I SEVERAL times made the explofion on the furface of fnow, when it always difperfed a confiderable quantity of it, making a hole near two inches deep, and almoft as broad as long; for it could not be made to pafs at a greater diftance than about three inches.

I WAS not a little furprifed to find that I could not make this electrical explofion pafs equally over the furface of fub-ftances which were conductors in nearly the fame degree; and for a long time imagined, that this property was peculiar to water, or to bodies that conducted by means of the water they contained. I could never make it pafs the furface of any kind of charcoal; though all the degrees of conducting power may be found in different pieces of it: and I was the more confirmed in my opinion, by obferving, that, though the explofion paffed perfectly well over the furface of a fmooth board, that had been juft wetted, and immediately wiped as clean as poffible; yet two hours after, when the board was quite dry, it would not pafs at all in the fame place. It alfo paffed with great violence over the furface of a bladder which had been moiftened about a quarter of an hour before, and then feem-ed to be quite dry; but would not pafs in the leaft degree two or three hours after. In the former cafe, the explofion had left a mark where it had paffed over the bladder, darker than the reft of the furface, a kind of polifh which was on it

being

being taken off: in the latter cafe, as the dry bladder conducted very imperfectly, the fire of the charge fpread in a
beautiful manner, covering a fpace of about an inch in diameter.

THIS electrical explofion would not pafs in the leaft degree,
over the furface of new glafs, notwithftanding its property of
diffufion above-mentioned feemed to promife that it might.
Neither would it pafs at all over the furface of alum, rock
falt, fal ammoniac, blue or green vitriol, or a piece of polifhed
agate ; though thefe are all conductors of a middle kind, like
water ; and feveral of them had very fmooth furfaces. It alfo refufed the furface of dry wood, and dry leather, even the
fmootheft cover of a book.

BUT I found that I had concluded too foon, that this paffage of the electrical explofion was peculiar to the furface of
water, by finding, firft, that it paffed over the furface of a
touch-ftone, and then over a piece of the beft kind of iron
ore, exceeding fmooth on fome of its fides. This piece is
about an inch thick, and about three inches in its other de-
menfions. The full charge of a jar of three fquare feet would
not enter it. It was diverting to obferve how the electrical
explofion would make a circuit round its angles, when it was
made in a place remote from the jar. It looked like a thing
invulnerable.

THIS electrical explofion paffed over the furface of oil of
vitriol with a dull found, and a red colour, which was the
only appearance of the kind that I have yet met with. In all
other cafes, if it paffed at all, it was in a bright flame, and
with a report peculiarly loud. It paffed over the furface of the
moft highly rectified fpirit of wine without firing it ; but
                                                        when

when I took too great a diftance, the electric fire entered the fpirit, and the whole plate was in a blaze in a moment.

I once fancied that the fluidity of water was in a great meafure the caufe of this phenomenon; but I found I could not make it pafs over the furface of quickfilver, or melted lead; though neither of the rods with which the difcharge was made touched the metals. A dark impreffion was made on the furfaces of both the quickfilver and the lead, of the ufual fize of the circular fpot; and remained very vifible, notwithftanding the ftate of fufion in which the metals were.

So far was the electrical explofion from paffing over the fur-face of any metal, that I obferved, if the diftance through the air, in order to a paffage through the metal, was ever fo little nearer than the diftance along the furfaces, it never fail-ed to enter the metal; fo that its entering the furface of the metal, and its coming out again feemed to be made with-out the leaft obftruction. If as much water was laid on a fmooth piece of brafs as could lie upon it, it would not go over the furface of the water, but always ftruck through the water into the metal. But if the metal lay at any confider-able depth under the water, it would prefer the furface. It even paffed over three or four inches of the furface of water, as it was boiling in a brafs pot over the fire, in the midft of the fteam and the bubbles, which feemed to be no hindrance to it.

Animal fluids, of all kinds that I have tried, feemed in a peculiar manner to favour the paffage of the electrical explo-fion over their furfaces, and the report of thofe explofions was manifeftly louder than when water was ufed in the expe-riment. This I remarked more particularly when I made

ufe

ufe of milk, the white and yolk of an egg, both frefh broken, and after it had ftood a day or two, and had contracted a hard pellicle. In all the experiments with the egg, it was obferved, that no peculiar impreffion was made in the place where the electric matter firft came upon the furface.

It was very remarkable, that the report made by all thefe explofions, in which the electric matter paffed over the furfaces, was confiderably louder than when the difcharge was made between two pieces of metal; and they were obferved by perfons at fome diftance out of the houfe, and in a neighbouring houfe, very much to refemble the fmart cracking of a whip; and indeed it would not be very eafy to diftinguifh them. But the found made by thefe explofions, though by far the loudeft that I ever heard of the kind, fell much fhort of the report made by a fingle jar, of no very great fize, of Mr. Rackftrow's; who fays that it was as loud as that of a piftol.

It was pretty evident, that the diftance at which the fire paffed over animal fubftances was greater than it could be made on the furface of water; particularly in the firft experiment of the leg of mutton. It alfo paffed about ten inches over the furface of a piece of fpinal marrow taken from an ox.

I was much ftruck with a beautiful appearance which occurred in the courfe of thefe experiments, though it was of a different nature from them. When the electrical explofion does not pafs over the furface of water, but enters the fluid, it makes a regular ftar upon it, confifting of ten or a dozen rays; and, what is moft remarkable, thofe rays which ftretch towards the brafs rod that communicates with the cutfide of the battery are always longer than the reft; and if the explofion be made at fuch a diftance, as to be very near

taking

taking the surface, those rays will be four or five times longer than the rest; and a line bounding the whole appearance will be a beautiful ellipsis, one of whose *foci* is perpendicularly under the brass knob with which the discharge is made.

IT will be in vain to attempt these experiments without a considerable force. Nothing at all, to any purpose, can be done with a common jar; since the explosion of it will hardly pass over the surface of any conductor farther than it will discharge through the air. The charge of a jar containing three square feet of coated glass will not make any considerable appearance upon the water; and as far as I can judge, the distance at which the explosion will pass along any surface is in proportion to the strength of the charge. For this reason I make no doubt but that I could have performed all the experiments above-mentioned to much greater advantage, if I had applied a greater force, but that would have required more time, and a moderate force was sufficient to ascertain the facts.

N. B. IN these experiments, I put the discharging rod through a handle of baked wood; by which means, I could with safety lay one end of it upon the wires of the battery, and make the explosion with the other, on what substances I pleased.

SECTION

SECTION XII.

EXPERIMENTS ON THE TOURMALIN.

FATIGUED with the inceſſant charging of the electrical battery, and ſtunned with the frequent report of its explosion; I was deſirous of ſome reſpite from thoſe labours, and with pleaſure took up the gentle and ſilent TOURMALIN. And I make no doubt but that my readers, who muſt have ſympathized with me, will be equally pleaſed with the change.

IT was in the month of Auguſt 1766, that, being in London, I received from Dr. Heberden, who is glad to encourage every attempt in philoſophical inquiries, his ſet of tourmalins; among which was that fine one which had paſſed through the hands of Mr. Wilſon and Mr. Canton, and of which a drawing and deſcription are given in the 51ſt. vol. of the Philoſophical Tranſactions, p. 316. But notwithſtanding I had this valuable ſtone ſo long in my poſſeſſion, it was not till the latter end of December that I began to make any ex-

4 U   periments

periments with it, having, in the mean time, been engaged in other electrical pursuits. At length, however, having brought my other experiments to the state in which the reader hath seen them, I was desirous of being an eye witness of the wonderful properties of this stone, and of pursuing a few hints which had occurred to me with respect to it. The result of my experiments I shall lay before the reader, after having informed him in what manner, and with what precautions they were made.

The methods I used to apply heat to the tourmalin were various, but they will be sufficiently explained in the particular experiments. To ascertain the kind of electricity, I always had near me a *stand of baked wood,* from the top of which projected various arms for different purposes. Three of them were of glass, to two of which were fastened threads of silk, as it comes from the worm, supporting light pieces of down; from the other hung a fine thread, about nine or ten inches long; while a brass arm supported a pair of Mr. Canton's pith balls. At the other extremity of this arm, which was pointed, I could place a charged jar, to keep the balls constantly and equably diverging, with positive or negative electricity. Sometimes I suspended the balls, not insulated, within the influence of large charged jars. And lastly, I had always at hand *a fine thread of trial* not insulated, and hanging freely, to observe whether the stone was electrical or not when I began any experiments, and sometimes to measure the strength of the power which it had acquired.

Before I began any experiments, I never failed to try how long my electrometers would retain electricity, and in what degree. If the thread would retain the virtue for a few

minutes,

minutes, I generally preferred it, when I wanted to communicate the electricity of the tourmalin; because it would catch it in a moment. If the thread would not retain the virtue long enough, or if I wanted a less variable degree of electricity than the thread could retain, I had recourse to the *feathers*, which never failed to retain the virtue that was communicated to them for several hours together. I have often found them pretty strongly electrified, after remaining untouched a whole night, though there had been no fire in the room. They might be touched without any sensible loss of their electricity; but they received the virtue very slowly.

THE reader must observe, that by the *positive* or *negative* side of a tourmalin, in the following experiments, I always mean the side which is positive or negative while the stone is cooling. Also, when I mention *the tourmalin* without any distinction, I always mean Dr. Heberden's large one, the convex side of which is positive in cooling, and the flat side negative.

THE consideration of Mr. Wilke's experiments on the production of spontaneous electricity, by melting one substance within another, first made me conjecture, that the tourmalin might collect its electricity from the neighbouring air. To ascertain this circumstance I made the following experiments, which seem to prove that my conjecture was just. It was with a view to this experiment that I first expressed a desire to have a tourmalin in my possession. I afterwards found that Mr. Wilson had made an experiment mentioned p. 319, which is, in part, favourable to this hypothesis, though he supposed the electricity to permeate the stone, so that one side might have been supplied from the other. But the following

4 U 2   lowing

lowing experiments will show, that the suppofition of the permeability of the tourmalin to the electric fluid is altogether unneceffary to account for any of the appearances it exhibits.

ON the ftandard bar of a moft excellent pyrometer made by Mr. Ellicott, I laid a part of a pane of glafs, and upon the glafs Dr. Heberden's large tourmalin. The bar was heated by a fpirit lamp placed underneath it; and I treated the tourmalin in this manner, to afcertain with exactnefs when the heat was increafing, decreafing, or ftationary. In this difpofition of my apparatus I obferved, that, whenever I examined the tourmalin, the glafs had acquired an electricity oppofite to that of the fide of the ftone which had lain upon it, and equally ftrong. If, for inftance, I prefented the flat fide of the ftone to a feather electrified pofitively, as the heat was increafing, it would repel it at the diftance of about two inches, and the glafs would attract it at the fame, or a greater diftance; and when the heat was decreafing, the ftone would attract it, and the glafs repel it at the diftance of four or five inches. It made no difference which fide of the glafs I prefented; both fides attracting or repelling the fame feather with equal ftrength. When I faftened a fhilling with fealing wax upon the glafs, the events were always the fame. The electricity of both the fhilling and the glafs was always oppofite to that of the ftone. I was furprifed to obferve how foon the electricity, both of the ftone and the glafs, would change when it came to the turn; for in lefs than a minute I have fometimes found them the reverfe of what they were before.

THERE was, however, in the cafes in which I laid the

convex

convex fide of the tourmalin upon the flat furface of the glafs, or fhilling, one exception to the rule above-mentioned; viz. that, in cooling, the glafs and fhilling were pofitive, as well as the ftone. This I imagined to be owing to the ftone touching the furface on which it lay in fo few points, that it collected its electricity from the air, and imparted it to the body on which it lay; and this fuppofition was confirmed by experiment. For getting a mold made for the convex fide of the ftone in plaifter of Paris, and heating the tourmalin in the mold, faftened to a flip of glafs, I always found the mold and the glafs poffeffed of the electricity contrary to that of the ftone, and equally ftrong. When they were cooling, the mold feemed fometimes to be more ftrongly negative than the ftone was pofitive; for, at one time, when the ftone repelled the thread at the diftance of about three inches, the mold attracted it at the diftance of near fix.

Having made the experiments above-mentioned with the tourmalin placed upon glafs, or conducting fubftances laid upon the glafs, I had the curiofity to try what would be the confequence of heating and cooling the ftone in contact with other fubftances, both electrics and conductors. And thefe experiments brought me gradually to the difcovery of a method of reverfing all the experiments that have hitherto been made upon the tourmalin, making that fide which is pofitive in heating or cooling to be negative, and that which is negative to be pofitive; fo that the kind of electricity fhall be juft what the operator fhall direct, by the application of proper fubftances to the ftone.

I began thefe experiments with fubftituting another tourmalin inftead of the piece of glafs above-mentioned; and
when

when only one of the tourmalins was heated, they were both affected juſt as the tourmalin and glaſs had been. If, for inſtance, the negative ſide of a hot tourmalin was laid upon the negative ſide of a cold one, this latter became poſitive, as a piece of glaſs would have been in the ſame circumſtances.

WHEN I heated both the tourmalins, though they were faſtened together with cement, they both acquired the ſame power that they would have done in the open air. In theſe caſes, as the ſtones could not be made to touch one another in a ſufficient number of points, nothing could be concluded from the experiments. The ſame objection lay againſt heating or cooling the tourmalin upon rough glaſs; when I always found them both to be affected as they would have been if the glaſs had been ſmooth.

THIS conſideration made me think of cooling the tourmalin in contact with *ſealing wax,* which might be made to fit the ſtone as exactly as poſſible, though it were ever ſo irregular. Accordingly I half buried the negative ſide of a tourmalin in hot ſealing wax; and when it was cold, turning it out of its waxen cell, found it poſitive (contrary to what it would have been in the open air) and the wax negative. The other ſide of the tourmalin, which was expoſed to the open air, was affected in the ſame manner as it would have been if the oppoſite ſide had been expoſed to the air too, ſo that both ſides were poſitive in cooling. As the negative ſide of the tourmalin became poſitive by cooling in wax, I had no doubt but that the poſitive ſide would be ſo, as I actually found it.

I WOULD have aſcertained the ſtate of the different ſides of

the

the tourmalin when it was *heating* in wax, but I found it extremely difficult to do it with fufficient certainty. It cannot be known exactly when the ftone begins to cool in thefe circumftances : befides, in this method of treatment, it muft neceffarily be fome time in the open air before it can be prefented to the electrometer ; and the electricity of the fides in heating is by no means fo remarkable as it is in cooling.   In the attempts I did make with the pofitive fide of the tourmalin buried in wax, I generally found it negative, but once or twice it feemed to be pofitive.

WHEN I cooled the tourmalin in *quickfilver*, contained in a china cup, it always came out pofitive, and left the quickfilver negative ; but this effect could not be concluded to be the confequence of the application of the one to the other, becaufe it is almoft impoffible to touch quickfilver with the tourmalin without fome degree of friction ; which never fails to make both fides ftrongly pofitive, though it be quite cold, and efpecially if the ftone be dipped deep into it.

IT then occurred to me, that the tourmalin would not be apt to receive any friction from fimple preffure againft the palm of my hand ; and this being a conducting fubftance communicating with the earth, the circumftances of the experiment would be new, and might poffibly produce new appearances.   The event more than anfwered my expectations: for in heating or cooling the tourmalin in contact with the palm of my hand, each fide of the ftone was affected exactly in a manner contrary to what it would have been if expofed to the open air.   In this cafe, though the pofitive appearances may be fufpected to be ambiguous, on account of the difficulty of avoiding fome fmall degree of friction, in removing

moving the stone from the hand; yet the negative appearances are, by that very circumstance, rendered the more indisputable, and therefore remove the objection from the positive ones. For the greater satisfaction of my reader, I shall relate these experiments exactly as they were made.

I FASTENED the convex side of Dr. Heberden's large tourmalin to the end of a stick of sealing wax, and when it was quite cold, I pressed the flat side of it pretty hard against the softest part of the palm of my hand. Immediately upon this, presenting it to an electrified feather, it appeared to be strongly negative, contrary to what it would have been if exposed to the open air; and it continued negative till it had acquired all the heat it could get from my hand, when its power decreased, though it was sensibly negative to the last. Perceiving no alteration, I let the stone cool in the open air; when, according to Mr. Canton's rule, it grew more strongly negative, till it was quite cold. Thus the same side of the stone was made negative both in heating and cooling.

HEATING the same flat side, by holding it near a red hot poker, and then just touching it with the palm of my hand (when I could not bear it to rest a moment) it became positive. Letting it cool in the air, it was negative, and touching it again with my hand it became positive. Thus I made the same side of the stone alternately positive and negative for a considerable time; and at length, when I could bear to keep it upon my hand, it acquired a strong positive electricity, which continued till it was brought to the heat of my hand.

To complete these experiments, I removed the wax from the convex side, and fastened it to the flat side of the stone. Then warming the convex side, by pressing it against the

palm

palm of my hand, it became pretty ftrongly pofitive, contrary to what it would have been if heated in the open air, and continued pofitive in a fmall degree after it had got all the heat it could from my hand.   Letting it cool in the open air, it grew, according to Mr. Canton's rule, more ftrongly pofitive, and continued fo till it was quite cold.   Thus the fame fide of the ftone was made pofitive both in heating and cooling.

I THEN heated the convex fide, by holding it near a red hot poker, and preffing it againft the palm of my hand, as foon as I could bear it, it became (contrary to what it would have been in the open air) pretty ftrongly negative; though it be extremely difficult to get a negative appearance from this fide.   It cannot always be catched when it is heating in the open air.   Care, however, muft be taken, left a flight attraction of the electrified feather, by a body not electrified, be miftaken for negative electricity.

HAVING made the above-mentioned experiments, to fee how the tourmalin would be affected by being heated or cooled in contact with various fubftances, to which only one of its fides was expofed at once; I made others in which the ftone was entirely furrounded by them.   It appeared very evident, from Mr. Canton's experiment, mentioned p. 323, that it could anfwer no purpofe to inclofe it in fubftances that were conductors: for though the two electricities fhould be generated, the equilibrium would inftantly be reftored between them.   I therefore made ufe of electric fubftances only, and began with *oil* and *tallow*, both covering the tourmalin with them when it was hot, and alfo heating it in boiling oil.   But this treatment produced no new appearance, the electricity of

4 X   the

the ftone being only a little leffened. The event was the fame when a tourmalin was covered with *cement* made of bees wax and turpentine.

At laft I made a fmall tourmalin very hot, and dropping melted fealing wax upon it, covered the ftone all over, to the thicknefs of about a crown piece; and found it to act nearly, if not quite as well through this coating of wax, as if it had been expofed to the open air. I take it for granted, that the infide of the cafe of wax next to the ftone was poffeffed of the electricity oppofite to that of the ftone, at the fame time that the outfide was the fame with it. A pretty deception may be made by means of this experiment: for if a tourmalin be concealed in a ftick of fealing wax, the wax will feem to have acquired the properties of the tourmalin.

Heating the ftone, or letting it cool in *vacuo* might eafily be imagined to have the fame effect as heating or cooling it in contact with conducting fubftances; I had the curiofity, however, to try the experiment, by letting it cool in an exhaufted receiver, in which I had a contrivance to bring a thread of trial near it, or withdraw it at pleafure. The ftone was fet upright on its edge, by means of bits of glafs which it touched but in a few points, The confequence was, that the virtue of the ftone feemed to be diminifhed about one half; owing, perhaps, to the vacuum not being fufficiently perfect. For the fame reafon, the tourmalin has but little virtue immediately upon being taken out of boiling water, or after being heated in flame.

One time I fixed a thin piece of glafs, with a fmall coating upon it, oppofite and parallel to the flat fide of the tourmalin, and at about a quarter of an inch diftance from it, in an

exhaufted

exhaufted receiver; to obferve whether the electricity would be tranfmitted from the glafs to the ftone through the vacuum: but though the glafs was electrified, it was fo flightly, that I could not be certain of what kind it was.

In order to afcertain the circumftances relating to the change of the electricity of the tourmalin with more exactnefs than could be done by heating and cooling the ftone in any of the ufual methods, I laid it upon the ftandard bar of the pyrometer, and communicated heat to it by a fpirit lamp placed underneath it. The refult of thefe experiments was in general agreeable to Mr. Canton's rules; but a few circumftances occurred in this method of treating it, which could not be determined in any other; and therefore it may be worth while juft to mention them. I generally heated the bar, which is of iron, eight inches long, till the index moved feventy degrees, each of which correfponds to one 7200th. part of an inch; and obferved, that whichever fide of the ftone lay uppermoft, it was extremely difficult to afcertain the nature of its electricity all the time the heat was increafing; though, in order to do it, I held over it an electrified thread, about two inches in length, faftened to a ftick of fealing wax, which juft fupported it in an horizontal fituation. It was evident, however, that it was electrified, by its attracting a thread of trial at the diftance of about a quarter of an inch; but if I took the ftone off the bar, and immediately prefented the fide that had lain upon it to an electrified thread or feather, I always perceived the convex fide to be negative, and the flat fide pofitive in the fame circumftances; but not half fo much as they were in the contrary ftate by cooling. In this cafe, the two powers were very

diftindiſtinguiſhable by the ſmall thread above-mentioned, as the ſtone lay upon the bar; and alſo by bits of down faſtened to ſilk threads. One of theſe, which had touched the convex ſide of the ſtone, as it lay uppermoſt upon the bar, could not be made to touch it again in leſs then five hours and a half.

To ſee what would be the effect of keeping the tourmalin in the very ſame degree of heat a conſiderable time together, I laid it upon the middle part of the bar, heated by two ſpirit lamps, one at each extremity, and making the index move forty five degrees, I kept it in the ſame degree of heat, without the leaſt ſenſible variation, for above half an hour together; and obſerved, that the upper ſide, which was the convex one, was always electrified to a ſmall degree, attract-ing a fine thread at the diſtance of about a quarter of an inch. If in that time I took it off the bar ever ſo quick, and preſented it to an electrified feather; the flat ſide, which lay upon the bar, was negative, and the upper ſide very ſlightly poſitive; as appeared by its only not attracting the feather. When I put a piece of glaſs betwixt the ſtandard bar and the tourmalin, and kept them likewiſe in the ſame degree of heat, for the ſame ſpace of time, the reſult was the ſame as before, and the glaſs was ſlightly electrified, in a kind oppo-ſite to that of the ſtone.

IN heating the tourmalin upon the pyrometer, one of its ſides was neceſſarily made much hotter than the other. This inconvenience I avoided in the following method of treat-ment, which, though not ſo accurate in ſome reſpects, has peculiar advantages in others. By means of two rough places in the ſtone, I tied it in a ſilk thread, which only touched

the

the extreme edge of it on both fides. Being in this manner perfectly infulated, I contrived to make it hang in the air, at any diftance from a fire, or candle, &c; and by twifting the ftring, I could make it prefent both its fides alternately, fo as to heat it very equally.

When, in this manner, I had made it fo hot, that I could hardly bear to handle it, I let it remain in the fame fituation a quarter of an hour, in order to be fure that it was heated equally throughout. Then, with a bundle of fine thread, held fome time before in the fame degree of heat, I took off the electricity which the ftone had acquired in heating, and continuing it in the fame fituation, I found it acquired extremely little, if any electricity. Sometimes, when I thought it had acquired a little (which might be occafioned by the variation of heat in the fire) it was fo fmall, that I could not determine of what kind it was. This fully fatisfied me of the juftnefs of Mr. Canton's obfervation, that it is not *heat*, but the circumftance of changing its degree of heat that gives electricity to this ftone.

If the ftone be heated pretty fuddenly, I have fometimes found that it may be handled, and preffed with the fingers feveral times before the electricity it acquires in heating will be changed, though it begins to cool the moment it is removed from the fire.

In this fame method of treatment, I verified Mr. Canton's obfervation, that when the tourmalin is heated, and fuffered to cool again, without either of its fides being touched, the fame fide will be pofitive or negative the whole time of the increafe and decreafe of the heat. But, as he obferves, p. 240, in his experiments on hot air, the ftone muft, in this cafe,

be

be heated only to a small degree. I also proved the converse of this proposition : for, beginning where I left the stone in the last experiment, and removing it farther from the fire, both sides acquired a strong electricity, as usual ; and bringing it again nearer to the fire, I observed that both the sides not only retained the electricity they had acquired in cooling, all the time it was heating, but a considerable time after it had remained in the same degree of heat.

I CANNOT, however, entirely acquiesce in the reason that Mr. Canton gives for this appearance : for if the surrounding air would conduct the electric fluid from the positive side of the stone to the negative, I should think it would be in the same situation as in the experiment Mr. Canton made upon it surrounded with water, p. 323, and that neither side would discover any electricity at all. When the heat is three or four times greater than is sufficient to change the electricity of the two sides, the virtue of the stone is the strongest, and appears to be so when it is tried in the very neighbourhood of the fire. In the very center of the the fire, the stone never fails to cover itself with ashes, attracted to it from all sides, and from this property it acquired its name in Dutch.

IT requires, indeed, some time for the electricity of the sides to change from one state to the other ; and therefore the time of the sensible change is not always at the time of its beginning to cool, but these two circumstances will be brought nearer together the hotter the stone is made, because then the efforts (of whatever kind they are) to acquire any particular species of electricity will be the most vigorous, and sooner produce their effect ; so as be to more able to overcome obstacles to it, such as must arise from the contrary
electricity

electricity with which the ftone is poffeffed. Thus, if either fide of the ftone be in a ftate to acquire either kind of electricity, and a quantity of the contrary electricity be communicated to it by friction, or *ab extra,* that foreign electricity will be either only weakened, or loft, or changed; and thefe in a longer or a fhorter fpace of time, according to the vigour, as we may fay, with which the ftone is made to exert itfelf to counteract that influence. But I have great reafon to fufpect my own opinion, when it is different from that of fo accurate and excellent a judge of this fubject as Mr. Canton.

It is a fact, however, that the ftone often changes its electricity very flowly; and the electricity it acquires in cooling never fails to remain many hours upon it, with very little diminution. It is even poffible that, in fome cafes, the electricity acquired by heating may be fo ftrong, as to overpower that which is acquired by cooling; fo that both fides may fhow the fame power in the whole operation. And I am very certain that, in my hands, both the fides of Dr. Heberden's large tourmalin have frequently been pofitive for feveral hours together, without any appearance of either of them having been negative at all. Perhaps the flat fide of this ftone, which is pofitive in heating, might continue fo according to Mr. Canton's obfervation; and the electricity of the convex fide might have changed, as it very often does, too foon for me to obferve it. This fact, however, has happened fo often with me, and is fo very remarkable, that I think I ought not to omit the mention of it, let the caufe be what it will.

This appearance happened fo conftantly when I firft began to make experiments with the tourmalin, that I had concluded the Duc de Noya had reafon to affert, contrary to

Æpinus,

Æpinus, that both sides of the tourmalin in all cases acquired positive electricity; and I should have acquiesced in that opinion, had it not been for the friendly remonstrances of Dr. Franklin and Mr. Canton; in consequence of which I renewed my experiments, and at length found other appearances. At the time above-mentioned, I generally heated the tourmalin by presenting each side alternately to a red hot poker, or a piece of hot glass held at the distance of about half an inch; and sometimes I held it in the focus of a burning mirror; but I have since found the same appearance when I have heated it in the middle of an iron hoop made red hot. The stone, in all these cases, was fastened by its edge to a stick of sealing wax. This appearance I have observed to happen the oftenest when the iron hoop has been exceeding hot, so that the outside of the stone must have been heated sometime before the inside; and I also think there is the greatest chance of producing this appearance when the convex side of the stone is made the hotter of the two. When I heat the large tourmalin in this manner, I seldom fail to make both sides positive till the stone be about blood warm. I then generally observe a ragged part of the flat side, towards one end of the stone, will become negative first, and by degrees the rest of the flat side; but very often one part of the flat side will, in this method of treatment, be strongly positive half an hour after the other part is become negative.

This account of the appearance is made the more probable by the manner in which the stone was affected when only one of its sides was heated at one time. For when the convex side only was heated, the stone often continued a long time with both its sides positive, generally till it was not sensibly

warm.

warm.   But, in this cafe, before the convex fide became pofitive, it would fometimes be negative two or three minutes.
On the other hand, when the flat fide only was heated, it
would be pofitive a long while, and the convex fide negative;
but the flat fide becoming negative a confiderable time before
the convex fide ceafed to be fo, both fides would continue
negative till the ftone was nearly cold.

EXTREMELY forry I am for the article with which I muft
clofe this feḉtion. In the firft of the above-mentioned courfes
of experiments, that fine tourmalin, which has been fo often
mentioned in the courfe of this work, flipped out of my
hands; and though it fell only from the height of my breaft,
upon a boarded floor, two pieces were broke off from one of
its ends. The ftone, however, is more disfigured than injured by the accident: for the larger of the fragments weighs
but ten grains, and the fmaller only one, while the reft of it
weighs four penny weights fixteen grains. I cannot perceive that its virtue is at all leffened. Mr. Wilfon obferves,
that there were feveral cracks in it; and for that reafon I had
been careful never to expofe it to any great degree of heat.

IT is broke with eight or ten different faces, each of which
have a moft exquifite polifh; but there is no appearance of
any *ftrata* or *laminæ* in the internal ftruḉture of the ftone.
A piece of glafs or pitch might be fuppofed to break in the
fame manner. The larger of the fragments has confiderable
power, and the two fides have the fame different powers that
they had when they were part of the entire ftone.

4 Y   SECTION

S E C T I O N   XIII.

MISCELLANEOUS EXPERIMENTS.

I. OBSERVATIONS ON THE ELECTRIC SPARK TAKEN THROUGH SEVERAL PIECES OF METAL.

MARCH the 24th. 1766. I obferved that an electric fpark taken from the prime conductor itfelf was not near fo ftrong and pungent, as one taken through a piece of metal infulated, and interpofed between my finger and the conductor.

THE effect was the fame whatever was the form of the interpofed piece of metal. And, in this manner, whatever was prefented received a full and ftrong fpark : whereas a great part is commonly diffipated, in pencils or ftars, even when pretty large brafs knobs are prefented to the prime conductor itfelf, if the excitation be very powerful; unlefs both the conductor and knob have one precife degree of convexity, adapted to one another.

ONE fingle brafs ball made the fpark as ftrong as the interpofition of a long piece of metal, or of many pieces.

WHETHER

WHETHER one, a few, or a great number of pieces were used, it seemed that the intervals taken together must be equal.

BUT these intervals taken together will be larger when the pieces are placed in a right line, than when they are laid in a curve.

WHETHER one body, or a number of them be interposed; if a spark be solicited, it will not strike the first, unless it can, at the same time, strike all the rest.

ALL these experiments succeed, in the same manner, with the explosion of a charged jar.

SOME of the above-mentioned circumstances, I afterwards found, had been taken notice of by Signior Beccaria.

II. A DECEPTION RELATING TO THE DIRECTION OF THE ELECTRIC SPARK

As I was once amusing myself with taking long sparks from a large prime conductor of polished copper, and considering the deceptions that electricians had fallen into with respect to the direction of the electric matter; I could not help being struck with one deception, which the evidence of my senses would never have rectified; and which showed very clearly, how little the evidence of the senses is to be depended upon in such cases. I observed, that, whether I made this large conductor give, or take the electric fire (for I could make it do either at pleasure, and with the same force) I still fancied that a spark taken with a brass ball above the conductor descended to it, and that a spark taken below it descended

from

from it; but sparks taken laterally seemed to have no one certain direction.

III. An experiment intended to ascertain whether electric substances, in their natural state, contain more of the electric fluid than conductors.

Thinking to ascertain Dr. Franklin's hypothesis, concerning the essential difference between conductors and non-conductors, I made a pretty large piece of glass red hot (in which state I had proved it to be a real conductor of electricity) and placed it upon a smooth piece of copper, insulated; supposing that, if electric substances had naturally a much greater share of the electric fluid than conductors, this piece of glass, in passing from a conducting to a non-conducting state, must exhaust the copper of its natural share of the electric fluid, and leave it electrified negatively. But I could perceive no kind of electricity, either in the copper, or the glass, during the whole time of its cooling.

Some time after, I found that J. F. Cigna had endeavoured to ascertain the same thing, by reducing ice into water; but ice and water are both conductors of electricity.

IV. The MUSICAL TONE of various discharges ascertained.

As the course of my experiments has required a great variety of electrical explosions I could not help observing a

great

great variety in the mufical tone made by the reports. This excited my curiofity to attempt to reduce, this variation to fome meafure. Accordingly, November the 17th. by the help of a couple of fpinets, and two perfons who had good ears for mufic, I endeavoured to afcertain the tone of fome electric explofions; and obferved, that every difcharge made feveral ftrings, particularly thofe that were chords to one another, to vibrate: but one note was always predominant, and founded after the reft. As every explofion was repeated feveral times, and three of us feparately took the fame note, there remained no doubt but that the tone we fixed upon was, at leaft, very near the true one. The refult was as follows.

A jar containing half a fquare foot of coated glafs founded F fharp, concert pitch. Another jar of a different form, but equal furface, founded the fame.

A jar of three fquare feet founded C, below F fharp. A battery, confifting of fixty four jars, each containing half a fquare foot, founded F below the C.

The fame battery, in conjunction with another of thirty one jars, each containing a fquare foot, founded C fharp. So that a greater quantity of coated glafs, always gave a deeper note.

Differences in the degree of a charge in the fame jar made little or no difference in the tone of the explofion; if any, a higher charge gave rather a deeper note.

From thefe experiments it will be eafy for any perfon to compare the quantity of fquare feet of coated glafs, with the lengths of mufical ftrings giving the fame note. For this purpofe, I could eafily have found more terms of the feries; but I am afraid philofophers in general will think it trifling
                                                                    enough

enough to have found so many. I do not expect that electrical explosions will ever be introduced into concerts of music; or that these experiments will be of any use to measure the extent of the clouds from which a clap of thunder proceeds. But true philosophers will not absolutely despise any new fact or observation, though it have no immediate, or apparent use.

V. Experiments on the Effects of giving a METALLIC TINGE to the Surface of Glass.

It has long been a question among electricians, where the electric matter that constitutes the charge of a plate of glass lies; whether within the pores of the glass, or only upon the surface; and some experiments I have made will perhaps be thought to throw some light on this difficult subject.

I considered that the common coating of a jar is not in actual contact with the glass, but that the metallic tinge, which is given to glass by an electric explosion of the metal upon its surface, is probably in contact with it, if not lodged in its pores. I therefore gave a coating of this kind to both sides of a plate of glass; and at first imagined that the glass coated in this manner did receive a charge, as well as if it had been coated in the common way; for it gave a real shock: but I very well remember, at that time, being a little surprised to see the electric fire run over the surface of that coating, a thing not possible in the common way. However, not sufficiently attending to that circumstance, I

was

was purfuing the experiment, and trying whether, by combining this piece of glafs with a large battery, and making the difcharge of both upon this metallic tinge, I could not melt part of it, and thereby fetch it out of the glafs; as that method would have melted, and abfolutely difperfed a confiderable part of the coating of a common jar. But I was prodigioufly furprifed to find, that, though the connection of this metallic tinge with the battery was complete, the difcharge could not be made by bringing the difcharging rod upon it; though within three quarters of an inch of another brafs rod, that formed the communication between this plate of glafs and the battery. This convinced me that the metallic tinge did not anfwer the purpofe of a coating; and I prefently fatisfied myfelf, that a piece of uncoated glafs would receive juft fuch a charge as the tinged glafs had done.

To afcertain this matter ftill further, I ftruck a tinge of this kind along two oppofite fides of a glafs tube, about half a yard in length; and holding it with my hand in contact with a part of this metallic tinge, found that it was excited juft like another tube: for when I difcharged the electricity of any part of the tube where the tinge was ftruck, it did not at all difcharge other parts of the tube, whither the fame tinge extended. Alfo the electric fnapping from the tinged part of the glafs could not be diftinguifhed from the fnapping at other places; except that, fometimes, where the gold lay thicker than ordinary, a denfer ftream of electric matter was vifible on its furface, and ran in feveral fmall ftreaks, in different directions, from the place where the fpark was taken.

This experiment feems to fhow, that a coating of metal exceedingly near the furface of the glafs is not at all affected

either

either by the excitation or charging of it; and seems to confirm the hypothesis of the electric fluid not entering the pores of the glass.

As the giving this metallic tinge to both sides of a plate of glass is not very easy, the reader will not, perhaps, be displeased to be informed in what manner I succeeded in it. After fatiguing myself a long time in endeavouring to strike a piece of leaf brass into the two sides of a plate of glass, to serve instead of a coating (having always broken the glass in fixing either the first or second coating) I at length put two other pieces of glass, one on each side of that to which I intended to give the tinge, with pieces of leaf brass between them both; and making one explosion through both of them at the same time, the upper and the lower piece of glass were shattered to pieces, but the middle piece (being equally affected on both sides) remained whole, and the coatings were nearly as I wished them.

VI. An experiment intended to ascertain whether FERMENTATION contributes to the production of electricity.

September the 3d. In order to determine whether any of the electric fluid was discharged from, or acquired by bodies in a state of *fermentation*; I hung a pair of pith balls at the extremity of a piece of wire communicating with a quantity of steel filings, fermenting with oil of vitriol, inclosed in a glass vessel. But they never separated in the least.

VII.

VII. AN EXPERIMENT INTENDED TO ASCERTAIN WHETHER EVAPORATION CONTRIBUTES TO THE PRODUCTION OF ELECTRICITY.

DECEMBER the 26th. I put a small quantity of water upon a thin piece of glaſs, and made it all ſuddenly evaporate by a red hot iron held under it; but the glaſs had acquired no degree of electricity. The weather was froſty.

VIII. AN EXPERIMENT INTENDED TO ASCERTAIN WHETHER FREEZING BE ACCELERATED OR RETARDED BY ELECTRIFICATION.

JANUARY the 6th. 1767. I expoſed two diſhes of water in the open air, while it was freezing intenſely, and electrified one of them pretty ſtrongly; but could perceive no difference in the time, either when it began to freeze, which was in about three minutes, or in the thickneſs of the ice when both had been frozen ſome time.

HAPPENING to caſt my eyes into the fields, out of the window, through which I had put the board which I uſed for the purpoſe of this experiment, I obſerved, on each ſide of the electrified wire, the ſame *dancing vapour*, which is ſeen near the ſurface of the earth in a hot ſummer's day, or near any heated body that occaſions an exhalation of vapours.

4 Z   IX.

IX. THE EXAMINATION OF A GLASS TUBE, WHICH HAD BEEN A LONG TIME CHARGED AND HERMETICALLY SEALED.

DECEMBER the 30th. I examined a glafs tube, about three feet in length, one half of which I had charged in the month of March preceding, and then fealed hermetically; but could not perceive that it was excited in the leaft degree, either by heating or cooling. The difference in the refult of this experiment from feveral of Mr. Canton's, related p. 296, I attribute to the thicknefs of the glafs of my tube. Mr. Canton charged fmall balls exceeding thin. I alfo obferved that there was no perceivable difference in the excitation of the charged or uncharged part of this tube, and that both parts acted exceeding well.

I AFTERWARDS opened this tube, and pouring a quantity of leaden fhot into it, found it to contain a very good charge. It gave me one confiderable fhock, and feveral fmall ones; as I made no ufe of an outward coating, but only difcharged it by grafping it in feveral places by my hand.

X. THE WEIGHT REQUISITE TO BRING SOME BODIES INTO CONTACT ASCERTAINED BY THE ELECTRICAL EXPLOSION.

IT is plain from optical experiments, and alfo from a variety of other confiderations, that bodies of no great weight, lying

lying upon one another, are not in actual contact. As the same thing is demonstrated by an electric spark being visible between pieces of metal lying upon one another, and other effects of electricity (particularly the fusion of the parts through which it goes out of one body and enters another, not actually in contact with it) I was desirous to determine, by this criterion, what weight was sufficient to bring bodies into actual contact. With these views, I began with laying twenty smooth shillings upon one another, and making the discharge of the battery through them ; thinking that the fusion would disappear, when the weight was sufficient to press them into contact. But I found that the whole column was not sufficient ; for every piece was melted on both its sides, so that every two contiguous sides had spots exactly corresponding to one another. The deepest impressions were made near the top of the column, but they did not diminish with exact regularity. Perhaps small particles of dust might prevent some of them from coming sufficiently near one another.

AFTERWARDS, I gradually increased my weights, till I found that about six pounds was sufficient for my purpose. The fusion was visible under that weight, but never under above half a pound more, though I repeated the experiment several times.

I HAD some suspicion, that the largeness of the explosion might have occasioned a momentary repulsion, separation, and consequent fusion of these pieces of metal, though pressed by such a weight, but I found I was not able to produce any fusion ; under a greater weight than that above-mentioned, though, instead of thirty two square feet of coated glass, I used above sixty.

4 Z 2   XI.

XI. THE EFFECT OF THE ELECTRICAL EXPLOSION TRANSMITTED THROUGH VARIOUS LIQUORS.

I BELIEVE it is generally fuppofed, that ale, and other liquors are turned four by lightning, and I was defirous of afcertaining whether that fact (if it be one) was owing to the liquors being properly ftruck with the lightning, or to the ftate of the air, &c. during the thunder ftorm. In order to this, I provided myfelf with a glafs tube, nine inches long, and about a quarter of an inch in diameter, and by inferting a wire into one end of it, which was ftopped with fealing wax, could eafily tranfmit an electrical fhock through any fubftances contained in it.

By this means, November the 13th. I began with difcharging the explofion of the battery through this tube, filled with *frefh fmall beer*, and obferved a confiderable quantity of fixed air, or fomething in the form of bubbles, to afcend in it; but when I tafted it, I could perceive no difference between it and that out of which it was taken. No doubt the efcape of fo much air would tend to make it grow ftale fomething fooner.

I THEN difcharged feveral large fhocks through a tube filled with *red wine*, but, after two or three days, could perceive no alteration in its tafte, or other fenfible qualities. In this difcharge, the electric matter did not immediately ftrike the wine, but a metal rod, which juft touched its furface: but I afterwards gave it two or three more fhocks, in which the wine itfelf was made to receive the explofion, but there was no variation in the effects.

I PASSED

I PASSED the shock through a tube filled with *milk*, in both the methods above-mentioned; but it was sweet three days after. Also a tube filled with *fresh ale* received several large shocks without undergoing any sensible change of properties.

IN all these explosions I held the tube in my hand, without feeling any thing of the shock.

I ALSO made the electric spark visible a great number of times in a small quantity of *syrup of violets*, without producing any change of colour, or other sensible qualities.

XII. OBSERVATIONS ON THE COLOURS OF ELECTRIC LIGHT.

FINDING it advanced in the writings of several electricians (who must have copied it from one another, without ever repeating the experiment, though it may be done so soon) that electric light contained no prismatic colours; I had the curiosity to try so extraordinary a fact, and immediately saw both the fallacy of the experiment when it was first made, and the cause of it. Holding a prism before my eyes, while the electric sparks were taken at the prime conductor, I observed as beautiful prismatic colours as any that are exhibited by the image of the sun; but when the light was a little diffused, as in those red or purple parts of a long spark, as it is called, the colours were not so vivid, and less easily distinguished from one another; and when the light was still more diffused, through a vacuum, the prism made no sensible alteration in the appearance of it. Thus the middle part of any large ob-

ject

ject appears of its natural colour through a prifm : for though the rays be really feparated, they are immediately confounded with others from different parts of the fame object; fo that its natural colour muft neceffarily be the refult.

As the flames of different bodies yield very different proportions of the prifmatic colours, I have often thought of attempting to afcertain the proportion of thefe colours in electric light, and compare it with the proportion of colours from light procured in various other ways. This may poffibly determine what that heterogeneous matter is that is roufed into action by the rapid paffage of the electric fluid, and which is the caufe of the light, the fmell, and perhaps other fenfible qualities of electricity; but I have not had leifure to purfue the inquiry.

I SHALL clofe this article with juft mentioning another deception, which fome perfons may poffibly lie under, with refpect to what is called the *length of the electric fpark.* When a jar is difcharged, it may be imagined, that a body of fire is feen extending from the infide to the outfide; whereas it is pretty certain, that that appearance is occafioned by the very rapid motion of a fingle ball of fire; in the fame manner as a lighted torch, with no greater motion than a man's arm can give to it, will feem to make an entire circle of fire. That the fire of an electrical explofion confifts of a ball, or cylinder, of no great length, feems pretty evident from one of the experiments with the circles, p. 667, in which the diameter of the circle was the fame, whether the explofion was taken at the diftance of half an inch, or of two inches; and alfo from the experiments of its paffage over furfaces, in which it was fometimes made twenty times longer than ufual, without any fenfible diminution of its thicknefs.

XIII.

XIII. OBSERVATIONS ON THE SMALL WIRES THAT COL-
LECT ELECTRICITY FROM THE EXCITED GLOBE.

HAVING made ufe of feveral brafs wires, about two inches
and a half long, to collect the electric matter from my globe;
I obferved, after a month or two, that about half an inch of
the ends of them, which touched the globe, had contracted
a blacknefs, particularly on the fide which lay next the globe.
I then took them off the ring to which they had hung, and rubbing them carefully, obferved, that the fame friction which
made the reft of the wire quite bright, made but little alteration in this acquired blacknefs. Recollecting, at the fame
time, S. Beccaria's theory of magnetifm, inftead of replacing
the wires, I hung two very fine needles in their place; and
December the 20th. after about two months, in which I had
made the moft ufe of the machine, I examined them, and
found that blacknefs at their points, but could not be fure
that they had acquired any degree of magnetifm. They had,
indeed, a very fmall degree; but I had not examined them
fo very accurately before I hung them on, as I did afterwards.
The experiment deferves to be repeated with more care, but
it requires a longer and more conftant ufe of an electrical machine than, it is probable, I fhall ever have an opportunity of
employing.

XIV.

XIV. EXPERIMENTS INTENDED TO ASCERTAIN THE DIFFERENCE IN THE CONDUCTING POWER OF DIFFERENT METALS.

IN a converſation I once had with Dr. Franklin, Mr. Canton, and Mr. Price, I remember aſking whether it was probable that there was any difference in the conducting power of different metals; and if there was, whether it was poſſible to aſcertain that difference; and as far as I can recollect, it was the Doctor who ſuggeſted the method which I have aſcribed to him, p. 490. Since that part of the work was printed, I have endeavoured to carry the ſcheme into execution, by tranſmitting the ſame exploſion of the battery through two wires at a time, of two different metals, and of the ſame thickneſs. They were hooked one to the other, and held faſt in hand viſes, after they were meaſured with a pair of compaſſes to exactly the ſame length. The experiments were much more pleaſing and ſatisfactory than I expected, but the reſult by no means correſponded to my ideas *a priori*.

I FIRST joined a piece of iron wire and a piece of copper wire. The exploſion totally diſperſed the iron, and left the copper untouched. The braſs likewiſe diſappeared when joined with the copper, and the iron when joined with the braſs.

So far the experiments were extremely eaſy; a ſingle charge of the battery ſufficing to determine the difference between any two; but when I came to compare the more perfect

fect metals, I found much more difficulty, and was obliged to try four or five charges of the battery upon every two: for, their conducting powers being nearly the same, I either made the charge too high, and difperfed them both; or too low, and touched neither of them. At length, I happened to hit upon fuch charges, that the copper vanifhed, and left both the filver and the gold; and the gold remained when the filver was difperfed. The hook, however, of the filver was melted off when the copper was difperfed, and the hook of the gold when the filver was difperfed: for the heat is always the greateft where the electric fire paffes from one body to another. Before the difperfion both of the copper and the filver, I had made explofions of fuch a ftrength, as though too fmall to melt them, gave them a bluifh tinge.

From thefe experiments it is eafy to fettle the order in which the metals above-mentioned are to be ranked, with refpect to the power of electricity to melt them. It is as follows. *Iron, brafs, copper, filver, gold.*

Not being able to get *lead* or *tin* drawn into wires, I got pieces of thofe metals rolled into plates equally thin, and taking fmall flips, of equal length and breadth, I tranfmitted the explofion through them; when the lead gave way the firft. I intended to have compared thefe plates with others of iron, brafs, &c. but had not an opportunity. I have little doubt but that tin would melt before iron; though indeed I had expected that tin would have melted before lead, and gold before filver. But according to Mr. Wilke's experiments, p. 335, lead is a worfe conductor than any of the other metals. My own experiments on the circular fpots, p. 662 made me expect that gold would have melted before filver.

5 A   It

IT is very remarkable, that when iron wire is melted by the electric explosion, *bright sparks* are generally disperfed about the room, in all directions; but that they are seldom, or never seen when wire of any other metal is used. If but a small residuum of a battery be taken between two iron rods, when the explosion is extremely little, a great number of small sparks will fly in all directions from the iron, to the distance of about an inch, and exhibit a beautiful appearance. Fewer of these sparks will be seen if one of the rods be brass, and, I think, none, in these small discharges, if they both be brass.

BEFORE any use can be made of these experiments, to determine the relative conducting power of the several metals, the order in which they melt with common heat should be compared with the order in which they melt with the electrical explosion; and I could not readily find a table of that kind. I think, however, that iron is said to require more heat to bring it into a state of fusion than any other metal, and it requires but a small force of electricity to do it; so that I think these two orders are not the same. Before this matter can be settled, it should likewise be found, how much more easily any of the metals will be melted before another, by transmitting shocks through wires of different lengths and thicknesses, which would be a very tedious business. I make no doubt but that an explosion which melts a copper wire of any given diameter would disperse an iron wire of twice the diameter, so that copper would be a much greater security, as a conductor to guard a building from lightning than iron, besides its being less liable to rust; but then it is more expensive.

XV.

XV. Experiments with an ELECTRIFIED CUP.

I SHALL close the account of my experiments with a small set, in which, as well as in the last, I have little to boast besides the honour of following the instructions of Dr. Franklin. He informed me, that he had found cork balls to be wholly unaffected by the electricity of a metal cup, within which they were held; and he desired me to repeat and ascertain the fact, giving me leave to make it public.

ACCORDINGLY, December the 21st. I electrified a tin quart vessel, standing upon a stool of baked wood; and observed, that a pair of pith balls, insulated by being fastened to the end of a stick of glass, and hanging entirely within the cup, so that no part of the threads were above the mouth of it, remained just where they were placed, without being in the least affected by the electricity; but that, if a finger, or any conducting substance communicating with the earth, touched them, or was even presented towards them, near the mouth of the cup, they immediately separated, being attracted to the sides; as they also were in raising them up, the moment that the threads appeared above the mouth of the cup.

IF the balls had hung in the cup a considerable time without touching it, and they were taken out immediately after the electricity of the cup was discharged, they were found to have acquired no degree of electricity.

IF they had touched any part of the cup, though they

showed

showed no electricity while they were within it; yet, upon being taken out, they appeared to have acquired some; which was more if they had touched a part near the edge of the cup, less if they had touched any part more remote from the edge, and least of all if they had touched the bottom only. If they had first touched the side near the top, and then the bottom, they came out with that small degree of electricity which they would have acquired, if they had touched the bottom only.

In any case, if the balls were taken out while the cup remained electrified, they necessarily acquired some degree of electricity, in passing the mouth of the cup.

To pursue this experiment a little further, I took a small coated phial, such as is represented upon the stool [c Pl. ii.] and observed, that when I held it by the wire, within the electrified cup, it acquired no charge, the electricity of the cup affecting both the inside and outside coating alike. If the external coating touched the bottom of the cup, the phial received a very small charge. If it was made to touch the side, it acquired a greater charge; and the nearer to the top it was held, the higher charge it received; the wire of the phial, which communicated with the inside coating, being further removed from the influence of the electricity of the cup.

May we not infer from this experiment, that the attraction of electricity is subject to the same laws with that of gravitation, and is therefore according to the squares of the distances; since it is easily demonstrated, that were the earth in the form of a shell, a body in the inside of it would not be attracted to one side more than another?

DOTH

DOTH it not follow from the experiments of the balls, compared with those with the phial, that no body can receive electricity in one place, unless an opportunity be given for its parting with it in another; at least, that a quantity must be repelled from any particular part before any more can enter; since a small body can no more receive electricity when all its sides are equally exposed to the action of an electrified body, than a phial can be charged when both its coatings are equally exposed to the same electricity?

Do not these experiments, likewise, favour the hypothesis of S. Beccaria, that there is no electrical attraction without a communication of electricity?

ADDITIONS

ADDITIONS AND CORRECTIONS.

PAGE 1. line 21. after *magnete* add, published in the year 1600.

P. 10. NOTE. Volume 2. in this place means the second volume of Jones's Abridgment. This was done inadvertently. Afterwards the abridgments of *Lowthorp*, *Jones*, *Eames*, and *Martin* are referred to as one work. There was no occasion to distinguish the *part* of the volume, because every thing that relates to the subject of electricity will be found together, among the *physiological papers*.

P. 37. l. 14. Instead of Mr. *Wheeler*, read Mr. *Du Faye*.

P. 53. ADD : Mr. Du Faye was the first who observed that electric substances attract the dew more than conductors. He observed that a glass vessel, placed on a metal cup, and set in the open air all day, will often be wet when the metal is dry. S. Beccaria accounts for this fact, by supposing that alterations in the electricity of the air easily produce correspondent alterations in the electricity of metals, in which the electric fluid moves with the utmost ease, but not in glass. Whenever, therefore, the state of the electric fluid in the air is altered, the glass is electrified *plus* or *minus*, and therefore attracts the vapours in the air. *

P. 145. ADD : Mr. Jallabert of Geneva carried the expe-

* Beccaria dell' elettricismo naturale et artificiale, p. 179.

riments

riments on plants further than the Abbé Nollet had done; and, by electrifying bottles in which the plants were growing in water, and placing in the same exposure other bottles, containing plants of the same kind; proved, in the clearest manner, that the electrified plants always grew faster, and had finer stems, leaves, and flowers than those which were not electrified, and consumed more of their water. *

P. 212. THE article relating to the conducting power of animal fluids was composed from my memory. I had written to Dr. Franklin for more accurate information, but his answer came too late. He informs me, that, of animal fluids, he had tried only *milk*, and that many years ago; but that Mr. Kinnersley, and others in America had since tried *blood* and *urine*, and also the sinews of animals newly killed; which, he understood, were all found to be exceeding good conductors, remarkably better than water.

P. 321. l. 2. ADD: Mr. Wilson observes, that when a tourmalin, which he had from Dr. Morton, was held between the eye and the light, and viewed in the direction through which the electric fluid is found to pass, it appears of a darker colour considerably than when it is viewed at right angles to the former direction. This appearance, he says, obtains in many other tourmalins, especially when they happen to be as conveniently shaped. †

P. 381. l. 2. READ, *If it will come on the full moon, you being at the change, observe these signs.* There was some mistake in the work from which I first copied the account, and I had made the best sense I could of it; but I have since ac-

* Ib. p. 125.   † Phil. Transf. Vol. 53, p. 448.

cidentally

cidentally met with the authority from which it was taken, which is Phil. Tranf. abridged. Vol. 2. p. 106.

P. 518. l. 28. ADD: The form of a coated plate of glafs is reprefented *b* plate ii.

P. 530. l. 24. ADD: Mr. Ramfden, mathematical inftrument maker in the Hay Market, I am informed, has lately conftructed an electrical machine, on a plan very different from all that I have yet heard of, in which, friction is not given to any kind of hollow glafs veffel whatever; but to a *circular plate of glafs*, about nine inches in diameter.

THIS plate turns vertically, and rubs againft four cufhions, each an inch and a half long, placed at the oppofite ends of the vertical diameter. The conductor is a brafs tube, and has two horizontal branches coming from it, reaching within about half an inch of the extremity of the glafs; fo that each branch, it is faid, takes off the electricity excited by two of the cufhions.

THE quantity of electricity excited by this machine is faid, by the maker, to be greater than can be produced from others. It is alfo faid to be much lefs affected by moift weather; and the mechanifm being very fimple, renders it lefs liable to be out of order.

NOT having had an opportunity of feeing this machine, I cannot affert any thing concerning it from my own knowledge; but the conftruction is original and ingenious.

A CATAL-

A CATALOGUE OF BOOKS WRITTEN ON THE SUBJECT OF ELECTRICITY, EXCLU-
SIVE OF PAPERS IN BOOKS OF PHILOSOPHICAL TRANSACTIONS, AND OTHER
MISCELLANEOUS WORKS; DISTINGUISHING [BY ASTERISMS] THOSE WHICH
THE AUTHOR HAD SEEN, AND MADE USE OF IN COMPILING THIS WORK.

* GILBERT de Magnete Magneticiſque corporibus, 1600, London, folio.
    * *Ottò de Guericke*'s experimenta nova Magdeburgica, 1672, Amſterdam, folio
    * *Hawkeſbee*'s phyſico mechanical experiments, 1709, 1719, London, octavo.
* *Deſaguliers*' diſſertation concerning electricity, 1742, London.
C. A. *Hauſenii* novi profectus in hiſtoria electricitatis, Leipſic, 1743.
Nollet's conjettures ſur les cauſes de l' elettricité des corps, Paris, 1745. Wilſon, p. 12.
    ———— eſſai ſur l' electricité des corps, 1746, 1754, Paris.
* *Wilſon*'s eſſay towards an explication of the phenomena of electricity deduced from
    the ether of Sir Iſaac Newton, 1746, London, octavo.
* *Watſon*'s experiments and obſervations, tending to illuſtrate the nature and properties
    of electricity, 3d. edi. 1746, London, octavo.
* ———— ſequel to the experiments and obſervations, tending to illuſtrate the nature
    and properties of electricity, 1746, 1747, London, octavo.
* *Freke*'s eſſay to ſhow the cauſe of electricity, 1746, London, octavo.
Frazenſtein's theoria electricitatis, more geometrico explicata, Hall, 1746.
* *Martin*'s eſſay on electricity, 1748, Bath, octavo.
Boze's recherche ſur la cauſe, et ſur la veritable theorie de l' electricité 1746.
Conjetture fiſiche intorna i fenomeni della machina elettrica, Rome, 1746, octavo.
    Hiſt. p. 168.
Franciſco Pivati's lettere della elletricita medica, Venice, 1747.
Boze's tentamina electrica, Wittemburg, 1747.
* *Rackſtrow*'s miſcellaneous obſervations, together with a collection of experiments on
    electricity, 1748, London, octavo.
Secondat's hiſtory of electricity, 1748. Daſibard's preface, p. 27.
* *Watſon*'s account of experiments, made to diſcover whether the electric power would
    be ſenſible at great diſtances, &c. 1748, London, octavo.
·Recueil des traités ſur l' electricité, traduits de l' alemand et de l' anglois, Paris, 1748,
    octavo, 3 vols.
* *Nollet*'s recherches ſur les cauſes particulieres des phenomenes electriques, 1749,
    Paris, twelves.
A latin diſſertation on electricity by *P. Francois Plata*, a jeſuit of Palermo, 1749. Hiſtoire.
* *Wilſon*'s treatiſe on electricity, 1750, 1752, London, octavo. N. B. *This book is
    quoted in this work by the title of* Wilſon's Eſſay.
* *Boulanger*'s traité de la cauſe et des phenomenes de l' electricité, 1750, Paris, twelves.
    N. B. *I have ſeen no more than one part of this treatiſe, which I had of Dr.* Watſon.
    How many more parts there are of the whole work, I have not been informed.
Joſ. Veratti obſervations ſur l' electricité, aux quelles on a joint les experiences faites
    a Montpelier pour guerir les paralytiques au moyen de l' electricité, a la Haye,
    1750, twelves.
Jallabert's experimenta electrica uſibus medicis applicata, 1750.
Electricorum effectuum explicatio, by *Father Bina*, 1751.
* Hiſtoire generale et particuliere de l' electricité, 1752. twelves.
J. H. Winkler's progr. de avertendi fulminis artificio ex doctrina electricitatis, Leipſic,
    1753.
* *Giambattiſta Beccaria* dell' elettriciſmo artificiale e naturale, 1753, Turin, quarto.
    N. B. *The copy of this treatiſe which I peruſed, and which was procured for me by
    Dr.* Franklin, *was defective; one ſheet being wanting, viz. from p. 182 to 191.*

        S. H.

S. H. Quelmalz diſſertatio de viribus electrices medicis, Leipſic, 1753, quarto.
* *Franklin*'s new experiments and obſervations on electricity, made at Philadelphia in America, part 1. 3d. ed. 1760.
------------------ part 2. 2d. ed. 1754.
------------------ part 3. 1754. London, quarto.
Brevis relatio de electricitate propria lignorum, authore *P. Wimaelino Ammerſin* De Lucerne Helvetiorum. Ord. Minim St. Franciſci de P. Conventual 1755. Nollet's letters, vol. 2. p. 235.
J. B. Landriani diſſertatio de nova electricitatis theoria. Milan. 1755. Wilke, 12.
J. A. Euleri, diſquiſitio de cauſa phyſica electricitatis, ab academia ſcientiarum imperiali petropolitana præmio coronata, Peterſburg, 1755. Wilke, 12.
* *Dalibard*'s hiſtoire abrigeé de l' electricité, and French verſion of Franklin's letters, 1766, Paris, 2 vols. twelves.
Lovet's ſubtile medium proved, 1756, London, octavo.
* *Johannes Carolus Wilke*'s diſputatio phyſica experimentalis de electricitatibus contrariis, 1757, Roſtock, quarto. N. B. *This admirable treatiſe conſiſts of four parts, but, to my great regret, the copy which I had contained only the three firſt of them. It was that which the author ſent to Dr.* Franklin, *before the remainder was printed.*
Butſchany diſſertatio de fulgure et tonitru ex phenomenis electricis Gottingen, 1757, quarto.
* *Giambattiſta Beccaria*'s lettere dell' elettriciſmo, 1758. Bologna, folio.
* *Wilſon* and *Hoadley*'s obſervations on a ſeries of electrical experiments, 2d. edi. 1759, London, quarto.
* *Æpinus*'s tentamen theoriæ electrcitatis et magnetiſmi, 1759, Peterſburg, quarto.
* *Nollet*'s lettres ſur l' electricité, tome 1. 1749, tome 2. 1760, Paris, twelves.
* *Du Tour*'s recherches ſur les differens mouvements de la maticre electrique, 1760, Paris, twelves.
* *Weſley*'s deſideratum, or electricity made plain aud uſeful, 1760, London, twelves.
Æpinus on the tourmalin, 1762.
Nollet's Leçóns de phyſique, tome 6th. 1764, Paris, twelves.
* *Johannes Franciſcus Cigna* de novis quibuidam experimentis electricis, from the memoirs of the academy of Turin for the year 1765, quarto.
* *Lovet*'s philoſophical eſſays in three parts, 1766, London, octavo.

OTHER BOOKS THE DATES OF WHICH I HAVE NOT FOUND.

Winkler's eſſay ſur la nature la cauſe et les effects de l' electricité. Hiſtoire, p. 33.
Boze's poem on electricity. Hiſtoire.
Obſervations ſur l' electricité par Mr.---chirurgien de la ſalpetre. Hiſt. p. 98.
Nouvelle diſſertation ſur l' electricité par un phyſicien de Chartres. Hiſtoire.
Mr. *Waitz* trailé de' l' electricité et de ſes cauſes. Nollet's recherches, p. 160.
Mr. *Krugen*'s meditations on electricity. Hall.
Tentamen de vi electrica ejuſque phenomenis, auctore Nic *Bammucaro*.
Laurentii Berand S. J. theoria electricitatis, Peterſburg. Wilke, p. 12.
Watkins on electricity.

As the reader will ſee by the aſteriſms what books I have had an opportunity of peruſing, he will ſee in what parts my hiſtory is moſt likely to be defective. And I ſhall think myſelf greatly obliged to any perſon, who will favour me with the uſe of any treatiſe which contains a diſcovery of importance. But I do not apprehend that any thing very material can have eſcaped me.

A N

A N

INDEX of THINGS.

5 B 2

it brought down by a kite, 353. fatal to professor Richman, 358.

Liquors, great explosions discharged through them, 724.

MACHINES ELECTRICAL, observations on their construction. 508, &c· various forms of them, 525, &c. the author's described, 530.

Magic picture, 563.

Magnetism, and electricity compared, 5, 6, 431, 727. not affected with electrification, 33. given by the electric explosion, 178, 351.

Maxims, practical, for the use of young electricians, 535.

Medical electricity, 408.

Medicated tubes, 146, &c.

Metallic conductors, their use in guarding buildings, 398. exemplified in the case of Newberry church in New England, 399. in the house of Mr. West, 400. in St. Bride's church, 401. the best construction of them, 405, &c.

Metals, melted by the electrical explosion, 292, 649. making some resistance to the passage of the electric fluid, 212. their calces, electrics, 238. calcined and revivified by the electric explosion, 294. tinge given by them to glass, 187, 649, 678. experiments relating to it, 718. the electric spark taken through pieces of them, 714, the difference of their conducting power, 728.

Musical tone of electrical explosions, 716.

OIL, shown to be a non-conductor, 614.

PAPER, quire of, pierced with the electrical explosion, 274.

Phosphorus, put upon pointed bodies electrified, 309.

Pointed bodies, their effect in electricity, 121, 144, 172, 173.

Propositions, a series of them comprising all the general properties of electricity, 433.

QUERIES, and hints calculated to promote further discoveries in electricity, 487.

RAIN, produced by electricity, 368.
Ribbons, electrified, 278.

SCIENCE, what branches of it are peculiarly useful to an electrician, 499.

Snow, produced by electricity, 371.
Spheroids, used instead of globes, 130.
Spider, electrical, 553.
Star, electrical, 75.
Steam, how affected by electricity, 215.
Stockings, electrical experiments with them, 267.
Stone, Portland a conductor when warm, 239. the conducting power of various kinds, 619.
Surfaces, of electric substances their different properties, 222. of conducting substances, the electrical explosion passing over them, 685.

THEORIES of electricity, 451. Mr. Wilson's, 450. Nollet's, 451. Du Tour's, 454. of positive and negative electricity, 455. of two electric fluids, 468.
Thunder clouds, described, 341. theory of them, 344.
Tourmalin, known to the antients, 314. the opposite electricity of its two sides, 316. the change of its electricity in heating and cooling, 321. its electrical properties found in other precious stones, 324. excited by small alteration of the heat of this atmosphere, 326. shown to collect its electricity from contiguous bodies, 699. the result of heating and cooling it in contact with various substances, 701.

VACUUM, experiments made in it, 41, 74, 124, 569. Torricellian, 571. made by the electrical explosion, 638.
Vanes, moving by electricity, 556.
Vegetables, lose their weight by being electrified, 129. and grow faster, 140.
Volcanos, exhibiting electrical appearances, 392.

WATER, an imperfect conductor of electricity, 209.
Water spouts, an electrical phenomenon, 377.
Wheel electrical, 565. self-moving, 566.
Wood, baked, a non-conductor. 196.

A N

A N

INDEX of NAMES.

INDEX OF NAMES.

Published by the Author of this Treatise, and sold by J. JOHNSON,
in Pater noster row. Price 3s. 6d. 8vo.

AN

ESSAY

ON A COURSE OF

LIBERAL EDUCATION

FOR

CIVIL AND ACTIVE LIFE.

WITH PLANS OF LECTURES ON

I.   The Study of History and general Policy.
II. The History of England.
III. The Constitution and Laws of England.

To which are added,

REMARKS ON A CODE OF EDUCATION,

Proposed by Dr. BROWN, in a late TREATISE, intitled,

THOUGHTS ON CIVIL LIBERTY, &c.

A short Account of a

CHART of BIOGRAPHY.

By JOSEPH PRIESTLEY, LL.D. F.R.S.

The Second Edition.

THE CHART of BIOGRAPHY, of which the Plate annexed exhibits a Specimen, is about three Feet in Length, and two Feet in Breadth. It reprefents the Interval of Time between the Year 1200 before Chrift, and 1800 after Chrift, divided, by an equal Scale, into Centuries. It contains about two thoufand Names of Perfons the moft diftinguifhed in the Annals of Fame; and the Length of their Lives is reprefented in it by Lines drawn in Proportion to their real Duration, and placed fo as to fhow, by Infpection, how long any number of Perfons were cotemporary, and how long one Life began before, or extended beyond another, with every other Circumftance which depends upon the Length of Lives, and the Relation they bear, both to one another, and to univerfal Time; Certainty being always reprefented by full Lines, and Uncertainty by Dots, or broken Lines. The Names are, moreover, diftributed into feveral Claffes by Lines running the whole Length of the Chart, and the Chronology is noted in one Margin by the Year before and after Chrift, and in the other by Succeffions of Kings.

As an Example of the Ufe of the Chart, let any Perfon but attend to the black Line which reprefents the Life of Sir Issac Newton; he will fee, by the Length and Situation of it, that that great Man was born before the Middle of the feventeenth Century, and lived till near the Middle of the eighteenth. He was born a few Years after the Death of Lord Bacon, and about as many Years before that of Descartes. He was a younger Man than Boyle, whom he outlived many Years; and Sir Hans Sloan, Montfaucon, Rollin, Bentley, and Le Clerc, lived to about his Age, and were his Cotemporaries the greateft Part of his Life. Almoft any Number of Lives may be compared with the fame Eafe, to the fame Perfection, and in the fame fhort Space of Time.

The Price of the Chart, together with a Book, containing a Description of it, a Continuation on a fmaller Scale, as high as the Creation, and a Catalogue of all the Names inferted in it, with the Dates annexed to them, is *Half a Guinea*.

Printed for the Author, and fold by himfelf in Warrington; by Mr. *J. Bowles*, in Cornhill; *T. Jefferyes*, at Charing-Crofs; and Mr. *R. Sayer*, in Fleet-Street, London where Specimens may be had gratis.

	1500		1600		1700		
50		50		50		50	

Historians &c.	Mirkhond · Baronius · Hale · Fitzherbert · Coke · Montfaucon · Machiavel · Brissonius · Dugdale · Guicciardin · Thuanus · Burnet
Critics &c.	I.C. Scaliger · Vossius · Rollin · W. Lilly · Casaubon · Temple · Politian · Turnebus · Selden · Bently
Poets &c.	Ariosto · Malherbe · Boileau · Holbein · Shakespeare · Dryden · Raphael · Tasso · Milton · Pope · Titian · Poussin · Handel
Mathemat. &c.	Paracelsus · Harvey · Boerhaave · Copernicus · Ld Bacon · Newton · Cardan · Descartes · Hans Sloan · C. Agrippa · T. Brahe · Boyle · Maclaurin
Divines &c.	Calvin · Pascal · Shaftesbury · Luther · Grotius · Le Clerc · Erasmus · Arminius · Tillotson · Beza · Locke
Statesmen	Francis 1st · Cromwel · Peter Gr · Columbus · Philip 2d · Turenne · Charles 12th · Albuquerque · Henry 4th · Lewis 14th · Charles 5th · Richlieu · Marlborough

| 50 | | 1500 | 50 | | 1600 | 50 | | 1700 | | 50 |

J. Priestley L.L.D. inv. et delin.t   J. Mynde Sc.t